Institute of Mathematical Statistics

LECTURE NOTES–MONOGRAPH SERIES

Volume 38

Model Selection

Edited by P. Lahiri

Institute of Mathematical Statistics
Beachwood, Ohio

Institute of Mathematical Statistics
Lecture Notes-Monograph Series

Series Editor:
Joel Greenhouse

The production of the *IMS Lecture Notes-Monograph Series* is managed by the IMS Business Office: Julia A. Norton, IMS Treasurer, and Elyse Gustafson, IMS Executive Director.

Library of Congress Control Number: 2001 135412

International Standard Book Number 0-940600-52-8

Printed in the United States of America

TABLE OF CONTENTS

PREFACE

This volume originates from a conference that I organized in Lincoln, Nebraska, U. S. A., during March 24-26, 1999. One of the main themes of the conference was the growing field of model selection. The conference was supported by the United States Postal Service, the National Center for Health Statistics, the Gallup Organization and the University of Nebraska-Lincoln. About eighty researchers including some of the contributors of this volume attended the conference.

Shortly after the conference, I discussed the possibility of publishing a volume on model selection with David Ruppert, the former editor of the IMS Lecture Notes-Monograph Series. We agreed that a volume containing a few long review papers on model selection would be of great interest to researchers. The intent was not to publish a volume of conference proceedings but to provide an overview of model selection from the perspective of a few experts in the field.

This volume contains four review papers with discussions and rejoinders. Since the papers are long, a table of contents at the beginning of each paper is provided.

We start the volume with the paper by C. R. Rao and Y. Wu, who survey various model selection procedures with reference to regression, categorical data and time series analysis. The paper also discusses the impact of carrying out statistical analysis after a model is selected.

The second paper by H. Chipman, E. I. George and R. E. McCulloch reviews the basics of Bayesian model selection and discusses at length two important practical issues associated with the implementation of Bayesian model selection: (i) specification of the prior distributions and (ii) the calculation of posterior distributions. These practical issues are then illustrated in variable selection for the linear models and CART model selection problems.

In the third paper, J. O. Berger and L. R. Pericchi motivate the Bayesian model selection and point out the need for objective Bayesian methods in model selection. The authors provide a comprehensive review of four objective Bayesian model selection approaches and extensively discuss criteria for evaluating different model selection methods. The objective Bayesian model selection methods are then critically examined and compared.

We conclude this volume with the paper by B. Efron and A. Gous who compare different model selection methods using hypothesis testing, a simple case of model selection. In one-dimensional case, the authors demonstrate how two apparently different approaches (Fisher's and Jeffreys') to hypothesis testing can be possibly reconciled by giving an interpretation of Fisher's scale in terms of Bayes factors. Similar arguments, however, fail in higher dimensions.

I would like to gratefully acknowledge the University of Nebraska-Lincoln, the Gallup Organization, the National Science Foundation and the University of Maryland at College Park for supporting the project.

I would like to take this opportunity to thank Joel Greenhouse, editor of the IMS Lecture Notes-Monograph Series, and David Ruppert for their encouragement and a number of constructive suggestions. I also thank Shijie Chen, Kristen Olson and Barbara Rolfes for their help in typing and editorial work and the Institute of Mathematical Statistics for publishing this volume.

I believe that the four long review papers along with their discussions and rejoinders provide an excellent overview of model selection. This volume will certainly be a valuable resource for researchers. I thank all the authors and discussants for their outstanding contributions in this volume.

P. Lahiri

Milton Mohr Distinguished Professor of Statistics

University of Nebraska-Lincoln

Adjunct Professor, Joint Program in Survey Methodology

University of Maryland at College Park

Senior Scientist, The Gallup Organization, Princeton

Model Selection
IMS Lecture Notes - Monograph Series (2001) Volume 38

On Model Selection

C. R. Rao and Y. Wu

Pennsylvania State University and York University

Abstract

The task of statistical model selection is to choose a family of distributions among a possible set of families, which is the best approximation of reality manifested in the observed data. In this paper, we survey the model selection criteria discussed in statistical literature. We are mainly concerned with those used in regression analysis and time series.

C.R. Rao is Eberly Professor, Department of Statistics,Pennsylvania State University, University Park, PA 16802, U.S.A; email: crr1@psu.edu. Y.Wu is Professor, Department of Mathematics and Statistics, York University, 4700 Keele Street, Toronto, Ontario, Canada M3J 1P3; email: wuyh@mathstat.yorku.ca. Rao was partially supported by Army Research Office: DAAD 19-01-1-0563 and Wu was partially supported by the Natural Sciences and Engineering Research Council of Canada.

Contents

1 Introduction

Let $\{M_\gamma,\ \gamma \in \Gamma\}$ be candidate models for the observations, where Γ is an index set. It is possible that the true model is not included in $\{M_\gamma\}$. Based on the data, we need to select a model from $\{M_\gamma,\ \gamma \in \Gamma\}$ through a suitable model selection criterion. Many model selection procedures have been proposed in the literature. Each one is designed for a particular application.

Model selection problems are encountered almost everywhere. In linear regression analysis, it is of interest to select the right number of nonzero regression parameters. With the smallest true model, statistical inferences can be carried out more efficiently. In the analysis of time series, it is essential to know the true orders of an ARMA model. In problems of clustering, it is important to find out the number of clusters. In the signal detection, it is necessary to determine the number of true signals, and so on.

In this paper, we survey the model selection criteria discussed in statistical literature. Almost all statistical problems can be considered as model selection problems, but in this paper, we will be mainly concerned with those used in regression analysis and time series. Some interesting examples can be found in Burnham and Anderson (1998), among others.

The paper is arranged as follows: In Section 2, model selection based on hypothesis testing is examined. In Section 3, selection of a model based on the prediction errors is surveyed. In Section 4, the information theoretic criteria are discussed. In Section 5, the role of cross-validation and bootstrap methods in model selection is covered. In Section 6, Baysian approaches to model selection are described. In Section 7, studies on robust model selection are examined. In Section 8, the results on order selection in time series are presented. In Section 9, model selection criteria in categorical data analysis are explored. In Section 10, the investigation on model selection in nonparametric regression is reviewed. In Section 11, data-oriented penalties are discussed. In Section 12, the effect of prior model selection based on data on the inferential statistical analysis of the same data is examined.

In the sequel, for an index set κ, $|\kappa|$ denotes the size of κ, $\boldsymbol{c}(\kappa)$ denotes the sub-vector containing the components of the vector $\boldsymbol{c}$ that are indexed by the integers in κ and $A(\kappa)$ denotes the sub-matrix containing the columns of a matrix A that are indexed by the integers in κ, P_A is used to denote the projection operator on the linear space generated by the column vectors of A, and $\boldsymbol{Y} \sim (\boldsymbol{\mu}, \Sigma)$ is used to represent the random vector $\boldsymbol{Y}$ distributed according to a multivariate distribution with mean vector $\boldsymbol{\mu}$ and covariance matrix Σ. For convenience, $\ell : m$ denotes the set of $\{\ell, \ell+1, \ldots, m\}$ in this paper, where $\ell \leq m$ are positive integers.

2 Selection of a model based on hypothesis testing

A model selection procedure can be constructed based on a sequence of hypothesis tests. Assume that there is an order in the set of candidate models $\{M_i,\ i = 1, 2, \ldots\}$ such that M_i is a preferable to M_{i+1}. A sequence of hypotheses, H_{io} : M_i *holds true* versus H_{ia} : M_{i+1} *holds true*, $i = 1, 2, \ldots$, can be tested sequentially. Once H_{io} is accepted, the test procedure stops and the model M_i is selected. Another such procedure is to replace H_{ia} by H'_{ia} : *One of* $\{M_j : j > i\}$ *holds true*. It can be seen that if all the tests in two test procedures have the same thresholds, the acceptance of M_i in the second test procedure will imply that of M_i in the first test procedure.

Now assume that there is a partial order in a finite index set. Using the partial order "$\prec$", Γ can be partitioned into equivalent classes Γ_i, $i = 0, 1, \ldots$. We further assume that some member of a subset Γ_0 has the smallest order. The problem is how to choose a model from the candidate models consisting of M_γ, $\gamma \in \Gamma$, by hypothesis testing. Suppose that the model M_{γ_1} is preferable to the model M_{γ_2} if $\gamma_1 \prec \gamma_2$. In this case, a model selection procedure can be constructed as follows: First, let $i = 0$. For each $\gamma \in \Gamma_0$, test the null hypothesis that M_γ holds true against the alternative hypothesis that one of the models $M_{\tilde{\gamma}}$, $\gamma \prec \tilde{\gamma}$ holds true. Find the largest $p-$value and if this value is less than the prechosen one, stop and select the model with that $p-$value. Otherwise, let $i = 1$, and repeat the above step with Γ_0 replaced by Γ_1. In general, if the procedure does not stop at the jth step, let $i = j + 1$ in the next step and repeat the previous step with Γ_j replaced by Γ_{j+1}.

For a better understanding of the procedures, let us consider the following linear model:

$$Y_i = \boldsymbol{x}_i'\boldsymbol{\beta} + \varepsilon_i, \quad i = 1, 2, \ldots, \tag{2.1}$$

where $\boldsymbol{x}_i$ are p-vectors, $\boldsymbol{\beta}$ is a p-vector parameter, and ε_i are random errors. Let κ denote a subset of $\{1, \ldots, p\}$. Based on the observations $(y_i, \boldsymbol{x}_i)$, $i = 1, \ldots, n$, we would like to decide whether the model (2.1) should be replaced by the following sub-model with a fixed κ:

$$Y_i = \boldsymbol{x}_i'(\kappa)\boldsymbol{\beta}(\kappa) + \varepsilon_i, \quad i = 1, 2 \ldots. \tag{2.2}$$

There are $2^p - 1$ such models.

Denote the model (2.2) by M_κ. If $\kappa_i = \{1, \ldots, i\}$, for convenience, write M_i for the model M_{κ_i}. For simplicity, consider the set of candidate models $\{M_i,\ i = 1, \ldots, p\}$. Hence, H_{io} is the hypothesis that $\beta_j = 0$ for $j > i$, H_{ia} is the hypothesis that $\beta_j = 0$ for $j > i + 1$ and $\beta_{i+1} \neq 0$, and H'_{ia} is the hypothesis that at least one of $\{\beta_j,\ j > i\}$ is not zero. Assume that the maximum likelihood estimators of unknown parameters can be worked out. Hence, the likelihood ratio statistics may be used to perform the tests. It is easy to see that the performance of the procedure is affected a great deal by the critical values of the tests. Since the exact distributions of the likelihood ratio statistics are usually unknown, their asymptotic distributions are used to compute these thresholds, whose acceptable accuracy would need a large sample. The advantage for such procedures is that the probability of overfitting is somewhat under control.

For giving a more clear view of such procedures, we further assume that ε_i, $i = 1, 2, \ldots$, are independently and identically $N(0, \sigma^2)$ distributed and $\boldsymbol{x}_i$, $i = 1, 2, \ldots$, are fixed, which can be relaxed to the condition that $\boldsymbol{\varepsilon}_n | X_n \sim N(\boldsymbol{0}, \sigma^2 I_n)$, where $\boldsymbol{\varepsilon}_n = (\varepsilon_1, \ldots, \varepsilon_n)'$ and $X_n = (\boldsymbol{x}_1, \ldots, \boldsymbol{x}_n)'$. For testing H_{io} against H_{ia}, the likelihood ratio statistic is equivalent to

$$F^{(i)} = \frac{\boldsymbol{Y}_n'[P_{X_n(1:i+1)} - P_{X_n(1:i)}]\boldsymbol{Y}_n}{\boldsymbol{Y}_n'[I_n - P_{X_n(1:i+1)}]\boldsymbol{Y}_n},$$

where $\boldsymbol{Y}_n = (Y_1, \ldots, Y_n)'$. Under H_{io}, $F^{(i)} \sim F_{1,n-(i+1)}$. The critical value is given by $F_{1,n-(i+1)}(\alpha)$, where $\text{Prob}\left(F_{1,n-(i+1)} > F_{1,n-(i+1)}(\alpha)\right) = \alpha$, and hence under M_i the probability of making a type I error, i.e. choosing M_{i+1}, is α. It can be seen that the probability of underfitting, i.e. choosing M_i when it is not a true model, depends on $\beta(1 : i+1)$. As commented in Shao and Rao (2000), the underfitting probability converges to 0 as the sample size n increases to ∞.

In practice, *forward selection*, *backward elimination* and *stepwise regression* are popular model selection methods in linear regression. They are available in almost all of statistical software packages. In their applications, the candidate models consist of all $2^p - 1$ submodels. Controlled by one or two thresholds, the model is selected based on statistical hypothesis testing. The details can be found in Krishnaiah (1982), Miller (1990) and some textbooks on linear regression. Some authors prefer backward elimination to forward selection for the economy of effort (see Mantel 1970). But on account of the simplicity of computation and stopping rules, forward selection is recommended. For choosing the significance levels required, the most widely used level is 10% or 5%. But the overall power as well as the type I error rate are unknown unless the order of entry of the variables into the model is specified explicitly before applying any method. It is easily seen that the order of entry differs with observations. To avoid such difficulties, Aitkin (1974) and McKay (1977) proposed an application of a simultaneous testing procedure, but it requires considerable computation to obtain a set of significance levels as shown by Shibata (1986a). Note that model selection procedures based on R^2 (the square of the multiple correlation), adjusted R^2 or its equivalent MSE are also available in many statistical software packages.

Thall, Russell, and Simon (1997) proposed an algorithm, *backward elimination via repeated data splitting* (BERDS), for variable selection in regression. Initially, the data are randomly partitioned into two sets E,V, and an exhaustive backward elimination (BE) is performed in E. For each $\alpha_{\text{stay}} \in (0, 1)$ used in BE, the corresponding fitted model from E is validated in V by computing the sum of squared deviations of observed from predicted values. This is repeated m times, and the α^*, which minimizes the sum of the m sums of squares, is used as α_{stay} in a final BE on the entire data set. BERDS is a modification of the algorithm BECV (BE via cross-validation) proposed by Thall, Simon, and Grier (1992). Their extensive simulation study showed that, compared to BECV, BERDS has a smaller model error and higher probabilities of excluding noise variables, of selecting each of several uncorrelated true predictors, and of selecting exactly one of two or three highly correlated true predictors. Thall, Russell, and Simon (1997) also showed that BERDS is superior to standard BE with $\alpha_{\text{stay}} = .05$ or $.10$, and this superiority increases with the number of noise variables in the data and the degree of correlation among true predictors.

While the log likelihood ratio tests are fairly good when nested parametric hypotheses are involved, it is a different story for testing non-nested parametric hypotheses. Cox (1961, 1962) initiated research on testing separate families of hypotheses and proposed non-nested test statistics, which, according to Bera (2000), can also be viewed as Rao's score tests. Later, Williams (1970a,b) gave a different approach by directly simulating the distribution of the log likelihood ratio on a computer. According to Loh (1985), both tests are not satisfactory. Loh (1985) proposed an alternative procedure by repeated application of the parametric bootstrap method over slowly shrinking confidence regions, which, as justified by the author, is promising. There are other solutions to this kind of problems in the literature, which are not limited to parametric hypotheses.

Since in practical situations, the assumed null hypotheses are only approximations and they are almost always different from the reality, the choice of the loss function in the test theory makes its practical application logically contradictory, as commented by Akaike (1974). Bera (2000) gives a very inspiring discussion on hypothesis testing with misspecified models, which includes a historical review as well as possible future direction of research.

3 Model selection based on prediction errors

Model selection based on statistical hypothesis testing described in the last section involves many restrictions and further the choice of thresholds are open. As commented in Akaike (1969), the main difficulty in applying this kind of procedures and their relatives stems from the fact that they are essentially formulated in the form of successive tests of null hypotheses against multiple alternative hypotheses. Actually one of the alternative hypotheses is just the model one is looking for and thus it is very difficult for one to get the feeling of the possible alternative hypotheses to set reasonable significance levels. To overcome this difficulty, Akaike (1969) suggested an alternative decision theoretic approach based on prediction errors. In Akaike (1969), the final prediction error (FPE) was defined, which is the mean squared prediction error when a model fitted to the present data is applied to another independent observation, or to make a one step prediction. Based on the final prediction error, the parameters in each candidate model are estimated so that the minimum final prediction error is attained for this model, and then a model, which has the minimum final prediction error within the candidate models, is selected. We call this an FPE procedure. Note that when the purpose of model selection is for prediction, it may be wiser to choose a model based on the prediction errors than using the model selection methods discussed in Section 2.

For example, consider the model (2.2). We assume that ε_i, $i = 1, 2, \ldots$, are independently and identically distributed with mean zero and variance σ^2 and $\boldsymbol{x}_i$, $i = 1, 2, \ldots$, are fixed, which can be relaxed to the condition that $\boldsymbol{\varepsilon}_n | X_n \sim (\mathbf{0}, \sigma^2 I_n)$. We also assume that X_n is of full rank. The least squares estimate of $\boldsymbol{\beta}(\kappa)$ for the model (2.2) is $[X_n(\kappa)' X_n(\kappa)]^{-1} X_n(\kappa)' \boldsymbol{Y}_n$. Let $\boldsymbol{Y}_0$ consist of n new observations and $\boldsymbol{Y}_0 = X_n(\kappa)\boldsymbol{\beta}(\kappa) + \boldsymbol{\varepsilon}_0$ with $\boldsymbol{\varepsilon}_0 \sim (\mathbf{0}, \sigma^2 I_n)$. Then the FPE, i.e. the mean squared prediction error, is given by

$$
\begin{aligned}
& \frac{1}{n} \mathrm{E}[(\boldsymbol{Y}_0 - \hat{\boldsymbol{Y}}_0)'(\boldsymbol{Y}_0 - \hat{\boldsymbol{Y}}_0)] \\
= \; & \frac{1}{n} \mathrm{E}[(\boldsymbol{Y}_0 - X_n(\kappa)'\boldsymbol{\beta}(\kappa))]'[(\boldsymbol{Y}_0 - X_n(\kappa)'\boldsymbol{\beta}(\kappa))] = \sigma^2(1 + k/n), \qquad (3.1)
\end{aligned}
$$

where $k = |\kappa|$. Let S_κ denote the residual sum of squares under the model M_κ, and $\hat{\sigma}_\kappa^2$ denote $S_\kappa/(n-k)$. Using the unbiased estimator $\hat{\sigma}_\kappa^2$ to replace σ^2 in (3.1), we get $\hat{\sigma}_\kappa^2(1 + k/n)$, which is denoted by $\mathrm{FPE}(\kappa)$. The selected model M_{κ^*} can be obtained by minimizing $\mathrm{FPE}(\kappa)$, i.e.

$$M_{\kappa^*} = \arg\min_{M_\kappa} \mathrm{FPE}(\kappa).$$

The FPE procedure was originally derived for autoregressive time series models. A similar procedure was developed by Davisson (1965) for analyzing signal-plus-noise data. By using FPE, Akaike (1970) suggested a way to decide on the thresholds for the procedures discussed in Section 2.

Define, for the model (2.2) with $p = \infty$,

$$\mathrm{SH}(\kappa) = \hat{\sigma}_\kappa^2 (n + 2k)(n - k)/n.$$

Shibata (1980) proposed a model selection criterion based on the expectation of prediction errors, which is given as follows:

$$M_{\kappa^*} = \arg\min_{M_\kappa} \mathrm{SH}(\kappa).$$

This criterion is equivalent to FPE and was shown to be asymptotically efficient.

In Akaike (1970), the modified version of the FPE procedure was proposed for improving the consistency of the FPE procedure. For the model (2.2), let

$$(\mathrm{FPE})^{\lambda}(\kappa) = S_{\kappa}(1 + k/n^{\lambda}),$$

where $0 < \lambda < 1$. A model is chosen to be the one which minimizes $(\mathrm{FPE})^{\lambda}(\kappa)$ within candidate models. We call such a procedure as the $(\mathrm{FPE})^{\lambda}$ procedure.

Even though the consistency of the $(\mathrm{FPE})^{\lambda}$ procedure is improved over the FPE procedure, it is still not consistent and the probability of overfitting is still greater than zero. Many authors have sought to modify the overfitting property by further adjusting the second term of the FPE procedure by multiplying it by δ and proposing the following FPE_{δ} procedure:

$$\min_{M_{\kappa}} \mathrm{FPE}_{\delta}(M_{\kappa}) = \hat{\sigma}_{\kappa}^{2}(1 + \delta k/n),$$

where δ may or may not depend on n.

The FPE_{δ} procedure or its equivalent was discussed by Akaike (1970, 1974), Atkinson (1980, 1981), Bhansali and Downham (1977), Shibata (1976, 1980, 1986a,b), Zhang and Krieger (1993) among others. Based on empirical evidence, some authors have suggested that the FPE_{δ} procedure with δ between 2 and 6 would do well in most situations, but such an ad hoc choice of δ seems to be lacking in theoretical justification. In view of the inconsistency of FPE and $(\mathrm{FPE})^{\lambda}$ procedures, it is unlikely that any finite δ would lead to a consistent procedure. To achieve consistency, it may be necessary to make δ dependent on sample size. Specially, it has been suggested that $\delta = \delta_n$ should satisfy $\delta_n \to \infty$ and $\delta_n/n \to 0$. See Bozdogan (1987), Nishii (1988), Rao and Wu (1987), Shao (1997), Zhao, Krishnaiah, and Bai (1986a,b) among others. By assuming normality, Venter and Steel (1992) studied the choice of the quantity δ in the FPE_{δ} procedure for selecting a member of a class of linear models having orthogonal structure. Two approaches are discussed, namely fixing the maximal estimation risk at a prescribed level and using minimax regret. In Zhang (1994) it was argued that a choice of $\delta \in [3, 4]$ would be adequate for most practical purposes, and by using decision theoretic properties of FPE_{δ}, it was shown that the incorrect models are sometimes preferable to the true model.

Mallows (1973) took a different approach to model selection criterion in a linear regression problem. Consider the model (2.2) and assume that the conditions made previously hold true. The fitted regression subset at the point $\boldsymbol{x}_i$ is given by

$$\hat{Y}_{i,\kappa} = \boldsymbol{x}_i(\kappa)\hat{\boldsymbol{\beta}}_{\kappa},$$

where $\hat{\boldsymbol{\beta}}_{\kappa}$ is the least squares estimate of $\boldsymbol{\beta}(\kappa)$ under the model (2.2). If $\mathrm{E}(\hat{Y}_{i,\kappa}) = \mu_{i,\kappa}$, then $\mu_{i,\kappa}$ generally differs from $\boldsymbol{x}_i(\kappa)'\hat{\boldsymbol{\beta}}_{\kappa}$ because of possible bias in the M_{κ}. Let $\mathrm{E}(Y_i) =$

θ_i. Then

$$\begin{aligned} \mathrm{E}[\hat{Y}_{i,\kappa} - \theta_i)^2] &= \mathrm{Var}(\hat{Y}_{i,\kappa}) + (\mu_{i,\kappa} - \theta_i)^2 \\ &= \sigma^2 \boldsymbol{x}_i(\kappa)'(X_n(\kappa)'X_n(\kappa))^{-1}\boldsymbol{x}_i(\kappa) + (\mu_{i,\kappa} - \theta_i)^2, \end{aligned}$$

which implies that

$$\begin{aligned} \Delta_\kappa &= \frac{1}{\sigma^2}\mathrm{E}\left[\sum_{i=1}^n (\hat{Y}_{i,\kappa} - \theta_i)^2\right] \\ &= \frac{1}{\sigma^2}\left[\sum_{i=1}^n \sigma^2 \boldsymbol{x}_i(\kappa)'(X_n(\kappa)'X_n(\kappa))^{-1}\boldsymbol{x}_i(\kappa) + \sum_{i=1}^n (\mu_{i,\kappa} - \theta_i)^2\right] \\ &= k + \frac{1}{\sigma^2}\sum_{i=1}^n (\mu_{i,\kappa} - \theta_i)^2. \end{aligned}$$

Since $\mathrm{E}(S_\kappa) = \sigma^2(n-k) + \sum_{i=1}^n (\mu_{i,\kappa} - \theta_i)^2$, Δ_κ can be estimated by

$$C_\kappa = \frac{S_\kappa}{\hat{\sigma}^2_{1:p}} + 2k - n.$$

However, the quantity C_κ is not an unbiased estimator of Δ_κ. Mallows (1995) suggested that any candidate model where $C_\kappa < k$ should be carefully examined as a potential best model. This procedure is called C_p criterion.

Shibata (1980) showed that FPE is an asymptotically efficient procedure. Since FPE and C_p procedures are asymptotically equivalent (see, e.g., Nishii 1984), C_p is also asymptotically efficient. Note that both procedures are not consistent.

Considering a sequence of models with kth model given by (2.2) for $\kappa = 1 : k$, Breiman and Freedman (1983) proposed the following criterion also based on the expectation of prediction errors:

$$M_{1:\hat{k}} = \arg \min_{M_{1:k},\, k \le n/2} \hat{\sigma}_{1:k}(1 + k/(n-1-k)).$$

This criterion was shown to be asymptotically efficient.

In a series of papers, Rissanen introduced minimum description length (MDL) principle as a process of searching for models and model classes with the shortest code length (see Rissanen 1989). The application of the MDL principle with predictive code length is called the predictive MDL (PMDL). Consider the linear model (2.2). Based on his PMDL principle, Rissanen (1986a,b,c) proposed a new criterion that selects the model which minimizes

$$PLS(\kappa) = \sum_{i=m+1}^{n} (y_i - \boldsymbol{x}_i'(\kappa)\hat{\boldsymbol{\beta}}_\kappa^{(i-1)})^2,$$

where $\hat{\boldsymbol{\beta}}_\kappa^{(j)}$ is the least squares estimate based on $\{y_i, \boldsymbol{x}_i(\kappa);\ i \le j\}$ and m is the first integer j so that $\hat{\boldsymbol{\beta}}_\kappa^{(j)}$ is uniquely defined. Since $(y_i - \boldsymbol{x}_i'(\kappa)\hat{\boldsymbol{\beta}}^{(i-1)}(\kappa))^2$ is the square of the prediction error at stage i, this criterion is called the predictive least squares (PLS) principle, which is strongly consistent. A drawback with PLS is that the data must be ordered, and the result may depend on the particular order selected. Hence, the symmetric PLS was proposed. See Rissanen (1989) for more details.

If one first uses a criterion based on data to select a set of regressors and then estimates the regression coefficients, such a popular strategy is called an s/e procedure. Foster and George (1994) proposed a measure for the evaluation of variable selection procedures in multiple regression. This measure, which is called risk inflation, is the maximum possible increase in risk of the consequent s/e procedure due to selecting rather than knowing the "correct" predictors. The risk inflation is obtained as the ratio of risk of a s/e estimator to the risk of the ideal (but unavailable) selection/estimation estimator which uses only the "correct" predictors. Consider the models (2.2). In the case of orthogonal predictors, the authors argued that compared to overall inclusion, AIC, C_p and BIC offer smaller risk inflation and hence they proposed a model selection procedure (RIC) as follows:

$$M_{\kappa^*} = \arg\min_{M_\kappa}[S_\kappa + k\hat{\sigma}^2(2\log p)],$$

which, as they stated in their paper, substantially improves on AIC, C_p and BIC and is close to optimal. For the general case, it is unfortunate that the model selection procedure based on the risk inflation depends on the correlation structure of the predictors. See Foster and George (1994) for details.

In a linear regression model, for attenuating possible excessive modelling biases, a large number of predictors are usually introduced at the initial stage of modelling. To enhance predictability and to select significant variables, one usually applies stepwise deletion, subset selection and ridge regression. While these three methods are useful in practice, they ignore stochastic errors inherited in the previous stages of variable selections (see, e.g., Fan and Li 2001). Tibshirani (1996) proposed a new approach, called least absolute shrinkage and selection operator (LASSO), which simultaneously selects variables and estimates parameters. By using the LASSO, some regression coefficients are shrinked and others are set to be zero. According to Tibshirani (1996), LASSO retains good features of both subset selection and ridge regression, and can be applied to generalized linear models, besides, the LASSO estimate is also a Bayes estimate. As a matter of fact, the LASSO is closely related to penalized likelihood with the L_1 penalty. Fan and Li (2001) generalized the LASSO method by proposing the penalized likelihood with a smoothly clipped absolute deviation (SCAD) penalty function along with a unified algorithm backed up by statistical theory, which resulted in an estimator with good statistical properties. Their approach includes the LASSO as its special case. Their simulation results showed that their method compared favorably with other approaches as an automatic variable selection technique. As shown in their paper, by the advantage of simultaneous selection of the variables and estimation of parameters, they were able to give a simple estimated standard error formula, which was tested to be accurate enough for practical applications.

Recently Breiman (1996) studied how to stabilize an unstable model selection procedure in linear regression. Such problems are very important and need further investigation.

4 Information theoretic criteria

Let $z_1, \ldots, z_n$ be n independent observations on a random vector $\boldsymbol{Z}$ with probability density function $g(z)$. Consider parametric family of density functions is $\{f_{\boldsymbol{\theta}}(z),\ \boldsymbol{\theta} \in \boldsymbol{\Theta}\}$ with a vector parameter $\boldsymbol{\theta}$ and parameter space $\boldsymbol{\Theta} \subset R^m$ for which the average log-likelihood is given by

$$\frac{1}{n}\sum_{i=1}^{n} \log f_{\boldsymbol{\theta}}(z_i), \tag{4.1}$$

where log denotes the natural logarithms. As n increases to infinity, this average tends to

$$S(g; f_{\boldsymbol{\theta}}) = \int g(z) \log f_{\boldsymbol{\theta}}(z)\, dz,$$

with probability 1, where the existence of the integral is assumed. The difference

$$K(g; f_{\boldsymbol{\theta}}) = S(g; g) - S(g; f_{\boldsymbol{\theta}})$$

is known as the Kullback-Leibler distance (information) between $g(z)$ and $f_{\boldsymbol{\theta}}(z)$ and takes positive values, unless $f_{\boldsymbol{\theta}}(z) = g(z)$ holds almost everywhere. Hence $S(g; f_{\boldsymbol{\theta}})$ is reasonable for defining a best fitting model by its maximization or, from the analogy to the concept of entropy, by minimizing $-S(g; f_{\boldsymbol{\theta}})$. Maximizing (4.1) with respect to $\boldsymbol{\theta}$ leads to the MLE $\hat{\boldsymbol{\theta}}$.

Consider the case that $g(z) = f_{\boldsymbol{\theta}_0}(z)$, where $\boldsymbol{\theta}_0 \in \boldsymbol{\Theta}$. When $\boldsymbol{\theta}$ is sufficiently close to $\boldsymbol{\theta}_0$,

$$K(f_{\boldsymbol{\theta}_0}, f_{\boldsymbol{\theta}}) \approx (\boldsymbol{\theta} - \boldsymbol{\theta}_0)' J (\boldsymbol{\theta} - \boldsymbol{\theta}_0)/2,$$

where J is the Fisher information matrix. When the MLE $\hat{\boldsymbol{\theta}}$ lies very close to $\boldsymbol{\theta}_0$, $K(f_{\boldsymbol{\theta}_0}, f_{\boldsymbol{\theta}})$ can be approximately measured by $(\hat{\boldsymbol{\theta}} - \boldsymbol{\theta}_0)' J (\hat{\boldsymbol{\theta}} - \boldsymbol{\theta}_0)/2$. Under certain regularity conditions $n(\hat{\boldsymbol{\theta}} - \boldsymbol{\theta}_0)' J (\hat{\boldsymbol{\theta}} - \boldsymbol{\theta}_0)$ is asymptotically distributed as chi-square with k degrees of freedom, and $\mathrm{E}[2nK(f_{\boldsymbol{\theta}_0}, f_{\hat{\boldsymbol{\theta}}})] \approx n(\boldsymbol{\theta} - \boldsymbol{\theta}_0)' J (\boldsymbol{\theta} - \boldsymbol{\theta}_0) + k$, where k is the number of independent parameters. By using

$$2\left(\sum_{i=1}^{n} \log f_{\boldsymbol{\theta}_0}(z_i) - \sum_{i=1}^{n} \log f_{\hat{\boldsymbol{\theta}}}(z_i)\right)$$

to approximate $n(\boldsymbol{\theta}-\boldsymbol{\theta}_0)'J(\boldsymbol{\theta}-\boldsymbol{\theta}_0)$, a correction is needed for the downward bias due to replacing $\boldsymbol{\theta}$ by $\hat{\boldsymbol{\theta}}$. Akaike (1973) added k as the correction and introduced the famous AIC criterion: Let

$$\text{AIC}(\boldsymbol{\theta}) = -2\log(\text{maximum likelihood}) + 2k, \tag{4.2}$$

where k is as defined above. The selected model is

$$M_{\boldsymbol{\theta}^*} = \arg\min_{M_{\boldsymbol{\theta}}} \text{AIC}(\boldsymbol{\theta}).$$

The justification of the correction k can be found in Akaike (1973), Linhart and Zucchini (1986) and Sakamoto, Ishiguro, and Kitagawa (1986) among others. Note that AIC is, in final analysis, based on the concept of minimizing the expected Kullback-Leibler distance (see, e.g., Sawa 1978, Sugiura 1978). It is worth mentioning that information theory (see, e.g., Guiasu 1977) has been a discipline only since the mid-1940s and covers a variety of theories and methods that are fundamental to many of the sciences.

For the model (2.2), assuming that the errors are $N(0,\sigma^2)$ distributed, AIC can be expressed as

$$\text{AIC}(\kappa) = n\log(S_\kappa/n) + 2k,$$

where S_κ is defined as before and $k = |\kappa|$.

Assuming that the errors have a multivariate normal distribution, Fujikoshi and Satoh (1997) proposed modified AIC and C_p for selecting multivariate linear regression models by reducing the bias of estimation of Akaike- and Mallows-type risks when the collection of candidate models includes both underspecified and overspecified models. Their simulation study showed that both modified AIC and C_p provided better approximations to their risk functions, and better model selection, than AIC and C_p.

For model selection in settings where the observed data are incomplete, Shimodaira (1994) proposed a natural extension of AIC, called predictive divergence for incomplete observation model criterion (PDIO). Cavanaugh and Shumway (1998) derived a variant of AIC based on the motivation provided by Shmodaira (1994), which can be evaluated using only complete-data tools, readily available through the EM algorithm and the supplemented EM algorithm. The authors compared their criterion with AIC and PDIO by simulation. The results showed that Cavanaugh and Shumway's criterion was less prone to overfitting than AIC and less prone to underfitting than PDIO.

Shibata (1980) has shown that AIC is asymptotically efficient. However, AIC is not consistent. Note that AIC, FPE and C_p are asymptotically equivalent (see, e.g., Nishii 1984). For small samples, many researchers have shown that AIC leads to overfitting (see, e.g., Hurvich and Tsai 1989). For improving on AIC, Sugiura (1978) and Hurvich and

Tsai (1989) derived AICc by estimating the expected Kullback-Leibler distance directly in regression models, where a second order bias adjustment was made, and the criterion is given as follows:

$$\text{AICc}(\boldsymbol{\theta}) = -2\log(\text{maximum likelihood}) + 2k\left(\frac{n}{n-k-1}\right) = \text{AIC}(\boldsymbol{\theta}) + \frac{2k(k+1)}{n-k-1}, \quad (4.3)$$

where k denotes the number of free parameters in the candidate model. The model for which AICc is smallest is chosen. From (4.3), it can be seen that AICc has an additional bias correction term. If n is large with respect to k, then the second order correction is negligible and AIC should perform well, which implies that AICc and AIC are asymptotically equivalent and hence AICc is asymptotically efficient but not consistent. Findley (1985) noted that the study of the bias correction is of interest in itself; the exact small sample bias correction term varies by type of models involved.

Denote a model selection criterion by MSC. Model A (with k variables) will be considered better than Model B (with $k+\ell$ variables) if $\text{MSC}(B) > \text{MSC}(A)$. Define the signal as $\text{E}[\text{MSC}(B) - \text{MSC}(A)]$ and the noise as the standard deviation of the difference denoted by $\text{sd}(\text{MSC}(B) - \text{MSC}(A))$. Then the signal-to-noise ratio is defined as $\text{E}[\text{MSC}(B) - \text{MSC}(A)]/\text{sd}(\text{MSC}(B) - \text{MSC}(A))$. See McQuarrie and Tsai (1998) for more details. AIC has a weak signal-to-noise ratio (see, e.g., McQuarrie and Tsai 1998) and hence it tends to overfit. In contrast, AICc's has better signal-to-noise ratio so that AICc should perform well regarding the overfitting. The performance of model selection criteria with weak signal-to-noise ratios could be improved if their signal-to-noise ratios could be strengthened. Unfortunately, there is no single appropriate correction for all criteria.

For the model (2.2), AICu was proposed by McQuarrie, Shumway, and Tsai (1997), where S_κ/n of AIC term in (4.3) was replaced by $S_\kappa/(n-k)$ and the other term remains the same, which provides better model choices than AICc for moderate to large sample sizes except when the true model is of infinite order.

For improving on the inconsistency of AIC criterion, Akaike (1978) and Schwarz (1978) introduced equivalent consistent model selection criteria conceived from a Bayesian perspective. Schwarz derived SIC for selecting models in the Koopman-Darmois family, while Akaike derived his model selection criterion BIC for the problem of selecting a model in linear regression. The two procedures introduced about the same time are equivalent. See McQuarrie and Tsai (1998) for more details. In Schwarz (1978), it was assumed that the observations come from a Koopman-Darmois family with density of the form

$$f(x, \boldsymbol{\theta}) = \exp(\boldsymbol{y}(x)'\boldsymbol{\theta} - b(\boldsymbol{\theta})),$$

where $\boldsymbol{\theta} \in \Theta$, a convex subset of R^p, and $\boldsymbol{y}$ is a p-dimensional sufficient statistic for $\boldsymbol{\theta}$.

Since the exact distribution of the prior need not be known for the sake of asymptotic nature of SIC, it suffices to assume that the prior is of the form $\sum \gamma_j \mu_j$, where γ_j is the prior probability for model M_j, and μ_j is the conditional prior of $\boldsymbol{\theta}$ given M_j. Further, Schwarz assumed a fixed loss for selecting the wrong model. As stated in Schwarz (1978), the Bayes solution consists of selecting the model with a high posterior probability. In large samples, this posterior probability can be approximated by a Taylor expansion. Schwarz found its first term to be the log of the MLE for the model M_j and its second term was of the form $k \log(n)$ where k is the dimension of the model and n is the sample size. The remaining terms in the Taylor expansion were shown to be bounded and hence could be ignored in large samples. The SIC is given as follows: Let

$$\mathrm{SIC}(\boldsymbol{\theta}) = -2\log(\text{maximum likelihood}) + k\log(n), \tag{4.4}$$

and choose the model for which SIC is smallest. It can be seen that the $2k$ term in AIC is replaced by $k\log(n)$ in SIC, which places a much stronger penalty for overfitting. When the parameters in SIC are estimated based on MDL principle, the resulting criterion is called MDL, which was derived in Rissanen (1978, 1983) under the assumption that there is no prior knowledge about $\boldsymbol{\theta}$.

BIC or SIC is strongly consistent but not asymptotically efficient. For small sample sizes, the chance of underfitting should not be overlooked. For improving the underfitting, it is natural to ask if $\log n$ can be replaced by a function of n which approaches infinity not as fast as $\log n$ when n tends to infinity. This function can not be constant. In Hannan and Quinn (1979), they argued, by applying the law of the iterated logarithm, that $\log n$ can be replaced by $c \log\log n$ with $c > 2$ in SIC without losing strong consistency. We call this new criterion HQ. When applying it, the underfitting is improved but does not vanish. It is unfortunate that a consistent model selection criterion usually tends to underfit when sample size is not large enough. All one can do is to find a consistent model selection criterion such that the underfitting is at its lowermost level. HQ meets such requirement.

For improving the small-sample performance of SIC, McQuarrie (1999) used the relationship between AIC and AICc to derive its small-sample correction denoted by SICc. He showed that SICc overfits less frequently than SIC, performs better in small samples and is asymptotically equivalent to SIC.

Consider the model (2.1). A framework is called prediction with repeated refitting if it allows model selection at each time, i.e., a model is chosen on the basis of the data available at time t, and the model selected is used to predict Y_{t+1}, while a framework is called prediction without refitting if a model is chosen on the basis of the training sample, and then the model selected is used to predict. Under the frame of finite-dimensional normal regression models, Speed and Yu (1993) compared model selection criteria according to prediction errors based upon prediction with refitting, and prediction without refitting and showed that Rissanen's accumulated prediction error and stochastic complexity criteria, AIC, SIC, and the FPE criteria achieve both low bounds for prediction with refitting and without refitting.

AIC was derived under the assumptions that (i) the estimation is by maximum likelihood and (ii) the parametric family of distributions includes the true model. Could these assumptions be somehow relaxed? Let

$$b(G) = \mathrm{E}_G \left[\frac{1}{n} \sum_{i=1}^{n} \log f_{\hat{\boldsymbol{\theta}}}(z_i) - \int \log f_{\hat{\boldsymbol{\theta}}}(\boldsymbol{u})\, dG(\boldsymbol{u}) \right],$$

where the expectation is taken over the true distribution G and $\hat{\boldsymbol{\theta}}$ is an estimate of $\boldsymbol{\theta}$. Without assuming that the true distribution belongs to the specified parametric family of probability distributions, $b(G)$ is asymptotically given by

$$b(G) = \frac{1}{n} \mathrm{tr}\{J(G)^{-1} I(G)\} + O(n^{-2}),$$

where $J(G)$ and $I(G)$ are defined by

$$J(G) = -\mathrm{E}_G \left[\frac{\partial^2 \log f_{\boldsymbol{\theta}}(z)}{\partial \boldsymbol{\theta} \partial \boldsymbol{\theta}'} \right] \quad \text{and} \quad I(G) = -\mathrm{E}_G \left[\frac{\partial \log f_{\boldsymbol{\theta}}(z)}{\partial \boldsymbol{\theta}} \frac{\partial \log f_{\boldsymbol{\theta}}(z)}{\partial \boldsymbol{\theta}'} \right].$$

Denote the bias corrected log likelihood by

$$\mathrm{TIC}(\boldsymbol{\theta}) = -2 \log(\text{maximum likelihood}) + 2\mathrm{tr}\{\hat{J}(G)^{-1} \hat{I}(G)\},$$

where $\hat{J}(G)$ and $\hat{I}(G)$ are respectively consistent estimates of $J(G)$ and $I(G)$, and choose the model for which TIC is smallest. This criterion is called TIC and was originally introduced by Takeuchi (1976) and also Stone (1977a), and later discussed extensively by Shibata (1989) and Konishi (1999). When the true model is included in the set of candidate models, $b(G)$ can be reduced to

$$b(G) = \frac{k}{n} + O(n^{-2}),$$

where k is the number of free parameters in the model, and TIC becomes AIC. If none of the candidate models is close to the true model, TIC is an alternative if sample size is large.

A generalized information criterion (GIC) was introduced in Konishi and Kitagawa (1996) by estimating the same Kullback-Leibler distance as in AIC while relaxing both the assumptions (i) and (ii). If the bias $b(G)$ can be estimated by appropriate procedures, then the bias corrected log likelihood is given by

$$\mathrm{GIC}(\boldsymbol{\theta}) = \frac{1}{n} \sum_{i=1}^{n} \log f_{\hat{\boldsymbol{\theta}}}(z_i) - \hat{b}(G),$$

where $\hat{\boldsymbol{\theta}}$ may be obtained by maximum likelihood, penalized likelihood or robust procedures. The estimated bias $\hat{b}(G)$ is generally given as an asymptotic bias and an approximation to $b(G)$. A model is selected such that the GIC is smallest. Konishi and Kitagawa (1996) employed a functional estimator $\hat{\boldsymbol{\theta}} = \boldsymbol{t}(\hat{G})$ with Fisher consistency and approximated $b(G)$ by a function of the empirical influence function of the estimator and the score function of the parametric model. They obtained the GIC in the form

$$\mathrm{GIC}(\boldsymbol{\theta}) = -2\sum_{i=1}^{n} \log f_{\hat{\boldsymbol{\theta}}}(z_i) + \frac{2}{n}\sum_{i=1}^{n} \mathrm{tr}\left\{ \boldsymbol{t}^{(1)}(z_i;\hat{G}) \frac{\partial \log f_{\boldsymbol{\theta}}(z_i)}{\partial \boldsymbol{\theta}'}\Big|_{\hat{\boldsymbol{\theta}}} \right\}.$$

Here $\boldsymbol{t}^{(1)}(z_i;\hat{G}) = (t_1^{(1)}(z_i;\hat{G}), \ldots, t_k^{(1)}(z_i;\hat{G}))'$ and $t_i^{(1)}(z_i;\hat{G})$ is the empirical influence function defined by

$$t_i^{(1)}(z_i;\hat{G}) = \lim_{\varepsilon\to 0} \frac{t_i((1-\varepsilon)\hat{G} + \varepsilon\delta_i) - t_i(\hat{G})}{\varepsilon}$$

with δ_i being a point mass at z_i. Note that AIC and TIC are special cases of GIC.

In Bozdogan (1987), CAICF (C denoting "consistent" and F denoting the use of Fisher information matrix) was proposed. Let

$$CAICF(\boldsymbol{\theta}) = -2\log(\text{maximum likelihood}) + k[\log(n) + 2] + \log|\hat{J}|.$$

CAICF criterion chooses a model for which CAICF is smallest. In Bozdogan (1988), an information theoretic measure of complexity called ICOMP for model selection for general multivariate linear and nonlinear structural models was proposed. The author claimed that ICOMP takes the spirit of AIC, but it is a different procedure than AIC in the sense that ICOMP is based on the entropic characterization of the measure of complexity of a model and that such a formulation provides a criterion of goodness of fit of a model. For a multivariate normal linear and nonlinear structural models, ICOMP is defined by

$$\begin{aligned} \mathrm{ICOMP}(\boldsymbol{\theta}) &= -2\log(\text{maximum likelihood}) \\ &+ \left\{ k\log\left(\frac{\mathrm{tr}\hat{\Sigma}_k}{k}\right) - \log|\hat{R}_k| \right\} + \left\{ n\log\left(\frac{\mathrm{tr}\hat{R}}{n}\right) - \log|\hat{\Sigma}_k| \right\} \qquad (4.5) \end{aligned}$$

where $\hat{\Sigma}_k$ is the estimated covariance matrix and $\hat{R}$ is the model residuals. A model with minimum ICOMP is chosen to be the best model among all candidate models. The author argued that minimization of ICOMP provides a trade-off between the accuracy of the estimated parameters, as measured by the interactions among the parameters, and the independent normal errors. The author asserted that ICOMP leads to a parsimonious description of the fitted model. As commented in Burnham and Anderson (1998), neither CAICF or ICOMP are invariant to 1-to-1 transformations of the parameters, and this feature would seem to limit their application. From (4.5), it can be seen:

1. The second term generally has the order of $\log n$. When the eigenvalues of the estimated covariance matrix are asymptotically proportionally identical with certain rate, it may tend to zero, which may cause serious overfitting. This happens in the balanced ANOVA.

2. The third term vanishes in case errors are homogeneous or may have order of n (e.g., $\mathrm{Var}(\varepsilon_{2i} = 2\mathrm{Var}(\varepsilon_{2i-1}))$, which may cause serious underfitting.

In Wei (1992) a model selection criterion FIC was proposed for linear regression based on Fisher information. Consider the linear model (2.2). Assume that $\varepsilon_i \sim N(0, \sigma^2)$ and that $\boldsymbol{x}_i$ is $\sigma(\varepsilon_1, \ldots, \varepsilon_{i-1})$ measurable. Then the conditional Fisher information matrix for $\boldsymbol{\beta}(\kappa)$ is $\sigma^{-2}\sum_{i=1}^n \boldsymbol{x}_i(\kappa)\boldsymbol{x}_i'(\kappa)$. The quantity $|\sum_{i=1}^n \boldsymbol{x}_i(\kappa)\boldsymbol{x}_i'(\kappa)|$ can be interpreted as the amount of information about $\boldsymbol{\beta}(\kappa)$. Denote

$$\mathrm{FIC}(\kappa) = n\hat{\sigma}_\kappa^2 + \hat{\sigma}^2 \log|\sum_{i=1}^n \boldsymbol{x}_i(\kappa)\boldsymbol{x}_i'(\kappa)|.$$

A model is selected for which FIC is the smallest. In FIC, the redundant information by introducing a spurious variable is used to represent its penalty. Compared with PLS (predictive least squares), the author argued that FIC is permutation invariant, easy to compute, no initialization problem is involved and is strongly consistent and, further, FIC seems to have better small sample performance.

A widely used procedure for inference about parameters of interest in the presence of nuisance parameters is based on the profile log-likelihood function. However, this procedure may give inconsistent or inefficient estimates. Since the profile log-likelihood itself is not a log-likelihood, Shi and Tsai (1998a) argued that one must consider conditional log-likelihood, modified profile log-likelihood, or marginal log-likelihood as alternative approaches. For simplicity, they proposed a model selection criterion based on marginal log-likelihood for linear regression. They first obtained an unbiased estimator of the expected Kullback-Leibler information of the marginal log-likelihood function of the fitted model and then derived the modified information criterion (MIC) based on it. Under some conditions, MIC is shown to be strongly consistent. Based on their simulation results, they indicated that MIC not only outperforms the efficient criteria AIC, AICc, FPE and C_p, but is superior (or comparable) to the consistent criteria BIC and FIC in both small and large sample sizes.

Consider the model (2.2). First let the set of all candidate models consist of $\{M_j\}$ where $M_j = M_{1:j}$. Denote $S_j = S_{1:j}$. Define

(1) $G_n^{(1)}(k) = S_k + kC_nS_p/(n-p), \quad k = 1, \ldots, p;$

(2) $G_n^{(2)}(k) = S_k + kC_n, \quad k = 1, \ldots, p;$

(3) $G_n^{(3)}(k) = n \log S_k + kC_n, \quad k = 1, \ldots, p.$

Rao and Wu (1989) and Bai, Rao, and Wu (1999) proposed the following selection rules based on $G_n^{(\ell)}$'s; the selected model is defined by $M_{\hat{k}_n}$ for which

$$G_n^{(\ell)}(\hat{k}_n) = \min_{1 \le k \le p} G_n^{(\ell)}(k). \tag{4.6}$$

The selection procedures defined above are called the general information criteria for linear regression (GIC-LR). Note that the equivalent criteria can be found in literature (e.g., Nishii 1984, Pötscher 1989, Shao 1997). Assume that C_n is a function of n satisfying the conditions

$$\frac{C_n}{n} \to 0, \quad \frac{C_n}{\log\log n} \to \infty. \tag{4.7}$$

It was shown in Rao and Wu (1989) and Bai, Rao, and Wu (1999) among others that under mild conditions, these criteria are strongly consistent.

Now consider the general situation where the set of all candidate models consist of all possible 2^p submodels. For each $1 \le i \le p$, denote

$$\boldsymbol{\beta}_{-i} = (\beta_1, \ldots, \beta_{i-1}, \beta_{i+1}, \ldots, \beta_p)'$$

and

$$X_{n,-i} = X_n(\{1 : i-1\} \bigcup \{i+1 : p\}).$$

Consider the model

$$y_n = X_{n,-i}\boldsymbol{\beta}_{-i} + \varepsilon_n. \tag{4.8}$$

Write the corresponding usual residual sum of squares by S_{-i}. In order to determine if the ith element of $\boldsymbol{\beta}$ is zero, we need only compare two models, one is the full model (2.1) and the other is the reduced model (4.8). Define, for $1 \le i \le p$,

(1) $G_n^{(1)}(-i) = S_{-i} - S_p - C_n S_p/(n-p);$

(2) $G_n^{(2)}(-i) = S_{-i} - S_p - C_n;$

(3) $G_n^{(3)}(-i) = n(\log S_k - \log S_p) - C_n.$

Then, choose the model as

$$\beta_i = 0 \;\text{ if }\; G_n^{(\ell)}(-i) \le 0 \;\text{ and }\; \beta_i \ne 0 \;\text{ if }\; G_n^{(\ell)}(-i) > 0$$
$$i = 1, \ldots, p.$$

Assume that C_n is chosen in accordance with the condition (4.7). It was shown in Rao and Wu (1989) and Bai, Rao, and Wu (1999) that under mild conditions, these criteria are strongly consistent. The advantage of such selection procedures is that it needs only the computation of $p+1$ residual sums of squares instead of 2^p residual sums of squares.

When p is large, this method can also be applied to the criteria discussed previously. Its disadvantage is that the underfitting may not be at nominal level. As an alternative method (Zheng and Loh 1995), the explanatory variables may be ordered using the t statistic. Using this method may need extra work in computing the t statistics and sorting. If all important predictors are significantly non-negligible, every method can give good detection of the smallest true model. The problem arises when some predictors are very critical. There are cases where a predictor is shown to be important when just this predictor is examined, and it becomes unimportant when the effect of some other predictors is eliminated (of course it is not for the limiting case). Therefore, a variable which is in fact more important than another may be excluded by wrong ordering. Both methods proposed by Zheng and Loh (1995) and Bai, Rao, and Wu (1999) respectively may face this problem. It needs deeper investigation to determine which performs better.

It is worth while mentioning that the assumption of normality and the assumption that the errors are identically distributed are not necessary for the criteria to be strongly consistent in this example. See Bai, Rao, and Wu (1999) for details.

In the problem of selecting a linear model to approximate the true unknown regression model, some necessary and/or sufficient conditions were established by Shao (1997) for the asymptotic validity of various model selection procedures such as AIC, C_p, FPE_α, BIC, SIC, GIC-LR, etc.. It was found that these selection procedures can be classified into three classes according to their asymptotic behavior. Under some fairly weak conditions, Shao (1997) showed that the selection procedures in one class are asymptotically valid if there exist fixed-dimension correct models; the selection procedures in another class are asymptotically valid if no fixed-dimension correct model exists. The procedures in the third class are compromises of the procedures in the first two classes.

Since the general information criterion for linear regression is consistent, it is of interest to know its convergence rate. In Shao (1998), some convergence rates for the error probabilities of the criterion, in terms of C_n and the order of the design matrix, were established. The author argued that the rates obtained there are sharper than the existing ones in the literature (e.g., Zhang 1993b) when the distribution of the response variable is non-normal.

5 Cross-validation, bootstrap and related model selection methods

Cross-validation is a method for model selection in terms of the predictive ability of the models. Suppose that n data points are available. A model is to be selected from a class of models. First, hold one data point and use the rest of $n-1$ data points to fit a model. Then check the predictive ability of the model in terms of the withheld data point. Perform this procedure for all data points. Select the model with the best average predictive ability. This procedure is described as the LOO (leave one out) method. For details, see Stone (1974, 1977a,b), Geisser (1975), Efron (1983, 1986), Picard and Cook (1984), Herzberg and Tsukanov (1986) and Li (1987) among others. Note that Allen's PRESS is equivalent to cross-validation (Allen 1974).

The cross-validation can be generalized as follows. Instead of choosing one data point for assessing the predictive ability, k data points are reserved for that purpose. The rest of $n_k = n - k$ data points are used to fit the model. There are ${}_nC_k$ different ways to partition the data set. The generalized cross-validation selects the model with the best average predictive ability among different ways of data splitting. It is easy to see that the computational complexity of this method increases as k increases.

Herzberg and Tsukanov (1986) did some simulation comparisons between the cross validation procedures with $k \equiv 1$ and $k \equiv 2$. They found that the leave-two-out cross-validation is sometimes better than the LOO cross-validation, although the two procedures are asymptotically equivalent in theory. When the number of predictors in any regression model under consideration is fixed, this type of cross-validation is not consistent and it can be shown that it is equivalent to Akaike information criterion. See also Geisser (1975), Burman (1989), and Zhang (1993a). It will not be the case if k is chosen to depend on n. For emphasizing this dependence, write k as $k(n)$. Shao (1993) showed that $k(n)/n \to 1$ as $n \to \infty$ is needed to guarantee that the cross-validation is asymptotically correct. When $k(n)$ is large, the amount of computation required to use the cross-validation seems impractical. Shao (1993) suggested several approaches (e.g., the balanced incomplete CV($k(n)$) and Monte Carlo CV($k(n)$)) to remedy it, and examined their performances in a simulation study. Wu, Tam, Li, and Zen (1999) have used Shao's method to estimate the number of super imposed exponential signals.

When the number of predictors in the regression model under consideration increases as n increases, the story is different. Li (1987) showed that under some conditions, the LOO cross-validation is consistent and is asymptotically optimal in some sense.

The bootstrap is a data resampling method for estimating or approximating the sampling distribution of a statistic and its characteristics. The general application of bootstrap method to model selection can be found in Linhart and Zucchini (1986). Recent developments in this area include bootstraping the mean squared prediction error (Shao 1996 and McQuarrie and Tsai 1998) and constructing bootstrapped estimate for the Kullback-Leibler discrepancy (Shibata 1997).

Breiman (1992) introduced a data-driven model selection based on the little bootstrap. Consider the model (2.2). Define the prediction error as

$$\mathrm{E}\|\boldsymbol{Y}_0 - X_n(\kappa)'\boldsymbol{\beta}(\kappa)\|^2 = N\sigma^2 + \|\mathrm{E}(\boldsymbol{Y}) - X_n(\kappa)'\boldsymbol{\beta}(\kappa)\|^2,$$

where $\boldsymbol{Y}_0$ denotes the vector of n new observations. Write

$$\mathrm{ME} = \|\mathrm{E}(\boldsymbol{Y}) - X_n(\kappa)'\boldsymbol{\beta}(\kappa)\|^2,$$

which is the error in fitting the true model. The following procedure describes how to get the little bootstrap ME estimate:

1. 1. Fit the full model (2.1), getting S_p and $\hat{\sigma}^2$. Do the variable selection, getting the sequence of subsets of indices $\kappa_0, \kappa_1, \ldots, \kappa_p$, and the values of S_{κ_j}, where $\kappa_0 = \emptyset$.

2. Generate $\{\varepsilon_{1i}\}$, $i = 1, \ldots, n$, as i.i.d. $N(0, t^2\hat{\sigma}^2)$ and form the new y data

$$\tilde{\boldsymbol{y}} = \boldsymbol{y} + \boldsymbol{\varepsilon}_1,$$

   where $\boldsymbol{\varepsilon}_1 = (\varepsilon_{11}, \ldots, \varepsilon_{1n})'$ and $t > 0$.

3. Using the data $(\tilde{y}_i, \boldsymbol{x}_i)$, find the subset sequence $\{\tilde{\kappa}_j\}$ following the same procedure as in Step 1, and compute the predictors $\hat{\boldsymbol{y}}$ and $\hat{\boldsymbol{y}}_{\tilde{\kappa}_j}$ based on the full model and the model $M_{\tilde{\kappa}_j}$.

4. Calculate

$$\frac{1}{t^2}\boldsymbol{\varepsilon}_1'(\hat{\boldsymbol{y}} - \hat{\boldsymbol{y}}_{\tilde{\kappa}_j}).$$

5. Repeat Steps 2, 3, and 4 a number of times and average the quantities computed in Step 4. Denote the average by $B_t(j)$.

6. The little bootstrap estimate is

$$\widehat{\mathrm{ME}}(\kappa_j) = S_{\kappa_j} - S_p + p\hat{\sigma}^2 - 2B_t(j).$$

Brieman proposed to select $M_{\tilde{\kappa}_{j*}}$ if

$$M_{\tilde{\kappa}_{j*}} = \arg\min_{M_{\tilde{\kappa}_{j*}}} \widehat{\mathrm{ME}}(\kappa_j).$$

In his paper, he commented on the choice of t and compared his method with C_p and replicate data method by simulation. His simulation results indicate that his method is better than C_p and he argued that all selection methods not based on data reuse give highly biased results and poor subset selection.

Recently the covariance inflation criterion (CIC) was proposed in Tibshirani and Knight (1999a), which adjusts the training error by applying the model selection procedure to permuted versions of the data set, to measure the covariance between the predicted values and the responses. In doing so, this criterion captures the inherent variability associated with an adaptive procedure, such as best subset regression. The CIC can be applied in the prediction problems such as regression and classification, and to nonlinear, adaptive prediction rules.

Consider the models (2.2) with squared error and fixed $\boldsymbol{x}_1, \ldots, \boldsymbol{x}_n$. Denote $\boldsymbol{z}_i = (y_i, \boldsymbol{x}_i)$, $i = 1, \ldots, n$, and $\boldsymbol{z}' = (\boldsymbol{z}_1, \ldots, \boldsymbol{z}_n)$. Let $\mu_i = \mathrm{E}(Y_i|\boldsymbol{x}_i)$, $\sigma^2 = \mathrm{Var}(Y_i|\boldsymbol{x}_i)$, and the conditional distribution of $Y_i|\boldsymbol{x}_i$ be F_{μ_i}. On the basis of $\boldsymbol{z}$ a model M_κ is chosen and the corresponding prediction rule $\eta_{\boldsymbol{z}}(\boldsymbol{x}, M_\kappa)$, indexed by a tuning parameter κ, is formulated.

The true error of the rule $\eta_{\boldsymbol{z}}(\boldsymbol{x}, M_\kappa)$ is

$$\mathrm{Err}(\kappa) = \frac{1}{n}\sum_{i=1}^{n} \mathrm{E}_{\mu_i}\{Y_i^* - \eta_{\boldsymbol{z}}(\boldsymbol{x}_i, M_\kappa)\}^2,$$

where $Y_i^* \sim F_{\mu_i}$ with the training set $\boldsymbol{z}$ fixed. This is sampling error and the training error (or apparent error) is

$$\overline{\mathrm{err}}(\kappa) = \frac{1}{n}\sum_{i=1}^{n} \mathrm{E}_{\mu_i}\{y_i - \eta_{\boldsymbol{z}}(\boldsymbol{x}_i, M_\kappa)\}^2.$$

Note that $\overline{\mathrm{err}}$ tends to be biased downwards as an estimate of Err because the training set $\boldsymbol{z}$ is used twice, both to construct the rule and to test it.

let $\hat{\sigma}^2$ be an estimate of the noise variance σ^2 and let

$$\hat{\sigma}_y^2 = \sum_{i=1}^{n}(y_i - \bar{y})/(n-1).$$

The covariance inflation criterion (CIC) is defined by

$$\mathrm{CIC}(\kappa) = \overline{\mathrm{err}}(\kappa) + \frac{2}{n}\frac{\hat{\sigma}^2}{\hat{\sigma}_y^2}\sum_{i=1}^{n} \mathrm{Cov}^0(Y_i^*, \eta_{\boldsymbol{z}^*}(\boldsymbol{x}_i, M_\kappa^*)) + \frac{2}{n}\hat{\sigma}^2.$$

A model is chosen to be the one which minimizes $\mathrm{CIC}(\kappa)$. The notation Cov^0 indicates covariance under the permutation distribution of $\boldsymbol{x}$ and y: $\boldsymbol{x}_i^* = (y_i^*, \boldsymbol{x}_i)$ with $y_1^*, \ldots, y_n^*$ a sample drawn without replacement from $y_1, \ldots, y_n$ and the $\boldsymbol{x}_i$ fixed. Here M_κ^* is the model, given a tuning parameter κ, chosen from the permuted data. Tibshirani and Knight (1999a) argued that the idea behind this definition is that, the harder the data are fitted, the more $\eta_{\boldsymbol{z}}(\boldsymbol{x}_i, M_\kappa)$ affects its own prediction, and hence the greater the optimism in $\overline{\mathrm{err}}(\kappa)$. Since in practice it is hard to compute all the permutations, the authors suggested taking a sample of them. Note that instead of sampling the responses without replacement, they can be sampled with replacement, and given an independent bootstrap distribution for the predictors and responses. See Tibshirani and Knight (1999a) for details. The authors commented that even if the little bootstrap procedure looks similar to the CIC in the context of linear regression, the uncertainty in the choice of t makes it difficult to generalize the little bootstrap to other settings. In contrast, the CIC can be defined for general prediction models for regression and classification, which was presented in Tibshirani and Knight (1999a). The CIC was also compared with AIC, BIC and RIC (the risk inflation criterion) in the paper.

Brieman (1996) showed how one can use the bootstrap for the more primary purpose of producing a better estimator. Breiman's bagging procedure applies a given estimator $\hat{\boldsymbol{\theta}}$ to each of B bootstrap samples, and then averages the B values to produce a new estimator $\tilde{\boldsymbol{\theta}}$. In a number of experiments involving trees, subset selection and ridge regression, Breiman showed that the bagged estimate $\tilde{\boldsymbol{\theta}}$ often has smaller mean squared error than the original $\hat{\boldsymbol{\theta}}$. The largest gains occurred for unstable estimators $\hat{\boldsymbol{\theta}}$, like subset selection and trees, for which small changes in the data can produce large changes in the estimate. The improvement in mean squared error is mostly due to a reduction in variance. As commented in Tibshirani and Knight (1999b), unfortunately the averaging process that produces the bagged estimate $\tilde{\boldsymbol{\theta}}$ also destroys any simple structure that is present in the original estimate $\hat{\boldsymbol{\theta}}$. A different use of the bootstrap was proposed in Tibshirani and Knight (1999b). They used bootstrap samples to provide candidate models for the model search. They argued that this has the advantage that it preserves the structure of the estimator while still inducing stability. Let $\boldsymbol{z} = (z_1, \ldots, z_n)$ be a training sample i.i.d. from a distribution F. Suppose that there is a model for the data that depends on a set of parameters $\boldsymbol{\theta}$. From the training sample, it is assumed that $\boldsymbol{\theta}$ is to be estimated by minimizing a target criterion

$$\hat{\boldsymbol{\theta}} = \arg\min_{\boldsymbol{\theta}} R(\boldsymbol{z}, \boldsymbol{\theta}).$$

Suppose also that there is a (possibly different) working criterion R_0 for which minimization is convenient. Tibshirani and Knight (1999b) proposed to estimate $\boldsymbol{\theta}$ by drawing bootstrap samples $\boldsymbol{z}^1, \ldots, \boldsymbol{z}^B$, estimating $\hat{\boldsymbol{\theta}}$ via R_0 from each sample

$$\hat{\boldsymbol{\theta}}^b = \arg\min_{\theta} R_0(\boldsymbol{z}^b, \boldsymbol{\theta}),$$

and then choosing $\hat{\boldsymbol{\theta}}$ as the value among the $\hat{\boldsymbol{\theta}}^b$ producing the smallest value of $R(\boldsymbol{z}, \theta)$:

$$\hat{\boldsymbol{\theta}}^B = \hat{\boldsymbol{\theta}}^{*b} \quad \text{where } \hat{\boldsymbol{\theta}}^{*b} = \arg\min_{b} R(\boldsymbol{z}, \hat{\boldsymbol{\theta}}^b).$$

As a convention, the original sample $\boldsymbol{z}$ is included among the B bootstrap samples. This procedure is called "Bumping" for Bootstrap Umbrella of Model Parameters. The value

$\hat{\boldsymbol{\theta}}^B$ is the bumping estimate of $\boldsymbol{\theta}$. The authors argued that the bumping provides a convenient method for finding better local minima, for resistant fitting, and for optimization under constraints.

6 Baysian approach to model selection

For the model (2.1), assume that $\boldsymbol{Y}|\boldsymbol{\beta},\sigma \sim N(X\boldsymbol{\beta},\sigma^2 I)$. Let $\boldsymbol{\tau} = (\tau_1,\ldots,\tau_p)'$ where $\tau_i = 0$ or 1 if β_i is small or large, respectively. The size of the $\boldsymbol{\tau}$th subset is denoted as $q_{\boldsymbol{\tau}} \equiv \boldsymbol{\tau}'\mathbf{1}$. Since the appropriate value of $\boldsymbol{\tau}$ is unknown, the uncertainty underlying variable selection is modelled by a hierarchical mixture prior $\pi(\boldsymbol{\beta},\sigma,\boldsymbol{\tau}) = \pi(\boldsymbol{\beta}|\sigma,\boldsymbol{\tau})\pi(\sigma|\boldsymbol{\tau})\pi(\boldsymbol{\tau})$. For this hierarchical setup, the marginal posterior distribution $\pi(\boldsymbol{\tau}|Y)$ contains the relevant information for variable selection. Based on the data Y, the posterior $\pi(\boldsymbol{\tau}|Y)$ updates the prior probabilities on each of the 2^p possible values of $\boldsymbol{\tau}$. Identifying each $\boldsymbol{\tau}$ with a submodel via $(\tau_i = 1) \Leftrightarrow (X(i)$ is included), those $\boldsymbol{\tau}$ with higher posterior probability $\pi(\boldsymbol{\tau}|Y)$ identify the more "promising" submodels, that is those supported most by the data and the prior distribution.

For identifying "promising" subsets of predictors for the model (2.1), a Bayesian procedure, called stochastic search variable selection (SSVS), was proposed in George and McCulloch (1993), which specifies a hierarchical Bayes mixture prior which uses the data to assign larger posterior probability to the more promising models. To avoid the overwhelming burden of calculating the posterior probabilities of all 2^p models, SSVS uses the Gibbs sampler to simulate a sample from the posterior distribution. The Gibbs sampler is effectively used to search for promising models rather than compute the entire posterior. The key to the potential of SSVS is the fast and efficient simulation of the Gibbs sampler.

George and McCulloch (1997) described a variety of approaches to Bayesian variable selection which includes SSVS as a special case. These approaches all use hierarchical mixture priors to describe the uncertainty present in variable selection problems. In the paper, hyperparameter settings which base selection on practical significance, and the implications of using mixtures with point priors were discussed. The authors showed that conjugate versions of these priors yield expressions for the posterior which can sometimes be sequentially computed using efficient updating schemes. According to the paper, when p is moderate (less than about 25), performing such sequential updating in a Gray Code order yields a feasible approach for exhaustive evaluation of all 2^p posterior probabilities, and for larger values of p, Markov chain Monte Carlo (MCMC) methods, such as the Gibbs sampler or the Metropolis-Hastings algorithms, can exploit such updating schemes to rapidly search for high probability models. The authors observed that estimation of normalization constants would provide improved posterior estimates of individual model probabilities and the total visited probability. In their paper, nonconjugate and conjugate MCMC implementations are compared on three simulated sample problems. They also illustrated the application of Bayesian variable selection to a real problem involving $p = 200$ potential regressors.

As discussed in George and McCulloch (1997), a variety of approaches for Bayesian model selection can be put into the following categories: 1. Prior specification and 2. Posterior computation. The prior specification corresponding to the removal of a predictor can be obtained by either a continuous distribution on β_i with high concentration at 0, or assigning an atom of probability to the event $\beta_i = 0$. Another distinguishing characteristic of prior specification is the difference between nonconjugate and conjugate forms for the coefficient priors. Nonconjugate forms offer the advantage of precise specification of a nonzero threshold of practical significance, and appear to allow for more efficient MCMC exploration with approximately uncorrelated predictors. Conjugate forms offer the advantage of analytical simplification which allows for exhaustive posterior evaluation in moderately sized problems (p less than about 25). In problems with large p where evaluation of posterior probabilities is not feasible, conjugate forms allow for exact calculation of relative posterior probabilities and estimates of total visited probability by MCMC posterior exploration. Furthermore, conjugate forms appear to allow for more efficient MCMC exploration with more correlated designs. For the purpose of posterior exploration, a large variety of MCMC algorithms can be constructed based on the Gibbs sampler and Metropolis-Hastings algorithms.

With prediction as the goal, Geisser (1993) considered it more appropriate to average predictions over the posterior distribution rather than using predictions from any single model. The potential of prediction averaging in the context of variable selection uncertainty has been nicely illustrated by Clyde, DeSimone, and Parmigiani (1996) and Raftery, Madigan, and Hoeting (1997) among others. In practice, there exists a situation where a single model is needed for prediction (e.g., Wakefield and Bennett 1996).

In Laud and Ibrahim (1995), a predictive Baysian viewpoint was advocated to avoid the specification of prior probabilities for the candidate models and the detailed interpretation of the parameters in each model. Consider probability models for the observable $\boldsymbol{y}$ conditioned on each model M_γ with the associated parameter vector $\boldsymbol{\theta}^{(M_\gamma)}$:

$$p(\boldsymbol{y}|M_\gamma, \boldsymbol{\theta}^{(M_\gamma)}), \quad \boldsymbol{\theta}^{(M_\gamma)} \in \boldsymbol{\Theta}^{(M_\gamma)}, \ \gamma \in \Gamma,$$

where $\boldsymbol{\Theta}^{(M_\gamma)}$ is the parameter space for the model M_γ and Γ is the index set. Suppose that a prior $\pi(\boldsymbol{\theta}^{(M_\gamma)}|M_\gamma)$ has been specified for each $\boldsymbol{\theta}^{(M_\gamma)}$, $\gamma \in \Gamma$. The posterior for $\boldsymbol{\theta}^{(M_\gamma)}$ under each model M_γ, given data $\boldsymbol{Y} = \boldsymbol{y}$, is given by $\pi(\boldsymbol{\theta}^{(M_\gamma)}|\boldsymbol{y},\ M_\gamma)$. Now envision replicating the entire experiment and denote by $\boldsymbol{Z}$ the vector of responses that might result. The predictive density for $\boldsymbol{Z}$ under model M_γ is

$$p(\boldsymbol{z}|M_\gamma, \boldsymbol{y}) = \int p(\boldsymbol{z}|M_\gamma, \boldsymbol{\theta}^{(M_\gamma)})\pi(\boldsymbol{\theta}^{(M_\gamma)}|\boldsymbol{y},\ M_\gamma)\, d\boldsymbol{\theta}^{(M_\gamma)}.$$

This density was called the predictive density of a replicate experiment (PDRE) in Laud and Ibrahim (1995). The replicative experiment is an imaginary device that puts the predictive density to inferential use, adapting the philosophy advocated in Geisser (1971). The imagined replication makes $\boldsymbol{y}$ and $\boldsymbol{Z}$ comparable, in fact exchangeable a priori. Moreover, the parameters in the model play a minimal role under replication. It seems clear that good models, among those under consideration, should make predictions close to what has been observed for an identical experiment. With this motivation, Laud and Ibrahim (1995) proposed three criteria. Consider

$$L_{M_\gamma} = \sqrt{\mathrm{E}\{(\boldsymbol{Z}-\boldsymbol{y})'(\boldsymbol{Z}-\boldsymbol{y})\}},$$

where the expectation is taken with respect to the PDRE. Good models will have small values of L_{M_γ}, which results in the first criterion. Based on PDRE itself, the second criterion is formulated considering that small values of $(\mathrm{PDRE})^{-1/n}$ indicate good models. The third criterion is based on the Kullback-Leibler distance between two predictive densities. Using these criteria, they implemented their methodology for three common problems arising in normal linear models: variable subset selection, selection of a transformation of predictor variables and estimation of a parametric variance function.

Suppose that we are considering two models, M_1 and M_2. Let $p(\boldsymbol{y}|M_i)$ and $p(M_i)$ be respectively the distribution of the data $\boldsymbol{Y}$ and the prior probability of the model M_i, $i = 1, 2$. The posterior probabilities of M_i, $i = 1, 2$, are given by

$$p(M_i|\boldsymbol{y}) = p(\boldsymbol{y}|M_i)p(M_i)/p(\boldsymbol{y}). \tag{6.1}$$

The posterior odds in favor of model M_1 over alternative M_2 are

$$\frac{p(M_1|\boldsymbol{y})}{p(M_2|\boldsymbol{y})} = \left(\frac{p(\boldsymbol{y}|M_1)}{p(\boldsymbol{y}|M_2)}\right)\left(\frac{p(M_1)}{p(M_2)}\right). \tag{6.2}$$

Let $\pi_i(\boldsymbol{\theta}_i)$ be the prior distributions of the d_i-dimensional parameter vector $\boldsymbol{\theta}_i$ under the models M_i, $i = 1, 2$. Expressing $p(\boldsymbol{y}|M_i)$ as the average of the usual likelihood $p(\boldsymbol{y}|\boldsymbol{\theta}_i)$ over the parameter space, we have

$$p(\boldsymbol{y}|M_i) = \int p(\boldsymbol{y}|\boldsymbol{\theta}_i)\pi_i(\boldsymbol{\theta}_i)\, d\boldsymbol{\theta}_i,$$

which, together with (6.1), implies that

$$p(M_i|\boldsymbol{y}) = \int p(\boldsymbol{y}|\boldsymbol{\theta}_i)\pi_i(\boldsymbol{\theta}_i)\, d\boldsymbol{\theta}_i p(M_i)/p(\boldsymbol{y}).$$

Hence, (6.2) can be expressed as

$$\frac{p(M_1|\boldsymbol{y})}{p(M_2|\boldsymbol{y})} = \left(\frac{\int p(\boldsymbol{y}|\boldsymbol{\theta}_1)\pi_1(\boldsymbol{\theta}_1)\, d\boldsymbol{\theta}_1}{\int p(\boldsymbol{y}|\boldsymbol{\theta}_2)\pi_2(\boldsymbol{\theta}_2)\, d\boldsymbol{\theta}_2} \right) \left(\frac{p(M_1)}{p(M_2)} \right).$$

Since it is often that the prior odds $p(M_1)/p(M_2)$ is 1, the ratio

$$B_{12} = \frac{\int p(\boldsymbol{y}|\boldsymbol{\theta}_1)\pi_1(\boldsymbol{\theta}_1)\, d\boldsymbol{\theta}_1}{\int p(\boldsymbol{y}|\boldsymbol{\theta}_2)\pi_2(\boldsymbol{\theta}_2)\, d\boldsymbol{\theta}_2}$$

is defined as the Bayes factor. B_{12} can be viewed as the "weighted" likelihood ratio of M_1 to M_2 and hence can be interpreted solely in terms of comparative support of the data for the two models (see Kass and Raftery 1995). Computing B_{12} requires specification of $\pi_i(\boldsymbol{\theta}_i)$, $i = 1, 2$. Often in Bayesian analysis, one can use noninformative (or default) priors. Commonly used priors are the "uniform" prior, the Jeffreys prior, and the reference prior (see, e.g., Berger and Bernardo 1992). Since the Bayes factor compares model M_1 to alternative M_2, it has been used for model selection.

Akaike (1983) mentioned that model comparisons based on the AIC are asymptotically equivalent to those based on Bayes factors. As commented in Kass and Raftery (1995), this is true only if the precision of the prior is comparable to that of the likelihood, but not in the more usual situation where prior information is small relative to the information provided by the data. In the latter more usual situation, the SIC indicates that the model with the highest posterior probabilities is the one that minimizes SIC($\boldsymbol{\theta}$) given in (4.4).

It is unfortunate that the Bayes factors typically depend rather strongly on the prior information and several problems arise using the Bayes factor when prior information is weak (see, e.g., Berger and Pericchi 1996a,b, De Santis and Spezzaferri 1997, Kass and Raftery 1995). As commented in De Santis and Spezzaferri (1997), assigning a diffuse proper prior to the parameters $\boldsymbol{\theta}_i$ is critical because the flatter the prior is, the more penalized the corresponding model M_i is. Furthermore, when the distribution $\pi_i(\boldsymbol{\theta}_i)$ is improper and defined only up to arbitrary constants, the Bayes factor itself is a multiple of these arbitrary constants. In this situation of weak prior information, several authors have proposed the use of partial Bayes factors (see, among others, Berger and Pericchi 1996a,b, O'Hagan 1995). The idea is to use part of the data as a training sample to update the prior distributions of the models and the remainder of the data to compute the Bayes factor.

Assume $\boldsymbol{y} = (\boldsymbol{y}(m)', \boldsymbol{y}(n-m)')'$, where $\boldsymbol{y}(m)$ is a proper training sample of size m, that is, a subsample such that $0 < \int p(\boldsymbol{y}(m)|\boldsymbol{\theta}_i)\pi_i(\boldsymbol{\theta}_i)\, d\boldsymbol{\theta}_i < \infty$, $i = 1, 2$. The training sample is minimal if it does not contain subsets that are proper training samples. The partial Bayes factor for model M_1 against model M_2 is then defined as

$$B_{12}(m) = \frac{\int p(\boldsymbol{y}(n-m)|\boldsymbol{\theta}_1)\pi_1(\boldsymbol{\theta}_1|\boldsymbol{y}(m))\, d\boldsymbol{\theta}_1}{\int p(\boldsymbol{y}(n-m)|\boldsymbol{\theta}_2)\pi_2(\boldsymbol{\theta}_2|\boldsymbol{y}(m))\, d\boldsymbol{\theta}_2}, \tag{6.3}$$

where $\pi_i(\boldsymbol{\theta}_i|\boldsymbol{y}(m))$ is the posterior distribution of the parameter $\boldsymbol{\theta}_i$, $i = 1, 2$. O'Hagan (1995) showed that the partial Bayes factor is less sensitive to the priors than the Bayes factor. By (6.3), it can be seen that the partial Bayes factor does not depend on absolute values of prior distributions but on their relative values and on the other hand, the partial Bayes factor depends on the specific $\boldsymbol{y}(m)$ chosen. As described in De Santis and Spezzaferri (1997), in finite sample, when the size m of the training sample increases, the sensitivity of the partial Bayes factor to prior distributions decreases, but at the same time its discriminatory power decreases. To eliminate the dependence of the partial Bayes factor on $\boldsymbol{y}(m)$ and to increase its stability, Berger and Pericchi (1996a,b) suggested averaging the partial Bayes factor corresponding to all the possible training samples and obtained intrinsic Bayes factor. A simple alternative that avoids averaging is the fractional Bayes factor proposed in O'Hagan (1995), which is given as

$$B_{12}^{\mathrm{F}}(m) = B_{12}\left(\frac{\int [p(\boldsymbol{y}|\boldsymbol{\theta}_2)]^{m/n}\pi_2(\boldsymbol{\theta}_2)\, d\boldsymbol{\theta}_2}{\int [p(\boldsymbol{y}|\boldsymbol{\theta}_1)]^{m/n}\pi_1(\boldsymbol{\theta}_1)\, d\boldsymbol{\theta}_1}\right).$$

The fractional Bayes factor has an asymptotic motivation: if m and n are both large, the likelihood based on $\boldsymbol{y}(m)$ is approximated by the one based on $\boldsymbol{y}$, raised to the power m/n. The comparison of the intrinsic Bayes factor and the fractional Bayes factor can be found in Berger and Pericchi (1996a,b) and De Santis and Spezzaferri (1997) among others.

7 Robust model selection

For the methods discussed in previous sections, it can be seen that there is an involvement of the distribution information of the models, direct, indirect, or as vehicle. How can we cope with the case when there are departures from the distributional assumptions or there exist outliers in the data at hand? Robust model selection criteria have been proposed for this purpose.

According to Qian and Künsch (1999), the following three issues should be taken into consideration when proposing a robust criterion. First, it should take into account the possibility that observations of both response and predictors may contain gross errors.

Therefore, the criterion should be somewhat robust to small number of outliers or small change in all of the data. Second, the criterion should be consistent if a finite-dimensional true model exists. Third, the effect of the signal-to-noise ratio on the empirical performance of the criterion needs to be taken care of.

Back to the literature, Ronchetti (1985) proposed a robust version of AIC, called AICR. In AICR, the first term of (4.2) is replaced by the sum of the discrepancy functions computed at the M-estimate and the second term is replaced by a product of α and the number of free parameters, where the choice of α follows from the asymptotic equivalence of the AIC given in Stone (1977a). It is easy to see that robustness of the AICR depends on the robustness of the M-estimation. Hampel (1983) suggested a modified version of it. Härdle (1987) investigated the properties of a selection criterion for regression which is asymptotically equivalent to the AICR and showed that it is asymptotically efficient. A similar idea was used by Martin (1980) and Behrens (1991) for autoregressive models. For general parametric models AICR was discussed in Ronchetti (1997).

Antoch (1986, 1987) introduced an algorithm to perform variable selection, where α-trimmed least squares estimators for parameters are computed for all possible submodels and then compared to the same estimator obtained in the full model. The submodels which lead to estimates whcih are "indistinguishable" from that of the full model are considered acceptable.

In Hurvich and Tsai (1990) a small sample criterion for the selection of least absolute deviations regression models was developed. Their criterion provides an exactly unbiased estimator for the expected Kullback-Leibler information when the error distribution is double exponential and the model is not underfitted. The selection procedure performs better than both AIC and AICR with the L_1-norm discrepancy function.

Recently, Ronchetti and Staudte (1994) presented a robust version of Mallow's C_p, denoted RC_p, which can be used with a large variety of robust estimators for the parameters, including M-estimators, bounded influence estimators, and one-step M-estimators with a high breakdown starting point. RC_p chooses the models that fit the majority of the data by taking into account the presence of outliers and possible departures from the normality assumption on the error distribution. Later, Sommer and Staudte (1995) implemented RC_p for Mallows-type estimators so that leverage points as well as outliers can be downweighted. Some examples can be found in both the papers to support the applications of RC_p. Another robust version of C_p can be derived from the Wald test as proposed by Sommer and Huggins (1996). Consider a set of candidate models $\{M_{\boldsymbol{\theta}},\ \boldsymbol{\theta} \in \boldsymbol{\Theta}\}$, where the candidate models are indexed by their parameter vector $\boldsymbol{\theta}$ and $\boldsymbol{\Theta}$ is the parameter vector space in $\mathcal{R}^p$. Let $\hat{\boldsymbol{\theta}}$ be an M-estimator of $\boldsymbol{\theta}$. Define $W_\kappa = W_{\boldsymbol{\theta}(\kappa)} - p + 2k$, where $W_{\boldsymbol{\theta}(\kappa)} = n\hat{\boldsymbol{\theta}}(\kappa)'\widehat{\Sigma}(\hat{\boldsymbol{\theta}})^{-1}(\kappa)\hat{\boldsymbol{\theta}}(\kappa)$, the Wald test statistic for testing the null hypothesis that $\boldsymbol{\theta}(\kappa) = \mathbf{0}$, and κ denotes a subset of $\{1, \ldots, p\}$. Then a model is selected for which W_κ is the smallest. An advantage of such a model selection criterion is that it is easily adapted to other nonadditive error model structures.

Using Shao's cross-validation methods (Shao 1993) for choice of variables as a starting point, a robust algorithm for model selection was proposed in Ronchetti, Field, and Blanchard (1997). Since Shao's techniques are based on least squares, they are sensitive to outliers. The authors developed their robust procedure using the same ideas of cross-validation as Shao but use estimators that are of optimal bounded influence for prediction. They demonstrated the effectiveness of their robust procedure in providing protection against outliers both in a simulation study and in a real example and contrasted the results with those obtained by Shao's method, demonstrating a substantial improvement in choosing the correct model in the presence of outliers with little loss of efficiency at the normal model.

A robust version of the Schwartz criterion was proposed in Machado (1993). It was shown in his paper that under some assumptions, the smallest true model would be selected with probability approaching one as $n \to \infty$. Consider the model (2.2). In Burman and Nolad (1995), an M-estimation-based Akaike-type criterion was presented, where the penalty term is the product of the number of free parameters and the estimate of C_ρ. The C_ρ is given by

$$C_\rho = \frac{\sum_{i=1}^{n} \operatorname{var}[\rho_1(Y_i - \boldsymbol{x}_i'\boldsymbol{\beta}_0)]}{\sum_{i=1}^{n} R_2[\mathrm{E}(Y_i) - \boldsymbol{x}_i'\boldsymbol{\beta}_0]},$$

where ρ is a convex discrepancy function with a unique minimum at 0, and twice differentiable in expectation, ρ_1 is the derivative of ρ, R_2 is the second derivative of $\mathrm{E}[\rho(\varepsilon + t)]$, and $\boldsymbol{\beta}_0$ is the minimizer of $\sum_{i=1}^{n} \mathrm{E}\rho(Y_i - \boldsymbol{x}_i'\boldsymbol{\beta})$. Many examples were given in the paper for the applications of the criterion. Based on the newly developed theory of stochastic complexity (Qian and Künsch 1998) in linear regression, Qian and Künsch (1999) proposed a model selection procedure. This criterion is itself a model selection procedure based on M-estimation, where the discrepancy function for the M-estimation was the Huber's function defined as

$$\rho_c(t) = \begin{cases} \frac{1}{2}t^2, & |t| < c, \\ c|t| - \frac{1}{2}c^2, & |t| \geq c. \end{cases}$$

Under some conditions, the criterion was shown to be strongly consistent in the paper. An extensive simulation study, which compares their method with several robust model selection procedures, can also be found there. By approximating the expected discrepancy functions, a criterion was suggested in general in Shi and Tsai (1998b). For linear regression, the authors proposed three criteria based on M-estimation, called AICR*, AICcR* and AICcR, respectively, where AICcR* was obtained by following the similar approaches as in the derivations of AICc. Consider the model (2.2). Define

$$R_n(\kappa) = \sum_{i=1}^{n} \rho(y_i - \boldsymbol{x}_i'(\kappa)\hat{\boldsymbol{\beta}}_\kappa) + q(k)C_n, \tag{7.1}$$

where $\hat{\boldsymbol{\beta}}_\kappa$ is the M-estimator for the model M_κ corresponding to the discrepancy ρ, i.e.

$$\sum_{i=1}^{n} \rho(y_i - \boldsymbol{x}_i'(\kappa)\hat{\boldsymbol{\beta}}_\kappa) = \min_{\boldsymbol{\beta}(\kappa)} \sum_{i=1}^{n} \rho(y_i - \boldsymbol{x}_i'(\kappa)\boldsymbol{\beta}(\kappa)), \tag{7.2}$$

and $q(k)$ is a strictly increasing function of k and C_n is a function of n. It can be seen that in (7.1), the first term is a generalization of a minimum negative log likelihood function and the second term is the penalty on the use of models involving more parameters. Wu and Zen (1999) introduced the selection criterion called Criterion R based on $R_n(\kappa)$ under which M_{κ^*} is selected such that

$$R_n(\kappa^*) = \min_{\kappa} R_n(\kappa), \tag{7.3}$$

where C_n is such that $n^{-1}C_n \to 0$ and $(\log\log n)^{-1}C_n \to \infty$ as $n \to \infty$. This criterion includes many classical model selection criteria as special cases, and it is shown to be strongly consistent in their paper.

A general form of M-estimation was proposed in Bai and Wu (1997), where the discrepancy functions may be convex functions or differences of convex functions. The model covered all linear and nonlinear regression models, AR time series, EIVR models, etc. as its special cases. Hence it is worth while to construct a model selection criterion based on it.

It is also of interest to examine methods to assess influence in model selection problems. Léger and Altman (1993) examined the use of "leave-one-out" measure of changes in predicted values to assess influence of individual observations in model building. They suggested this measure considering multicollinearity among the independent variables. It seems to us that other measures can also be proposed and studied.

8 Order selection in time series

Consider the following autoregressive model of order p:

$$\boldsymbol{x}_t - \Phi_1 \boldsymbol{x}_{t-1} - \cdots - \Phi_p \boldsymbol{x}_{t-p} = \boldsymbol{z}_t, \tag{8.1}$$

where $\Phi_1, \ldots, \Phi_p$ are real $m \times m$ matrices, and $\{z_t\}$ is the error process.

The autoregressive models are the popular models for time series data. In previous sections, we have discussed many model selection methods and some of them, e.g., FPE, $(\text{FPE})^\lambda$, FPE_α and HQ, were originally derived for autoregressive models. In the analysis of autoregressive models, it is of interest to know the order of the optimal autoregressive model. Hence, a criterion is needed to fulfill this task. Generally speaking, most model selection methods, which work for linear regression, apply to the selection of the order of an autoregressive model. In the framework of stationary autoregressive models, Hannan and Quinn (1979) proved the strong consistency of the order estimators obtained by HQ for the univariate case, and Quinn (1980) obtained a similar result for the multivariate case. For the nonstationary autoregressive models with independently and identically distributed errors, weak consistency of the order estimators was established independently by Paulsen (1984) and Tsay (1984). Paulsen (1984) also discussed the multivariate case. The nonstationarity considered in both papers arises from the fact that the characteristic polynomial is allowed to have roots not only outside but also on the unit circle. Paulsen and Tjøstheim (1985) also discussed the case of nonstationarity where autoregressive scheme is stable but the error process is allowed to have a nonconstant variance. Pötscher (1989) gave strong consistency results for order estimation in nonstationary autoregressive models under the assumptions weaker than those employed in Paulsen (1984), Tsay (1984) and Paulsen and Tjøstheim (1985). He assumed the error process to be a martingale difference with respect to $\{\mathcal{F}_t\}$, where $\mathcal{F}_t$ is the σ-algebra generated by $\{x_s,\ s \le t\}$ in the model (8.1) with $m = 1$.

Using asymptotic efficiency as the criterion, Hurvich and Tsai (1989, 1993) studied the order estimation in the autoregression models without assuming the bound of the possible orders. They proposed a bias-corrected version of AIC (AICc), which works well in this case. The correction is of particular use when the sample size is small, or when the number of fitted parameters constitutes a large fraction of the sample size. The corrected method is asymptotically efficient if the true model is infinite dimensional. Furthermore, when the true model is of finite dimension, the method is found to provide better model order choices than any other asymptotically efficient method. Applications to nonstationary autoregressive and mixed autoregressive moving average models are also discussed there. Hurvich, Shumway, and Tsai (1990) suggested another order estimator which provides somewhat better model selection than the previous one if none of the candidate model dimensions exceeds one-half the sample size and provides a much better model selection than the previous one if some of the candidate models have large dimension and the sample size is small, when the autoregression models are estimated by maximum likelihood.

An order selection procedure based on the subsampling method can be found in Fukuchi (1999). He proposed to select a time series model empirically from a set of possibly nonnested and misspecified models by using estimated risk of prediction as a selection criterion. The author argued that compared with information theoretic criteria, his procedure was free of the problem of penalty selection. However the choice of subsample size will affect the performance of the procedure. According to Fukuchi (1999), the method of subset selection of stochastic regressors based on cross validation by Yao and Tong (1994) seems to be extendable to the model selection problem considered in the paper.

Suppose the goal is to make long range prediction, e.g., h-step forecasts, where we need to predict x_{n+h} from the time series $x_1, \ldots, x_n$. A simple method can be given based on the "plug-in" method (see Box and Jenkins 1970), in which an initial k-th order autoregression is chosen with k itself selected by an order selection criterion, and the multistep forecasts are obtained from this initial model fitted to the time series by repeatedly iterating the model and replacing the unknown future values by their own forecasts. Whittle (1963) observed that the plug-in method is optimal in a least-squares sense if the fitted model coincides with that generating the time series, or in a somewhat restricted sense, for prediction only one step ahead. Since all fitted models may be incorrect in practice, this observation suggests that for multistep prediction the plug-in method may not work well. A different approach may be desirable.

There has been much interest recently in the question of using lead-time (h) dependent estimates or model for multistep prediction of a time series. It is easy to see that such study involves solving an order selection problem, which is essential for forecasting. Earlier references advocating lead-time dependent model selection and/or parameter estimation for multistep forecasting include Findley (1983), Tiao and Xu (1993) and Lin and Granger (1994). In Bhansali (1996), a direct procedure involving a linear least-squares regression of x_{t+h} on $x_t, \ldots, x_{t-k+1}$ was used for estimating the prediction constants, with $k = \tilde{k}_h$, say, treated as a random variable whose value is selected anew for each h by an order selection criterion. He showed that the order selection by suitable h-step generalizations of the AIC and FPE criteria or their equivalents are asymptotically efficient for h-step prediction as the bound is attained in the limit if $\tilde{k}_h$ is selected by any of these criteria. The comparison between the plug-in method and the direct procedure can be found in Bhansali (1997).

In Hurvich and Tsai (1997), a version of the corrected AIC (AICc) was developed for the selection of an h-step-ahead linear predictor for a weakly stationary time series in discrete time. A motivation for this criterion was provided in terms of a generalized Kullback-Leibler distance which is minimized at the optimal h-step predictor, and which is equivalent to the ordinary Kullback-Leibler distance when $h = 1$. In their simulation study, it was found that if the sample size is small and the predictor coefficients are estimated by Burg's method (Burg 1978), AICc typically outperforms both the ordinary AIC and FPE for h-step prediction. Note that Chen, Davis, Brockwell, and Bai (1993) presented simulation results that for a finite order autoregressive process, the Burg estimator can perform very poorly in small samples if the model order used in the estimator greatly exceeds the true order: they found that the Yule-Walker estimator performs much better at these high orders. Hurvich and Tsai (1996) argued that the reason for them to use Burg's estimator is that if the AICc is used it is rare that a large model order is selected. They presented evidence to indicate that Burg estimation can produce much better selected predictors than Yule-Walker estimation.

Liu (1996) investigated the simultaneous multistep forecasts. First, a univariate autoregressive model is translated into a constrained multivariate regression model. Based on this transformation, it was shown that the model selection procedures derived from one-step ahead forecasts also keeps some optimality in the sense of multistep forecasts. The author obtained the multistep versions of BIC, FIC, and C_p.

Now consider the following autoregressive moving average (ARMA) model:

$$\boldsymbol{x}_t - \Phi_1 \boldsymbol{x}_{t-1} - \cdots - \Phi_p \boldsymbol{x}_{t-p} = \boldsymbol{z}_t + \Psi_1 \boldsymbol{z}_{t-1} + \cdots + \Psi_q \boldsymbol{z}_{t-q}, \tag{8.2}$$

where $\Phi_1, \ldots, \Phi_p$ and $\Psi_1, \ldots, \Psi_q$ are real $m \times m$ matrices, and $\{\boldsymbol{z}_t\}$ is the error process. The orders of this model are p and q.

The monograph by Choi (1992) gives a comprehensive survey and an extensive bibliography on the ARMA model identification. As pointed out by Choi, the FPE, AIC, BIC, SIC, HQ, MDL, and PLS and similar procedures can be used to select the orders of the ARMA models. Lai and Lee (1997) expanded the list and they extended Rissanen's accumulated prediction error criterion and Wei's FIC from linear to general stochastic regression models, which includes ARMA models as its special case. They showed that these criteria are strongly consistent under certain conditions.

Zhang and Wang (1994) proposed the order determination quantity (ODQ) as a new way to solve order estimation problems for the model (8.2) with $m = 1$. The ODQ is defined as

$$\text{ODQ}_n(p, q) = n\hat{\sigma}_n^2(p, q) - n\hat{\sigma}_n^2(p^*, q^*) - a_n$$

where $\hat{\sigma}^2(\cdot, \cdot)$ denotes an estimate of the common variance of the noise sequence; $0 < p \leq p^*$, $0 < q \leq q^*$; (p^*, q^*) is an upper bound of the unknown true order (p_0, q_0), which can be arbitrarily large but fixed and is supposed to be known *a priori*; n is the sample size; and $a_n > 0$ is a data-dependent constant such that $a_n/n \to 0$ and $a_n/(\log n)^\gamma \to \infty$ almost surely, where $\gamma = 1$ for pure autoregressive models and $\gamma \geq 1$ is a nonrandom constant to be specified in the future for general ARMA models. $(\hat{p}, \hat{q})$ is determined to satisfy

$$\mathrm{ODQ}_n(\hat{p}-1, \hat{q}) > 0, \ \ \mathrm{ODQ}_n(\hat{p}, \hat{q}-1) > 0 \text{ and } \mathrm{ODQ}_n(\hat{p}, \hat{q}) < 0,$$

instead of by minimizing $\mathrm{ODQ}(p, q)$ over (p, q). The authors argued that theoretical analysis and simulation showed that the ODQ has higher identifiability for unknown true orders, provides clear separation points and requires less computational effort than the order estimation criteria such as AIC, BIC, HQ, PLS, etc. Under certain conditions, it was shown that ODQ is strongly consistent for unstable autoregressive processes.

Note that if an ARMA model is invertible, it can be approximated by an autoregressive model of order m for large m. If the set of candidate models consists of autoregressive models while the true model is an ARMA model, an efficient order selection criterion is recommended against a consistent order selection criterion. A simulation study for the case that the true model is a moving average model while the set of candidate models consists of autoregressive models can be found in McQuarrie and Tsai (1998).

The following nonlinear time series model was studied in Chen, McCulloch, and Tsay (1997):

$$\begin{aligned} x_t &= f(x_{t-1}, \cdots, x_{t-p}; a_{t-1}, \ldots, a_{t-q}; \boldsymbol{\beta}_f) + a_t, \\ a_t &= g_t z_t, \\ g_t &= g(x_{t-1}, \ldots, x_{t-u}; a_{t-1}, \ldots, a_{t-v}; g_{t-1}, \ldots, g_{t-w}; \boldsymbol{\beta}_g), \end{aligned} \tag{8.3}$$

where x_t is a univariate time series, $f(\cdot)$ and $g(\cdot)$ are two known functions with finite dimensional parameter vectors $\boldsymbol{\beta}_f$ and $\boldsymbol{\beta}_g$, respectively, p, q, u, v, and w are non-negative integers, and $\{z_i\}$ is a sequence of independent and identically distributed random variables with mean zero and variance one. The function $g(\cdot)$ is assumed to be positive; it governs the evolution of the volatility of the innovational series a_t. Examples of the model so defined were presented there. In that paper, they claimed that there was little discussion of model selection across different classes of nonlinear models and that much work on model selection in the literature focuses on nested models for which the traditional maximum likelihood ratio tests or Rao's score tests or information criterion functions apply. It is easy to see that for non-nested models, model discrimination becomes much more involved, especially when the competing models are nonlinear. Li (1993) adopted the idea of separate families of hypotheses of Cox (1962) and proposed a test statistic for discriminating between bilinear and threshold models. Chen, McCulloch, Tsay (1997) argued that Li's test was closely related to the method of selecting a model with smaller residual variance and was not applicable to other nonlinear models. They proposed an approach that is, as they asserted, widely applicable in univariate time series analysis for linear or nonlinear models, and can discriminate between non-nested nonlinear models.

Their approach is based on Gibbs sampling and in particular they treated starting values of the time series and the innovational series as parameters and considered the conditional likelihood function of a parameter given the others. The approach also requires some prior specification, which is the probability that an individual observation is generated by a specified model given that both observations adjacent in time are generated by the same model. The drawback is that it may take substantial computing time in some applications.

The order determination criteria can also be used to test the hypotheses of white noise model against autoregressive models (see Pukkila and Krishinaiah 1988a, b) for details). Based on this idea, a procedure for identifying ARMA models was proposed in Pukkila, Koreisha, and Kallinen (1990).

The selection of a model when the candidate models are some general stochastic models needs some investigation. It is certain that the model selection methods, e.g., AIC, BIC and HQ, can be adopted to this case, but their performances may not be satisfactory. Better selection procedures need to be explored.

9 Model selection in categorical data analysis

Data collected in the social sciences for measuring attitudes and opinions on various issues and demographic characteristics such as gender, race, and social class and in biomedical sciences to measure such factors as severity of an injury, degree of recovery from surgery, and stage of a disease are categorical. Categorical data also arises in other sciences.

For categorical data, the problems such as checking independence of attributes, selection of optimal explanatory variables, and selection of an optimal categorization are of special interest. Appropriate procedures to solve these problems by AIC can be found in Sakamoto (1991) among others. Sakamoto (1991) also proposed ABIC, an extension of AIC, for evaluating Baysian binary regression models.

Generalized linear models were introduced by Nelder and Wedderburn (1972). This family contains important models for categorical data such as logit and probit models for quantal responses, loglinear models and multinomial response models for counts, as well as linear regression and analysis of variance models for continuous response variables. A generalized linear model is specified by three components: a random component, a systematic component, and a link. For such a model, the deviance or Pearson's chi-square goodness-of-fit statistic may be used to measure the fit.

When a generalized linear model selection problem is about selecting optimal explanatory variables, it is not hard to see that the model selection methods discussed in Sections 2-4 can adjust themselves to serve generalized linear model selection problems (e.g., Agresti 1990, Bai, Krishnaiah, Sambamoorthi, and Zhao 1992, Christensen 1997, Hosmer, Jovanovic, and Lemeshow 1989, McCullagh and Nelder 1989, Pregibon 1979). For example, consider a set of loglinear models. In this case, stepwise procedures can be performed by starting with an initial model and then using rules for adding or deleting terms to arrive at a final model. Note that a model selection procedure may be improved or modified to adapt to the situation. As an example, the backward elimination is modified for controlling the experimentwise error rate (Aitkin 1978, 1979).

The generalized linear models usually do not include a dispersion parameter. McCullagh and Nelder (1989) suggested that it is often wise to assume that a dispersion parameter is present in the model unless the data or prior information indicate otherwise. Hurvich and Tsai (1995) generalized the AICc to an extended quasi-likelihood model, which includes the generalized linear model with a dispersion parameter as a special case.

Qian, Gabor, and Gupta (1996) considered the problem of selecting a model with the best predictive ability in a class of generalized linear models. A predictive least quasi-deviance criterion was proposed to measure the predictive ability of a model. This criterion is obtained by applying the idea of the predictive minimum description length principle and the theory of quasi-likelihood functions. The resulting predictive quasi-deviance function is an extension of the predictive stochastic complexity of the model. Under rather weak conditions the authors showed that the predictive least quasi-deviance method is consistent. Also, the authors showed that the selected model converges to the optimal model in expectation. The method was then modified for finite sample applications. Examples and simulation results were presented in the paper. There is still much work to be done in this direction.

Random effect models are useful for explanations of overdispersion, correlation and subject-specific inference. Hence generalized linear models with random effects are very desirable in practice. The choice of a model in such cases needs some study.

10 Model selection in nonparametric regression

In nonparametric regression, local polynomial, kernel and smoothing spline methods among others have been used to construct nonparametric estimates of smooth regression functions (see, e.g., Fan and Gijbels 1996 and Simonoff 1996). These estimators use a smoothing parameter to control the amount of smoothing performed on a given data set, where the parameter is chosen using a selection criterion. Many methods have been proposed for selecting the parameter.

As commented in Loader (1999), the methods for bandwidth selection for procedures such as kernel density estimation and local regression can be divided into two broad classes. One of the classes includes classical methods such as cross-validation, Mallows' C_p, AIC, etc. while the other class contains plug-in methods. In a plug-in method, the bias of an estimate $\hat{f}$ is written as a function of the unknown f, and usually approximated through Taylor series expansions, and a pilot estimate of f is then "plugged in" to derive an estimate of the bias and hence an estimate of the mean integrated squared error, and then the optimal bandwidth minimizes this estimated measure of fit. According to Loader (1999), substantial "evidence" has been collected to establish superior performance of modern plug-in methods in comparison to methods such as cross validation; this has ranged from detailed analysis of rates of convergence, to simulations, to superior performance on real datasets. Loader (1999) took a detailed look at some of this evidence, looking into the sources of differences. He argued that his findings challenge the claimed superiority of plug-in methods on several fronts. First, plug-in methods are heavily dependent on arbitrary specification of pilot bandwidths and fail when this specification is wrong. Second, the often-quoted variability and undersmoothing of cross validation simply reflects the uncertainty of bandwidth selection; plug-in methods reflect this uncertainty by oversmoothing and missing important features in complicated situations. Third, in terms of the asymptotic theory, plug-in methods use available curvature information in an inefficient manner, resulting in inefficient estimates. Asymptotically, the plug-in based estimates are beaten by their own pilot estimates.

Recently an interesting approach for selecting the smoothing parameters of nonparametric regression estimators was proposed in Hart and Yi (1998). Their method was based on one-sided cross-validation instead of ordinary cross-validation. The authors argued that by using one-sided cross-validation their method retains the nature of ordinary cross-validation and has much better statistical properties. It was shown that statistical properties of their method are comparable to those of a plug-in methods.

However, due to great variability and a tendency to under smooth, the "classical" criteria, such as generalized cross-validation and AIC are not ideal for selecting the smoothing parameter. Hurvich, Simonoff, and Tsai (1998) addressed these problems by proposing a nonparametric version of their AICc criterion. The authors argued that AICc, unlike plug-in methods, can be used to choose smoothing parameters for any linear smoother, including local quadratic and smoothing spline estimators, and AICc is competitive with plug-in methods for choosing smoothing parameters, and also performs well when a plug-in approach fails or is unavailable. Since in some applications neither parametric nor nonparametric estimation may give a reasonable fit to the data, Shi and Tsai (1999) and Simonoff and Tasi (1999) obtained AICc for semiparametric regression models.

Consider the selection of a hard wavelet threshold for recovery of a signal embedded in additive Gaussian white noise, a problem closely related to that of selecting a subset model in orthogonal normal linear regression. The existing approaches, such as AIC, Donoho and Johnstone's universal method (Donoho and Johnstone 1994), Nason's cross-validatory method (Nason 1996), etc. were presented in McQuarrie and Tsai (1998). A computationally efficient algorithm for implementing Nason's method can also be found there. Hurvich and Tsai (1998) proposed a data-dependent method of hard threshold selection based on a cross-validatory version of AICc, which, like universal thresholding and Nason's method, can be implemented in $O(n \log n)$ operations (where n is the sample size). The simulation results presented in McQuarrie and Tsai (1998) showed that both of the cross-validatory methods outperform universal thresholding.

As another approach for using wavelet decompositions to select a regression model, Antoniadis, Gijbels, and Grégoire (1997) suggested the determination of the number of nonzero coefficients in the vector of wavelet coefficients based on the idea of MDL. They pointed out that the class of functions tested by their criterion allowed them to approximate quite efficiently alternatives composed by complicated functions with inhomogeneous smoothness.

In the theory of linear models, the concept of degrees of freedom plays an important role. This concept is often used for measurement of model complexity, for obtaining an unbiased estimate of the error variance, and for comparison of different models. A concept of generalized degrees of freedom (GDF) that is applicable to complex modelling procedures was developed in Ye (1998). The definition is based on the sum of the sensitivities of each fitted value to perturbation in the corresponding observed value. The concept is nonasymptotic in nature and does not require analytic knowledge of the modelling procedures. The concept of GDF offers a unified framework under which complex and highly irregular modelling procedures can be analyzed in the same way as classical linear models. Besides, there is an interesting connection between the GDF and the half-normal plot.

Consider a response vector $\boldsymbol{Y} = (Y_1, \ldots, Y_n)'$,

$$\boldsymbol{Y} \sim N(\boldsymbol{\mu}, \sigma^2 I),$$

where σ^2 is assumed to be known and $\boldsymbol{\mu} = (\mu_1, \ldots, \mu_n)'$ is an $n \times 1$ mean vector. Define a modelling procedure $\mathcal{M}$ as a mapping from R^n to R^n that produces a set of fitted values $\hat{\boldsymbol{\mu}} \equiv \hat{\boldsymbol{\mu}}(\boldsymbol{Y})$ from $\boldsymbol{Y}$. Note that $\boldsymbol{\mu}$ often depends on some observed covariates, and so does the modelling procedure $\mathcal{M}$. The GDF for modelling $\mathcal{M}$ are given by $D(\mathcal{M}) = \sum_{i=1}^{n} h_i^{\mathcal{M}}(\boldsymbol{\mu})$, where

$$h_i^{\mathcal{M}}(\boldsymbol{\mu}) = \frac{\partial \mathrm{E}_{\boldsymbol{\mu}}[\hat{\mu}_i(\boldsymbol{Y})]}{\partial \mu_i}.$$

In classical linear models, the GDF reduces to the standard degrees of freedom. If $\mathcal{M}$ is a linear smoother, then $D(\mathcal{M})$ reduces to the trace of the smoothing matrix. Efron (1986) obtained the concept of "expected optimism" by using the average of the covariance form of Y_i and $\hat{\mu}_i$. Ye (1998) argued that the covariance form is less intuitive and more difficult to analyze and estimate. The author pointed out that $\mathrm{E}_{\boldsymbol{\mu}}[\hat{\mu}_i(\boldsymbol{Y})]$ is an infinitely differentiable function of $\hat{\boldsymbol{\mu}}$, which has three implications: 1. GDF is defined even when $\hat{\mu}_i$ is highly irregular or even discontinuous. 2. Because $h_i^{\mathcal{M}}(\boldsymbol{\mu})$ is also infinitely differentiable, it can be estimated with its value $h_i^{\mathcal{M}}(\boldsymbol{Y})$. 3. Because $\mathrm{E}_{\boldsymbol{\mu}}[\hat{\mu}_i(\boldsymbol{Y})]$ can be viewed as a smoothing of the fitted value $\hat{\mu}_i$, what is important is the global behavior of $\hat{\mu}_i$, not the local behavior. For estimating $D(\mathcal{M})$, an algorithm, which is based on Monte Carlo method, was provided in Ye (1998). The algorithm is given as follows:

1. Repeat $t = 1, \ldots, T$;
2. Generate $\boldsymbol{\Delta}_t = (\delta_{t1}, \ldots, \delta_{tn})$ for the density $\prod(1/\tau^n)\phi(\delta_{ti}/\tau)$;
3. Evaluate $\hat{\mu}_i(\boldsymbol{Y} + \boldsymbol{\Delta}_t)$ based on the modelling procedure $\mathcal{M}$;
4. Calculate $\hat{h}_i$ as the regression slope from

$$\hat{\mu}_i(\boldsymbol{Y} + \boldsymbol{\Delta}_t) = \alpha + \hat{h}_i \delta_{ti}, \quad t = 1, \ldots, T,$$

$D(\mathcal{M})$ is estimated by $\hat{D}(\mathcal{M}) = \sum_i \hat{h}_i$. The parameter T determines the number of perturbations. It was suggested that $T \geq n$ in Ye (1998). $\hat{D}(\mathcal{M})$ depends on a tuning parameter τ, which the author referred to as the perturbation size. According to Ye (1998), it can be shown that $\hat{D}(\mathcal{M}) \to D(\mathcal{M})$ if $\tau \to 0$. This approach is similar to the little bootstrap of Brieman (1992).

Based on Ye (1998), the concept of GDF allows complex modelling procedures to be analyzed in a way similar to the analysis of classical linear models, and is independent of the sample size constraint and the complexity of the modelling procedure. Let $\{\mathcal{M}_\gamma, \gamma \in \Gamma\}$ be a set of modelling procedures, where Γ is an index set. He proposed two model selection procedures:

1. Let

$$A_e(\mathcal{M}_\gamma) = \|\boldsymbol{Y} - \hat{\boldsymbol{\mu}}_\gamma\|^2 - n\sigma^2 + 2D(\mathcal{M})\sigma^2,$$

and minimize $A_e(\mathcal{M}_\gamma)$ with respect to $\mathcal{M}_\gamma$. This is called the extended AIC (EAIC).

2. Let

$$\text{YGCV}(\mathcal{M}_\gamma) = \|\boldsymbol{Y} - \hat{\boldsymbol{\mu}}_\gamma\|^2/(n - D(\mathcal{M}))^2,$$

and minimize YGCV($\mathcal{M}_\gamma$) with respect to $\mathcal{M}_\gamma$. This is the analog of GCV in Graven and Wahba (1979). It is referred to as YGCV. The advantage of this criterion is that it does not assume a known σ^2.

When σ^2 is unknown, an estimate of it may be obtained by

$$s^2 = \|\boldsymbol{Y} - \hat{\boldsymbol{\mu}}\|^2/(n - D(\mathcal{M})).$$

Based on Ye (1998), EAIC and YGCV can be used to compare the performance of a nonparametric regression to a linear model as a way of diagnosing the adequacy of the linear model, or of selecting the most suitable nonparametric regression procedures among various alternatives, such as classification and regression trees, projection pursuit regression, and artificial neural networks. The criteria can also be used for selection of variables in nonparametric regression settings (Bickel and Zhang 1992 and Zhang 1991) by treating variable selection as a special case of selecting modelling procedures.

11 Data-oriented penalty

Let a model selection criterion be the sum of two components, where the first component measures the model fit and the other component is used for penalizing the model complexity. Among them, many have a data-independent penalty fixed or as an unbounded function of the sample size. In the first case, a procedure tends to overfit and yields less prediction error and in the later case, a procedure is quite stable with the penalty falling in some interval varied with the sample size. The choice of the penalty will affect the performance of such a model selection criterion. A theoretical study on the effect of a penalty of a classical linear model selection criterion can be found in Shao (1997). Hence, it is needed to find a good data-oriented penalty so that a procedure with its use will perform well. Such efforts can be found in Rao and Wu (1989), Wei (1992) and Shao (1998) among others.

Recently, a promising model selection procedure has been proposed by J. S. Rao-Tibshirani by searching for an optimal penalty in terms of minimizing an objective function via methods such as the cross-validation (see J. S. Rao 1994). Bai, Rao, and Wu (1999) proposed a method of constructing a data-oriented penalty. For selecting a model from candidate models given in (2.2), consider the model selection criterion $G_n^{(1)}$ defined in (4.6). A data-oriented procedure to select the penalty C_n proposed in Bai, Rao, and Wu (1999) is as follows:

1. Compute any consistent estimate $\widetilde{\boldsymbol{\beta}}_n = (\widetilde{\beta}_{1n}, \ldots, \widetilde{\beta}_{pn})'$ of $\boldsymbol{\beta}$ and the mean squared error $\widetilde{\sigma}_p^2$. For instance, $\widetilde{\boldsymbol{\beta}}_n$ can be chosen to be the least square estimate of $\boldsymbol{\beta}$.

2. Compute $\hat{\varepsilon}_n = \boldsymbol{y}_n - X_n \widetilde{\boldsymbol{\beta}}_n$.

3. Let $\overline{\boldsymbol{\beta}}_n = (\overline{\beta}_{1n}, \ldots, \overline{\beta}_{pn})'$ be defined as follows:

$$\overline{\beta}_{in} = \begin{cases} \widetilde{\beta}_{in}, & \text{if } |\widetilde{\beta}_{in}| \geq \eta, \\ \eta \operatorname{sign}(\widetilde{\beta}_{in}), & \text{if } |\widetilde{\beta}_{in}| < \eta, \end{cases} \qquad \text{for } i = 1, \ldots, p,$$

   where the constant η is a suitable chosen threshold value.

4. Let

$$\boldsymbol{u}_n(h) = X_n(h)\overline{\boldsymbol{\beta}}_n(h) + \hat{\varepsilon}_n, \quad h = 1, \ldots, p.$$

   Compute

$$D_n(q,h) = \overline{S}_q(h) - \overline{S}_h(h), \quad q = 0, 1, \ldots, p$$

   where $\overline{S}_q(h) = (\boldsymbol{u}_n(h))'(I - P_q)\boldsymbol{u}_n(h)$. It can be shown that $\overline{S}_p(h) = S_p$ if $\overline{\boldsymbol{\beta}}_n = \widetilde{\boldsymbol{\beta}}_n$ Define

$$\Delta_h = \min_{q<h} \left\{ \frac{D_n(q,h)}{(h-q)\widetilde{\sigma}_p^2} \right\}.$$

5. Define

$$C_n^{(R)} = \frac{\text{average of } \{\Delta_h,\ h = 1, \ldots, p\}}{1 + \sqrt{\lfloor 0.01 n \rfloor}},$$

   where $\lfloor b \rfloor$ denotes the integer part of b.

Then choose $C_n^{(R)}$ as the penalty C_n in $G_n^{(1)}$. It was shown in Bai et al. (1992) that $C_n^{(R)}$ asymptotically satisfies the conditions given in (4.7), while it works well for small to moderate sample sizes. Similar results are true for the other two criteria $G_n^{(\ell)}$, $\ell = 2, 3$, defined in (4.6). Based on the same idea, Wu (2001) proposed a data-oriented penalty for Criterion R (7.3).

Shao and J. S. Rao (2000) combined hypothesis testing and model selection together and proposed a particular choice of the penalty parameter C_n for the model selection criterion $G_n^{(1)}$ defined in (4.6). The authors demonstrated that the new model selection procedure inherits good properties from both approaches, i.e., its overfitting and underfitting probabilities converge to 0 as $n \to \infty$, and when n is fixed, its overfitting probability is controlled to be approximately under a pre-assigned level of significance.

Consider the candidate models of form (2.2). Let $\mathcal{M}_c$ denote the set of true models in $\{M_\kappa\} = \mathcal{M}$ and $M_{\kappa 0}$ denote the optimal model, which is the model in $\mathcal{M}_c$ with the smallest dimension. If M_{κ^*} be the model selected by the model selection criterion $G_n^{(1)}$, $p_1 = P(M_{\kappa^*} \in \mathcal{M}_c)$ is the error probability of overfitting and $p_2 = P(M_{\kappa^*} \in \mathcal{M} - \mathcal{M}_c)$ is the error probability of underfitting, which depend on C_n. Let δ be a pre-assigned level of significance. Shao and J. S. Rao (2000) proposed the choice of C_n so that $p_1 \leq \min\{\delta, 1/\sqrt{n}\}$ (approximately) holds and otherwise as small as possible to minimize p_2. Simulation studies were presented in the paper. A real data analysis can also be found there.

There are still many interesting open problems. For example, how can one find a good data-oriented penalty if the number of available models is infinite? Another problem is how to find a good data-oriented penalty when the true model is not included in a given class of models. Certainly, we can still proceed as if the data were from one of the available models. Then what is the consequence of such a procedure? Since the data are also used to choose the penalty, what is its impact on the inferences performed on such selected model?

12 Statistical analysis after model selection

Once a model is selected, it is usual that statistical analysis is carried out based on the selected model and often the analysis will be done as if the selected model is the true model. It is easy to see that any statistical analysis based on the selected model is affected by the nature of the true model, candidate models and the model selection procedure.

It is already well known that the model selection procedure can severely affect the validity of standard regression procedures. Rencher and Pun (1980) demonstrated that a model selected by the best subset regression method tends to have an inflated value of R^2. Miller (1990) showed that, if one starts with a model selected from the data, then regression estimators may be biased and standard hypothesis tests may not be valid. Breiman (1992) has shown that models selected by classical data-driven methods can produce strongly biased estimates of mean squared prediction error, while both little bootstrap and cross-validation can produce relatively unbiased estimates of mean squared prediction error for data-selected submodels. Reviews of some of the difficulties induced by variable selection can be found in Bankroft and Han (1977), Burnham and Anderson (1998), Chatfield (1995), Cohen and Sackrowitz (1987) and Miller (1990) among others.

Cohen and Sackrowitz (1987) studied the problem of inference following a model selection and proposed a decision-theory approach for it, where the loss function includes components for model selection as well as for inference and allows for flexibility in emphasis on one or the other. Assuming that the data are normally distributed, Hurvich and Tsai (1990) presented Monte Carlo results on the coverage rates of confidence regions for the regression parameters, conditional on the selected model order. Their findings showed that conditional coverage rates are much smaller than the nominal coverage rates, obtained by assuming that the model is known in advance. In a very general context, Pötscher (1991) established the asymptotic properties of estimators of parameters based on a statistical model which has been selected via a model selection procedure. The asymptotic distributions of the estimators are obtained and the effects of the model selection process are illustrated numerically using the example of a distributed lag model. An important potential application of such results is to the generation of confidence regions for the parameters of interest. Kabaila (1995) demonstrated that a great deal of care must be exercised in any attempt at such an application. The author examined the effect of model selection procedures on confidence and prediction regions and emphasized that consistent estimation of the order of the model need not necessarily lead to confidence and prediction regions with optimal properties. As Pötscher (1991) noted, the asymptotic properties are of little value unless they hold for realized sample sizes. Pötscher and Novák (1998), by using simulation, studied the small sample distribution of estimators of parameters based on a statistical model which has been selected via a model selection procedure, and, in particular, evaluated the accuracy of the approximation provided by the asymptotic distribution in small samples. Zhang (1992) gave related results for the linear regression model, mainly concentrating on first and second moment properties of the estimators. Note that in Pötscher (1991), model selection procedures under investigation were testing procedures for variable selection while in Zhang (1992), FPE_α was used instead. In Pötscher and Novák (1998), both procedures were considered.

Based on Ye (1998), EAIC and YGCV can be used to evaluate the effect induced by various selection procedures, such as variable selection in linear models, and bandwidth selection in nonparametric regression. The author applied the proposed framework to measure the effect of variable selection in linear models, leading to corrections of selection bias in various goodness-of-fit statistics.

It is often that in statistical practice, the same data are used for formulating, fitting and checking a model, which may lead to inaccurate summaries and overconfident decisions. A Bayesian view of the problem can be found in Draper (1995). By virtue of recent computational advances, he discussed a Baysian approach to solving this problem and examined its implementation in some examples. But the approach introduced another problem of how to choose a prior, which is associated with Bayesian approaches in general.

The problem is actually very profound. Whether or not an answer is satisfactory depends on one's belief. Some people believe in the existence of a true model and others regard it as a fiction. The related discussion can be found in Chatfield (1995, 1996), and Draper (1995) among others.

Based on the above investigation, it can be seen that extensive research on the impact of model selection on statistical analysis is in great demand. Much work needs to be done on this direction.

13 Conclusions

In this paper, numerous model selection procedures are discussed. They are developed based on hypothesis testing, prediction errors, information measurement, MDL, resampling methods, Bayesian approach, etc.. Besides, the impact of carrying out statistical analysis after model selection is surveyed.

As demonstrated in this paper, the research on model selection is of great importance from both theoretical and practical points of view. It is not hard to see that the area of model selection is rich in problems, which are waiting to be solved. Some of them are already mentioned in previous sections.

In the end, we wish to emphasize that the model we use to analyze a data set depends on the specific questions to be answered. There are instances where different models may have to be used on the same data to answer different questions. Also, it is a good statistical practice to analyze the data under different possible models to answer a specific question to see how different or robust the answers are.

Finally, the model we select by using any of the methods described in the paper will depend on the sample size. Take for instance a simple regression problem for predicting a response measurement y using a set of p predictor variables $\boldsymbol{x}$. We may have a sample , $(y_1, \boldsymbol{x}_1), \ldots, (y_n, \boldsymbol{x}_n)$, of n observations and the use of a selection method may indicate that a subset of $\boldsymbol{x}$ may be sufficient to predict y. If we have a larger sample, the same selection method may indicate that a larger subset of $\boldsymbol{x}$ will provide better prediction (see Rao 1987).

REFERENCES

Agresti, A. (1990). *Categorical Data Analysis.* Wiley, New York.

Aitkin, M. A. (1974). Simultaneous inference and the choice of variable subsets in multiple regression. *Technometrics* **16** 221-227.

Aitkin, M. A. (1978). The analysis of unbalanced cross-classifications. With discussion. *J. Roy. Statist. Soc. Ser. A* **141**, 195-223.

Aitkin, M. (1979). A simultaneous test procedure for contingency tables. *Appl. Statist.* **28**, 233-242.

Allen, D. M. (1974). The relationship between variable selection and data augmentation and a method for prediction. *Technometrics* **16**, 125-127.

Akaike, H. (1969). Fitting autoregressive models for prediction. *Ann. Inst. Statist. Math.* **21**, 243-247.

Akaike, H. (1970). Statistical predictor identification. *Ann. Inst. Statist. Math.* **22**, 203-217.

Akaike, H. (1973). Information theory and an extension of the maximum likelihood principle. In *2nd Internat. Symp. on Information Theory* (Petrov, B.N. and Czàki, F., eds.) 267–281, Akademiai Kiadò, Budapest.

Akaike, H. (1974). A new look at the statistical model identification. *IEEE Trans. Automatic Control* AC-19, 716-723.

Akaike, H. (1978). A Bayesian analysis of the minimum AIC procedure. *Ann. Inst. Statist. Math.* **30** 9-14.

Akaike, H. (1983). Information measures and model selection. *Bull. Int. Statist. Inst.* **50**, 277-290.

Antoch, J. (1986). Algorithmic development in variable selection procedure. *Proc. of COMPSTAT* 83-90, Physica-Verlag, Heidelberg.

Antoch, J. (1987). Variable selection in linear models based on trimmed least squares estimator. In *Statistical Data Analysis Based on the L_1-norm and Related Methods* (Y. Dodge, ed.) 231-245, North Holland.

Antoniadis, A, Gijbels, I. and Grégoire, G. (1997). Model selection using wavelet decomposition and applications. *Biometrika* **84**, 751-763.

Atkinson, A. C. (1980). A note on the generalized information criterion for choice of a model. *Biometrika 67*, 413-418.

Atkinson, A. C. (1981). Likelihood ratios, posterior odds and information criteria. *J. Econometrics* **16**, 15-20.

Bera (2000) ?

Bai, Z. D., Krishnaiah, P. R., Sambamoorthi, N. and Zhao, L. C. (1992). Model selection for log-linear models. *Sankhyā Ser. B* **54**, 200-219.

Berger, J. O. and Bernardo, J. M. (1992). On the development of the reference prior method. In *Bayesian Statistics 4* (J. M. Bernardo, J. O. Berger, A. P. Dawid and A. F. M. Smith, eds.) 25-44, Oxford Univ. Press.

Berger, J. O. and Pericchi, L. R. (1996a). The intrinsic Bayes factor for model selection and prediction. In *Bayesian Statistics 5* (J. M. Bernardo, J. O. Berger, A. P. Dawid and A. F. M. Smith, eds.) 35-60, Oxford Univ. Press.

Berger, J. O. and Pericchi, L. R. (1996b). The intrinsic Bayes factor for model selection and prediction. *J. Amer. Statist. Assoc.* **91**, 109-122.

Bhansali, R. J. (1996). Asymptotically efficient autoregressive model selection for multistep prediction. *Ann. Inst. Statist. Math.* **48**, 577-602.

Bhansali, R. J. (1997). Direct autoregressive predictors for multistep prediction: Order selection and performance relative to the plug in predictors. *Statist. Sinica* **7**, 425-450.

Bhansali, R. J. and Downham, D. Y. (1977). Some properties of the order of an autoregressive model selected by a generalization of Akaike's FPE criterion. *Biometrika* **64**, 547-551.

Bickel, P. and Zhang, P. (1992). Variable selection in nonparametric regression with categorical covariates. *J. Amer. Statist. Assoc.* **87**, 90-97.

Box, G. E. P. and Jenkins, G. M. (1970). *Time Series Analysis: Forecasting and Control*, Holden Day, San Francisco.

Bozdogan, H. (1987). Model selection and Akaike's information criterion (AIC): The general theory and its analytical extensions. *Psychometrika* **52**, 345-370.

Bozdogan, H. (1988). ICOMP: A new model selection criterion. In *Classification and Related Methods of Data Analysis* (Hans H. Bock, ed.) 599-608, Amsterdam, North Holland.

Breiman, L. (1992). The little bootstrap and other methods for dimensionality selection in regression: X-fixed prediction error. *J. Amer. Statist. Assoc.* **87**, 738-754.

Breiman, L. (1996). Heuristics of instability and stabilization in model selection. *Ann. Statist.* **24**, 2350-2383.

Breiman, L. and Freedman, D. (1983). How many variables should be entered in a regression equation? *J. Amer. Statist. Assoc.* **78**, 131-136.

Burg, J. P. (1978). A new analysis technique for time series data. In *Modern Spectrum Analysis* (D. G. Childers, ed.) 42-48, IEEE Press, New York.

Burman, P. (1989). A comparative study of ordinary cross-validation, v-hold cross-validation, and repeated learning-testing methods. *Biometrika* **76**, 503-514.

Burman, P., Nolan, D. (1995). A general Akaike-type criterion for model selection in robust regression. *Biometrika* **82**, 877-886.

Burnham, K. P. and Anderson D. R. (1998). *Model Selection and Inference A practical Information-Theoretic Approach.* Springer-Verlag, New York.

Cavanaugh, J. E. and Shumway, R. H. (1998). An Akaike information criterion for model selection in the presence of incomplete data. *J. Statist. Plann. Inference* **67**, 45-65.

Chatfield, C. (1995). Model uncertainty, data mining and statistical inference (with discussion). *J. Roy. Statist. Soc. Ser. A* **158**, 419-466.

Chatfield, C. (1996). Model uncertainty and forecast accuracy. *J. Forecasting* **15**, 495-508.

Chen, C. H., Davis, R. A., Brockwell, P. J. and Bai, Z. D. (1993). Order determination for autoregressive processes using resampling methods. *Statist. Sinica*, **3**, 481-500.

Chen, C. W. S., McCulloch, R. E. and Tsay, R. S. (1997). A unified approach to estimating and modelling linear and nonlinear time series. *Statist. Sinica*, **7**, 451-472.

Choi, B. S. (1992). *ARMA Model Identification.* Springer-Verlag, New York.

Christensen, R. (1997). *Log-Linear Models and Logistic Regression*, 2nd ed. Springer-Verlag, New York.

Clyde, M. A., DeSimone, H. and Parmigiani, G. (1996). Prediction via orthogonalized model mixing. *J. Amer. Statist. Assoc.* **91**, 1197-1208.

Cohen, A. and Sackrowitz, H. B. (1987). An approach to inference following model selection with applications to transformation-based and adaptive inference. *J. Amer. Statist. Assoc.* **82**, 1123-1130.

Cox, D. R. (1961). Tests of separate families of hypotheses. In *Proc. Fourth Berkeley Symp. Math. Statist. Probab.* **1**, 105-123.

Cox, D. R. (1962). Further results on tests of separate families of hypotheses. *J. Roy. Statist. Soc. Ser. B* **24**, 406-424.

Craven, P. and Wahba, G. (1979). Smoothing noisy data with spline functions. *Numerical Mathematics* **31**, 377-403.

Davisson, L. D. (1965). The prediction error of stationary Gaussian time series of unknown covariance. *IEEE Trans. Inform. Theory* **11**, 527-532.

De Santis, F. and Spezzaferri, F. (1997). Alternative Bayes factors for model selection.

Canad. J. Statist. **25**, 503-515.

Donoho, D. L. and Johnstone, I. M. (1994). Ideal spatial adaptation by wavelet shrinkage. *Biometrika* **81**, 425-455.

Draper, D. (1995). Assessment and propagation of model uncertainty (with discussion). *J. Roy. Statist. Soc. B* **57**, 45-97.

Efron, B. (1983). Estimating the error rate of a prediction rule: improvement on cross-validation. *J. Amer. Statist. Assoc.* **78**, 316-331.

Efron, B. (1986). How biased is the apparent error rate of a prediction rule? *J. Amer. Statist. Assoc.* **81**, 461-470.

Fan, J. and Gijbels, I. (1996). *Local Polynomial Modelling and Its Applications.* Chapman and Hall, London.

Fan, J. and Li, R. (2001). Variable selection via penalized likelihood. *J. Amer. Statist. Assoc.* To appear.

Findley, D. F. (1983). On the use of multiple models for multi-period forecasting. In *Proc. Bus. Econ. Statist. Sect.* pp. 528-531. Washington, D. C., Amer. Statist. Assoc.

Findley, D. F. (1985). On the unbiasedness property of AIC for exact or approximating linear stochastic time series models. *J. Time Ser. Anal.* **6** 229-252.

Foster, D. and George, E. (1994). the risk inflation criterion for multiple regression. *Ann. Statist.* **22**, 1947-1975.

Fujikoshi, Y. and Satoh, K. (1997). Modified AIC and C_p in multivariate linear regression. *Biometrika* **84**, 707-716.

Fukuchi, J.-I. (1999). Subsampling and model selection in time series analysis. *Biometrika* **86**, 591-604.

Geisser, S. (1971). The inferential use of predictive distributions. In *Foundations of Statistical Inference* (V. P. Godambe and D. A. Sprott, eds.) 456-469, Holt, Rinehart and Winston, Toronto.

Geisser, S. (1975). The predictive sample reuse method with applications. *J. Amer. Statist. Assoc.* **70**, 320-328.

Geisser, S. (1993). *Predictive Inference: an Introduction.* Chapman and Hall, New York.

George, E. I. and McCulloch, R. E. (1993). Variable selection via Gibbs sampling. *J. Amer. Statist. Assoc.* **88**, 881-889.

George, E. I. and McCulloch, R. E. (1997). Approaches for Bayesian variable selection. *Statist. Sinica* **7**, 339-373.

Guiasu, S. (1977). *Information Theory with Applications.* McGraw-Hill, New York.

Hampel, F. R. (1983). Some aspects of model choice in robust statistics. *Proc. Int. Statist. Inst., 44th Session*, Book 2, 767-771, Madrid.

Hannan, E. J. and Quinn, B. G. (1979). The determination of the order of an autoregression. *J. Roy. Statist. Soc. Ser. B* **41**, 190-195.

Härdle, W. (1987). An effective selection of regression variables when the error distribution is incorrectly specified. *Ann. Inst. Statist. Math.* **39**, 533-548.

Hart, J. D. and Yi, S. (1998). One-sided cross-validation. *J. Amer. Statist. Assoc.* **93**, 620-631.

Herzberg, A. M. and Tsukanov, A. V. (1986) The design of experiments for model selection. *Proc. 1st World Congress Bernoulli Soc.* **2**, 175-178.

Hosmer, D. W., Jovanovic, B. and Lemeshow, S. (1989). Best subsets logistic regression. *Biometrics* **45**, 1265-1270.

Hurvich, C. M., Shumway, R. and Tsai, C. L. (1990). Improved estimators of Kullback-Leibler information for autoregressive model selection in small samples. *Biometrika* **77**, 709-719.

Hurvich, C. M., Simonoff, J. S. and Tsai, C. L. (1998). Smoothing parameter selection in nonparametric regression using an improved Akaike information criterion. *J. R. Stat. Soc. Ser. B* **60**, 271-293.

Hurvich, C. M. and Tsai, C. L. (1989). Regression and time series model selection in small samples. *Biometrika* **76**, 297-307.

Hurvich, C. M. and Tsai, C. L. (1990). Model selection for least absolute deviations regression in small samples. *Statist. Probab. Lett.* **9**, 259-265.

Hurvich, C. M. and Tsai, C. L. (1993). A corrected Akaike information criterion for vector autoregressive model selection. *J. Time Ser. Anal.* **14**, 271-279.

Hurvich, C. M. and Tsai, C. L. (1995). Model selection for extended quasi-likelihood in small samples. *Biometrics* **51**, 1077-1084.

Hurvich, C. M. and Tsai, C. L. (1996). The impact of unsuspected serial correlations on model selection in linear regression. *Statist. Probab. Lett.* **27**, 115-126.

Hurvich, C. M. and Tsai, C. L. (1997). Selection of a multistep linear predictor for short time series. *Statist. Sinica*, **7**, 395-406.

Hurvich, C. M. and Tsai, C. L. (1998). A crossvalidatory AIC for hard wavelet thresholding in spatially adaptive function estimation. *Biometrika* **85**, 701-710.

Kabaila, P. V. (1995). The effect of model selection on confidence regions and prediction regions. *Econometric Theory* **11**, 537-549.

Kass, R. E. and Raftery, A. (1995). Bayes factors. *J. Amer. Statist. Assoc.* **90**, 773-795.

Konishi, S. (1999). Statistical model evaluation and information criteria multivariate analysis. In *Design of Experiments, and Survey Sampling* (S. Ghosh, ed.), 369-399, Marcel Dekker, Inc.

Konishi, S. and Kitagawa, G. (1996). Generalized information criteria in model selection. *Biometrika* **83**, 875-890.

Krishnaiah, P. R. (1982). Selection of variables under univariate regression models. In *Handbook of Statistics II* (P. R. Krishnaiah, ed.), 805-820, North-Holland, Amsterdam.

Lai, T. L. and Lee, C. P. (1997). Information and prediction criteria for model selection in stochastic regression and ARMA models. *Statist. Sinica*, **7**, 285-310.

Laud, P. W. and Ibrahim, J. G. (1995). Predictive model selection. *J. Roy. Statist. Soc. Ser. B* **57**, 247-262.

Léger, C. and Altman, N. (1993). Assessing influence in variable selection problems. *J. Amer. Statist. Assoc.* **88**, 547-556.

Li, K. C. (1987). Asymptotic optimality for C_p, C_L, cross-validation and generalized cross-validation: discrete index set. *Ann. Statist.* **15**, 958-975.

Li, W. K. (1993). A simple one degree of freedom test for non-linear time series model discrimination. *Statist. Sinica* **3**, 245-254.

Lin, J.-L. and Granger, C. W. J. (1994). Forecasting from non-linear models in practice. *J. Forecasting* **13**, 1-9.

Linhart, H. and Zucchini, W. (1986). *Model Selection.* Wiley.

Liu, S. I. (1996). Model selection for multiperiod forecasts. *Biometrika* **83**, 861-873.

Loader, C. R. (1999). Bandwidth selection: classical or plug-in? *Ann. Statist.* **27**, 415-438.

Loh, W. (1985). A new method for testing separate families of hypotheses. *J. Amer. Statist. Assoc.* **80**, 362-368.

Machado, J. A. F. (1993). Robust model selection and M-estimation. *Econometric Theory* **9**, 478-493.

Mallows, C. L. (1973). Some comments on C_p. *Technometrics* **15**, 661-675.

Mallows, C. L. (1995). More comments on C_p. *Technometrics* **37**, 362-372.

Mantel, N. (1970). Why stepdown procedures in variable selection? *Technometrics* **12** 621-625.

Martin, R. D. (1980). Robust estimation of autoregressive models. In *Direction in Time Series* (D. R. Brillinger and G. C. Tiao, eds.), 228-262, IMS, Hayward, CA.

McCullagh, P. and Nelder, J. A. (1989). *Generalized Linear Models*, 2nd ed. Chapman and Hall, London.

McKay, R. J. (1977). Variable selection in multivariate regression: an application of simultaneous test procedures. *J. R. Statist. Soc. B* **39** 371-380.

McQuarrie, A. (1999). A small-sample correction for the Schwarz SIC model selection criterion. *Statist. Probab. Lett.* **44**, 79-86.

McQuarrie, A., Shumway, R. and Tsai, C. L. (1997). The model selection criterion AICu. *Statist. Probab. Lett.* **34**, 285-292.

McQuarrie, A. and Tsai, C. L. (1998). *Regression And Time Series Model Selection.* World Scientific, Singapore.

Miller, A. J. (1990). *Subset Selection in Regression.* Chapman and Hall, London.

Nason, G. P. (1996). Wavelet regression by cross-validation. *J. Roy. Statist. Soc. Ser. B* **58**, 463-479.

Nelder, J. and Wedderburn, R. W. M. (1972). Generalized linear models. *J. Roy. Statist. Soc. Ser. A* **135**, 370-384.

Nishii, R. (1984). Asymptotic properties of criteria for selection of variables in multiple regression. *Ann. Statist.* **12**, 758-765.

Nishii, R. (1988). Maximum likelihood principle and model selection when the true model unspecified. *J. Multivariate Anal.* **27**, 392-403.

O'Hagan, A. (1995). Fractional Bayes factors for model comparison (with discussion). *J. Roy. Statist. Soc. Ser. B* **57**, 99-138.

Paulsen, J. (1984). Order determination of multivariate autoregressive time series with unit roots. *J. Time Ser. Anal.* **5**, 115-127.

Paulsen, J. and Tjøstheim, D. (1985). Least squares estimates and order determination procedures for autoregressive processes with a time dependent variance. *J. Time Ser. Anal.* **6**, 117-133.

Picard, R. R. and Cook, R. D. (1984). Cross-validation of regression models. *J. Amer. Statist. Assoc.* **79**, 575-583.

Pötscher, B. M. (1989). Model selection under nonstationarity: autoregressive models and stochastic linear regression models. *Ann. Statist.* **17**, 1257-1274.

Pötscher, B. M. (1991). Effects of model selection on inference. *Econometric Theory* **7**, 163-185.

Pötscher, B. M. and Novák, A. J. (1998). The distribution of estimators after model selection: large and small sample results. *J. Statist. Comput. Simulation* **60**, 19-56.

Pregibon, D. (1979). Data analytic methods for generalized linear models. Ph. D. dissertation, Univ.. of Toronto, Canada.

Pukkila, T., Koreisha, S. and Kallinen, A. (1990). The identification of ARMA models. *Biometrika* **77**, 537–548.

Pukkila, T. and Krishnaiah, P. R. (1988a). On the use of autoregressive order determination criteria in univariate white noise tests. *IEEE Trans. Acoust., Speech and Signal Processing* **36**, 764-774.

Pukkila, T. and Krishnaiah, P. R.(1988b). On the use of autoregressive order determination criteria in multivariate white noise tests. *IEEE Trans. Acoust., Speech and Signal Processing* **36**, 1396-1403.

Qian, G., Gabor, G. and Gupta, R. P. (1996). Generalized linear model selection by the predictive least quasi-deviance criterion. *Biometrika* **83**, 41-54.

Qian, G. and Künsch, H. R. (1998). On model selection via stochastic complexity in robust linear regression. *J. Statist. Plann. Inference* **75**, 91-116.

Quinn, B. G. (1980). Order determination for multivariate autoregression. *J. Roy. Statist. Soc. Ser. B* **42**, 182-185.

Raftery, A. E., Madigan, D. and Hoeting, J. A. (1997). Bayesian model averaging for linear regression models. *J. Amer. Statist. Assoc.* **92**, 179-191.

Rao, C. R. (1987). Prediction of future observations in growth curve models. *Statist. Sci.* **2**, 434-471.

Rao, C. R. and Wu, Y. (1989). A strongly consistent procedure for model selection in a regression problem. *Biometrika* **76**, 369-374.

Rao, J. S. (1994). Adaptive subset selection via cost-optimization using resampling methods in linear regression models. Ph.D. dissertation, Univ. of Toronto, Canada.

Rencher, A. C. and Pun, F. C. (1980). Inflation of R^2 in best subset regression. *Technometrics* **22**, 49-53.

Rissanen, J. (1978). Modelling by shortest data description. *Automatica* **14**, 465-471.

Rissanen, J. (1983). A universal prior for integers and estimation by minimum description length. *Ann. Statist.* **11**, 416-431.

Rissanen, J. (1986a). Stochastic complexity and modelling. *Ann. Statist.* **14**, 1080-1100.

Rissanen, J. (1986b). A predictive least squares principle. *IMA J. Math. Control. Inform.* **3**, 211-222.

Rissanen, J. (1986c). Order estimation by accumulated prediction errors. In *Essays in Time Series and Allied Processes* (J. Gani and M. B. Priestly, eds.). *J. Appl. Probab.* **23A** 55-61.

Rissanen, J. (1989). Stochastic Complexity in Statistical Inquiry. World Scientific, Singapore.

Ronchetti, E. (1985). Robust model selection in regression. *Statist. Prob. Lett.* **3**, 21-23.

Ronchetti, E. (1997). Robustness aspects of model choice. *Statist. Sinica* **7**, 327-338.

Ronchetti, E., Field, C. and Blanchard, W. (1997). Robust linear model selection by cross-validation. *J. Amer. Statist. Assoc.* **92**, 1017-1023.

Ronchetti, E. and Staudte, R. G. (1994). A robust version of Mallows's C_p. *J. Amer. Statist. Assoc.* **89**, 550-559.

Sakamoto, Y. , Ishiguro, M. and Kitagawa, G. (1986). *Akaike Information Criterion Statistics.* KTK Scientific Publishers, Tokyo.

Sakamoto, Y. (1991). *Categorical Data Analysis by AIC.* KTK Scientific Publishers, Tokyo.

Sawa, T. (1978). Information criteria for discriminating among alternative regression models. *Econometrica* **46**, 1273-1291.

Schwarz, G. (1978). Estimating the dimension of a model. *Ann. Statist.* **6**, 461-464.

Shao, J. (1993). Linear model selection by cross-validation. *J. Amer. Statist. Assoc.* **88**, 486-494.

Shao, J. (1996). Bootstrap model selection. *J. Amer. Statist. Assoc.* **91** (1996), 655-665.

Shao, J. (1997). An asymptotic theory for linear model selection. *Statist. Sinica,* **7**, 221-264.

Shao, J. (1998). Convergence rates of the generalized information criterion. *J. Nonparametric Statist.* **9**, 217-225.

Shao, J. and Rao, J. S. (2000). The GIC for model selection: A hypothesis testing approach. *J. Statist. Plann. Inference* **88**, 215-231.

Shi, P. and Tsai, C. L. (1998a). On the use of marginal likelihood in model selection. Technical Report, Graduate School of Management, Univ. California, Davis.

Shi, P. and Tsai, C. L. (1998b). A note on the unification of the Akaike information criterion. *J. R. Stat. Soc. Ser. B* **60**, 551-558.

Shi, P. and Tsai, C. L. (1999). Semiparametric regression model selections. *J. Statist. Plann. Inference* **77**, 119-139.

Shibata, R. (1976). Selection of the order of an autoregressive model by Akaike's information criterion. *Biometrika* **63**, 117-126.

Shibata, R. (1980). Asymptotic efficient selection of the order of the model for estimating parameters of a linear process. *Ann. Statist.* **8**, 147-164.

Shibata, R. (1986a). Selection of regression variables. In *Encyclopedia of Statistical Sciences VII* (S. Kotz and N. L. Johnson, eds.), 709-714, John Wiley & Sons.

Shibata, R. (1986b). Selection of the number of regression variables; a minimax choice of generalized FPE. *Ann. Inst. Statist. Math.* **38**, 459-474.

Shibata, R. (1989). Statistical aspects of model selection. In *From Data to Model* (J. C. Willems, ed.), 215-240, Springer-Verlag.

Shibata, R. (1997). Bootstrap estimate of Kullback-Leibler information for model selection. *Statist. Sinica* **7**, 375-394.

Shimodaira, H. (1994). A new criterion for selecting models from partially observed data. In *Selecting Models from Data: Artificial Intelligence and Statistics IV, Lecture Notes in Statistics* (P. Cheeseman and R. W. Oldford, eds.), 21-29, Springer, New York.

Simonoff, J. S. (1996). *Smoothing Methods in Statistics.* Springer-Verlag.

Simonoff, J. S. and Tsai, C. L. (1999). Semiparametric and additive model selection Using an Improved AIC Criterion. *J. Comput. Graph. Statist.* **8**, 22-40.

Sommer, S. and Huggins, R. M. (1996). Variable selection using the Wald test and a robust C_p. *Appl. Statist.* **45**, 15-29.

Sommer, S. and Staudte, R. G. (1995). Robust variable selection in regression in the presence of outliers and leverage points. *Austral. J. Statist.* **37**, 323-336.

Speed, T. P. and Yu, B. (1993). Model selection and prediction: normal regression. *Ann. Inst. Statist. Math.* **45**, 35-54.

Stone, M. (1974). Cross-validatory choice and assessment of statistical prediction. *J. Roy. Statist. Soc. Ser. B.* **36**, 111-133.

Stone, M. (1977a). An asymptotic equivalence of choice of model by cross-validation and Akaike's criterion. *J. Roy. Statist. Soc. Ser. B.* **39**, 44-47.

Stone, M. (1977b). Asymptotics for and against cross-validation. *Biometrika* **64**, 29-38.

Sugiura, N. (1978). Further analysis of the data by Akaike's information criterion and the finite corrections. *Comm. Statist.- Theory Meth.* **7** 13-26.

Simonoff, J. S. and Tsai, C. L. (1999). Semiparametric and additive model selection Using an Improved AIC Criterion. *J. Comput. Graph. Statist.* **8**, 22-40.

Takeuchi, K. (1976). Distribution of informational statistics and a criterion of model fitting. *Suri-Kagaku* (Mathematic Sciences) **153**, 12-18. (In Japanese).

Thall, P. F., Simon, R., and Grier, D. A. (1992). Test-based variable selection via cross-validation. *J. Comput. Graph. Statist.* **1**, 41-61.

Thall, P. F., Russell, K. E. and Simon, R. M. (1997). Variable selection in regression via repeated data splitting. *J. Comput. Graph. Statist.* **6**, 1-34.

Tiao, G. C. and Xu. D. (1993). Robustness of MLE for multi-step predictions: the exponential smoothing case. *Biometrika* **80**, 623-641.

Tibshirani, R. (1996). Regression shrinkage and selection via the LASSO. *J. Roy. Statist. Soc. Ser. B* **58**, 267-288.

Tibshirani, R. and Knight, K. (1999a). The covariance inflation criterion for adaptive model selection. *J. Roy. Statist. Soc. Ser. B.* **61**, 529-546.

Tibshirani, R. and Knight, K. (1999b). Model search and inference via bootstrap bumping. *J. Comput. Graph. Statist.* **8**, 671-686.

Tsay, R. S. (1984) Order selection in nonstationary autoregressive models. *Ann. Statist.* **12**, 1425-1433.

Venter, J. H. and Steel, S. J. (1992) Some contributions to selection and estimation in the normal linear model. *Ann. Inst. Statist. Math.* **44**, 281-297.

Wakefield, J. C. and Bennett, J. E. (1996). The Bayesian modelling of covariates for population pharmacokinetic models. *J. Amer. Statist. Assoc.* **91**, 917-927.

Wei, C. Z. (1992). On predictive least squares principles. *Ann. Statist.* **20**, 1-42.

Whittle, P. (1963). *Prediction and Regulation by Linear Least-Square Methods.* English Univ. Press, London.

Williams, D. A. (1970a). Discrimination between regression models to determine the pattern of enzyme synthesis in synchronous cell cultures. *Biometrics* **28**, 23-32.

Williams, D. A. (1970b). Discussion of "A method for discriminating between models," by A. C. Atkinson. *J. Roy. Statist. Soc. Ser. B* **32**, 350.

Wu, Y. (2001). An M-estimation-based model selection criterion with a data-oriented penalty. *J. Statist. Comput. Simulation.* To appear.

Wu, Y., Tam, K. W., Li, F. and Zen, M. M. (1999). A note on estimating the number of super imposed exponential signals by the cross-validation approach. Technical Report, Center for Multivariate Analysis, Penn. State Univ.

Wu, Y. and Zen, M. (1999). A strongly consistent linear model selection procedure based on M-estimation. *Probability Theory and Related Fields* **113**, 599-625.

Yao, Q. and Tong, H. (1994). On subset selection in non-parametric stochastic regression. *Statist. Sinica* **4**, 51-70.

Ye, J. (1998). On measuring and correcting the effects of data mining and model selection. *J. Amer. Statist. Assoc.* **93**, 120-131.

Zhang, H. M. and Wang, P. (1994). A new way to estimate orders in time series. *J. Time Ser. Anal.* **15**, 545-559.

Zhang, P. (1991). Variable selection in nonparametric regression with continuous covariates. *Ann. Statist.* **19**, 1869-1882.

Zhang, P. (1992). Inference after variable selection in linear regression models. *Biometrika* **79**, 741-746.

Zhang, P. (1993a). Model selection via multifold cross validation. *Ann. Statist.* **21**, 299-313.

Zhang, P. (1993b). On the convergence rate of model selection criteria. *Commun. Statist.-Theory Meth.*, **22**, 2765-2775.

Zhang, P. (1994). On the choice of penalty term in generalized FPE criterion. In *Selecting Models from Data* (P. Cheeseman and R. W. Oldford, eds.), 41-49, Springer-Verlag, New York.

Zhang, P. and Krieger, A. M. (1993) Appropriate penalties in the final prediction error criterion: a decision theoretic approach. *Statist. Probab. Lett.* **18**, 169–177.

Zhao, L. C., Krishnaiah, P. R. and Bai, Z. D. (1986a). On detection of the number of signals in presence of white noise. *J. Multivariate Anal.* **20**, 1-25.

Zhao, L. C., Krishnaiah, P. R. and Bai, Z. D. (1986b). On determination of the number of signals when the noise covariance matrix is arbitrary. *J. Multivariate Anal.* **20**, 26-49.

Zheng, X. and Loh, W. (1995). Consistent variable selection in linear models. *J. Am. Statist. Assoc.* **90**, 151-156.

DISCUSSION

Sadanori Konishi

Kyushu University

I would like to begin by congratulating Drs. Rao and Wu for very concise review and clear exposition of model selection. The paper will stimulate future research in statistical model selection and evaluation problems.

Needless to say, model selection and evaluation are essential and of great importance in modelling process in various fields of natural and social sciences. Akaike (1973) introduced an information criterion as an estimator of the Kullback-Leibler measure of discriminatory information between two probability distributions, and a number of successful applications of AIC in statistical data analysis have been reported. Schwarz (1978) proposed a model selection criterion called BIC (Bayesian information criterion) from a Bayesian viewpoint. AIC and BIC are the most widely used model selection criteria in practical applications.

Now by taking advantage of fast computers, we may construct complicated nonlinear models for analyzing data with complex structure. Nonlinear models are generally characterized by a large number of parameters. We know that the maximum likelihood methods yield unstable parameter estimates and lead to overfitting. In such cases the adopted model is estimated by the maximum penalized likelihood method, Bayes approach, etc.

It might be noticed that the criteria AIC and BIC, theoretically, cover only models estimated by the maximum likelihood methods. The problem is: "Can AIC and BIC be applied to a wider class of statistical models?" Konishi and Kitagawa (1996) proposed an information-theoretic criterion GIC which enables us to evaluate various types of statistical models. By extending Schwarz's basic ideas, I will introduce a criterion to evaluate models estimated by the maximum penalized likelihood method.

Suppose we are interested in selecting a model from a set of candidate models M_1, $\cdots$, M_r for a given observation vector $\boldsymbol{y}$ of dimension n. It is assumed that each model M_k is characterized by the probability density $f_k(\boldsymbol{y}|\boldsymbol{\theta}_k)$, where $\boldsymbol{\theta}_k \in \Theta_k \subset R^k$. Let $\pi_k(\boldsymbol{\theta}_k|\lambda)$ be the prior distribution for parameter vector $\boldsymbol{\theta}_k$ under model M_k, where λ is a hyperparameter. Then the posterior probability of the model M_k for a particular data

Sadanori Konishi is Professor, Graduate School of Mathematics, Kyushu University, 6-10-1 Hakozaki, Higashi-Ku, Fukuoka 812-8581, Japan; email: konishi@math.kyushu-u.ac.jp.

set $\boldsymbol{y}$ is given by

$$\Pr(M_k|\boldsymbol{y}) = \frac{\Pr(M_k)\int f_k(\boldsymbol{y}|\boldsymbol{\theta}_k)\pi_k(\boldsymbol{\theta}_k|\lambda)d\boldsymbol{\theta}_k}{\sum_{\alpha=1}^{r}\Pr(M_\alpha)\int f_\alpha(\boldsymbol{y}|\boldsymbol{\theta}_\alpha)\pi_\alpha(\boldsymbol{\theta}_\alpha|\lambda)d\boldsymbol{\theta}_\alpha},$$

where $\Pr(M_k)$ is the prior probability for model M_k.

The Bayes approach for selecting a model is to choose the model with the largest posterior probability among a set of candidate models for a given value of λ, which is equivalent to choose the model that maximizes

$$\Pr(M_k)\int f_k(\boldsymbol{y}|\boldsymbol{\theta}_k)\pi_k(\boldsymbol{\theta}_k|\lambda)d\boldsymbol{\theta}_k = \Pr(M_k)\int \exp\left\{\log f_k(\boldsymbol{y}|\boldsymbol{\theta}_k) + \log \pi_k(\boldsymbol{\theta}_k|\lambda)\right\} d\boldsymbol{\theta}_k.$$

We now specify the prior distribution $\pi_k(\boldsymbol{\theta}_k|\lambda)$ on the parameters of each model to be

$$\pi_k(\boldsymbol{\theta}_k|\lambda) = (2\pi)^{-(k-q)/2}(n\lambda)^{(k-q)/2}|D|_+^{1/2}\exp\left\{-\frac{n\lambda}{2}\boldsymbol{\theta}_k' D\boldsymbol{\theta}_k\right\},$$

where D is a $k \times k$ known matrix of rank $k-q$ and $|D|_+$ is the product of nonzero eigenvalues of D. Then, using Laplace's methods for integrals in the Bayesian framework developed by Tierney and Kadane (1986) and Kass, Tierney and Kadane (1990), we have, under equal prior probabilities Pr (M_k), an asymptotic approximation

$$\begin{aligned}\mathrm{GBIC}(\lambda) &= -2\log\left\{\int f_k(\boldsymbol{y}|\boldsymbol{\theta}_k)\pi_k(\boldsymbol{\theta}_k|\lambda)d\boldsymbol{\theta}_k\right\} \\ &= -2\log f_k(\boldsymbol{y}|\hat{\boldsymbol{\theta}}_k) + n\lambda\hat{\boldsymbol{\theta}}_k' D\hat{\boldsymbol{\theta}}_k + q\log n \\ &\quad + \log|J_\lambda(\hat{\boldsymbol{\theta}}_k)| - \log|D|_+ - q\log 2\pi - (k-q)\log\lambda,\end{aligned}$$

where $J_\lambda(\boldsymbol{\theta}_k) = -n^{-1}\,\partial^2\left\{\log f_k(\boldsymbol{y}|\boldsymbol{\theta}_k)\right\}/\partial\boldsymbol{\theta}_k\partial\boldsymbol{\theta}_k' + \lambda D$ and the estimate $\hat{\boldsymbol{\theta}}_k$ is given by maximizing the penalized log-likelihood function

$$\ell_\lambda(\boldsymbol{\theta}_k) = \log f_k(\boldsymbol{y}|\boldsymbol{\theta}_k) - \frac{n\lambda}{2}\boldsymbol{\theta}_k' D\boldsymbol{\theta}_k.$$

Optimal value of λ is obtained as the minimizer of GBIC(λ) for each model, and then we choose a statistical model for which the value of the criterion GBIC is minimized over a set of competing models. The GBIC may be applied for evaluating various types of nonparametric regression models estimated by the maximum penalized likelihood method.

ADDITIONAL REFERENCES

Akaike, H. (1973). Information theory and an extension of the maximum likelihood principle. *2nd Internat. Symp. on Information Theory* (Petrov, B.N. and Csaki, F., eds.) 267–281, Akademiai Kiado, Budapest. (Reproduced in *Breakthroughs in Statistics,* **1**, 1992 (S. Kotz and N.L.Johnson, eds.) 610–624, Springer-Verlag, New York.)

Kass, R. E., Tierney, L. and Kadane, J. B. (1990). The validity of posterior asymptotic expansions based on Laplace's method. In *Bayesian and Likelihood Methods in Statistics and Econometrics*, (S. Geisser, J. S. Hodges, S. J. Press and A. Zellner, eds.), North-Holland, New York.

Tierney, L. and Kadane, J. B. (1986). Accurate approximations for posterior moments and marginal densities. *J. Amer. Statist. Assoc.* **81**, 82 - 86.

Rahul Mukerjee
Indian Institute of Management

I begin by congratulating the authors, Professors Rao and Wu, on this very illuminating and scholarly piece of work which will inspire future researchers in this area. They have done an enormous job of which we are the beneficiaries.

Considerable attention has been given in this paper on the important problem of selecting an appropriate sub-model starting from the linear model (2.1). I, therefore, find it relevant to briefly discuss some related issues in design of experiments. The discussion will be focussed primarily on discrete designs. Incidentally, experimental design problems under model uncertainty have been of substantial interest in recent years (Dey and Mukerjee, 1999; Wu and Hamada, 2000).

To motivate the ideas, consider the setup of a 2^n factorial experiment, a situation where there are n factors each at two levels. Suppose interest lies in identifying the active factors, i.e., the ones with nonzero main effects, under the absence of all interactions. A *factor screening experiment* is one which can achieve this. Interpreting the factors as regressors, the problem here is the same as that initiated by (2.1) and (2.2). The model (2.1) now consists of the general mean and the main effects of the two-level factors, each main effect being represented by a single parameter. Clearly, then at least $n + 1$ observations are needed to examine (2.1) and all possible sub-models thereof.

Rahul Mukerjee is Professor,Indian Institute of Management, Post Box No. 16757, Calcutta 700 027, India; email: rmuk1@hotmail.com

Unfortunately, in many practical situations, especially in exploratory studies, n can be quite large and even $n+1$ observations are not affordable. This poses additional concerns. A way out is possible via consideration of the phenomenon of effect sparsity. Often it is known that among the n factors at most k are active where the known quantity k is small compared to n. However, there is no knowledge about which factors are active. The problem is then to identify the active factors, which are at most k in number, and to estimate the corresponding main effects as well as the general mean.

The notion of search designs, pioneered by Srivastava (1975), plays a crucial role in handling problems of this kind. To highlight the underlying ideas, we consider a linear model $\mathrm{E}(\boldsymbol{Y}) = X_1\beta_1 + X_2\beta_2$, $\mathrm{Cov}(\boldsymbol{Y}) = \sigma^2 I_N$, where $\boldsymbol{Y}$ is the $N \times 1$ observational vector, $X = [X_1, X_2]$ is the known design matrix, β_1 is an unknown parametric vector, σ^2 is the common variance of the observations and I_N is the $N \times N$ identity matrix. The parametric vector β_2 is partially known. It is known that at most k its elements are nonzero, where k is small compared to the dimension of β_2. No knowledge is, however, available about which elements of β_2 are possibly nonzero and what their values are. Interest lies in estimating β_1 and searching and estimating the possibly nonzero elements of β_2. Following Srivastava (1975), an experimental design d (associated with a choice of X), that enables one to achieve this, is called a *search design* with resolving power (β_1, β_2, k). Observe that if β_1 is taken as the scalar representing the general mean and β_2 is taken as the vector of the main effects of the factors then the search design problem reduces to the factor screening problem introduced above. Of course, the idea of search designs applies to many other situations. For example, one can consider more complex factorials including asymmetric factorials and entertain interactions too in addition to the general mean and main effects.

The following result, due to Srivastava (1975), is a fundamental tool in the study of search designs.

Theorem 1. *For a design to have resolving power (β_1, β_2, k) in the above setup, it is necessary that for every $N \times 2k$ submatrix X_{20} of X_2, the matrix $[X_1, X_{20}]$ has full column rank. Furthermore, in the noiseless case $\sigma^2 = 0$, this condition is also sufficient for to have resolving power (β_1, β_2, k).*

For $\sigma^2 = 0$, if the rank condition of Theorem 1 is satisfied then the true model can be identified with probability unity. Further discussion on the actual identification of the true model is available in Srivastava (1975). The design problem here consists of finding a design such that the rank condition holds. The construction of such experimental plans and related combinatorics have received much attention in the literature and we refer to Gupta (1990) and Ghosh (1996) for informative reviews. The latter article also discusses the important issue of sequential experimentation in this and related contexts.

For $\sigma^2 > 0$, even under the rank condition, detection of the true model is not possible with probability unity. Discussion on the search procedure in this case is available in Srivastava (1975, 1996). The design problem here consists of ensuring the rank condition as well as attaining a high probability of correct search. Significantly new grounds have been broken in this direction by Shirakura et al. (1996). The actual construction of optimal designs for the case $\sigma^2 > 0$ deserves further attention.

With reference to the problem of factor screening, *supersaturated designs* have also been of much interest. With a 2^n factorial, under the absence of all interactions, let X be the $N \times (n+1)$ design matrix where N is the number of observations and the columns of X correspond to the general mean and the n main effects. As before, let $N < n+1$, and suppose the objective is to identify the active factors under effect sparsity. Since $N < n+1$, one cannot achieve orthogonality in the sense of making $X^T X$ a diagonal matrix, where X^T is the transpose of X. The idea of supersaturated designs, which dates back to Booth and Cox (1962) and has witnessed a revival of interest in recent years, essentially aims at choosing X such that the sum of squares of the off-diagonal elements of $X^T X$ is minimized. Li and Wu (1997) describe several other criteria for choosing such designs. Data analytic techniques for the use of supersaturated designs in the identification of the correct model, via detection of the nonzero main effects, have been discussed in Lin (1993, 1995); see Dey and Mukerjee (1999) and Wu and Hamada (2000) for more references on supersaturated designs.

Cheng et al. (1999) consider another approach towards the design problem for the study of model robustness and model selection. In connection with regular fractions of symmetric factorials, i.e., the ones specified by a system of linear equations over a finite field, they introduce the notion of *estimation capacity* which is a criterion of model robustness. The objective here is to retain full information on the main effects and as much information as possible on the two-factor interactions in the sense of entertaining the maximum possible model diversity, under the absence of interactions involving three or more factors. Cheng and Mukerjee (1998) report further theoretical results on designs with maximum estimation capacity.

Turning to continuous experimental designs, an important design problem in model selection is the one where an objective is the identification of the appropriate degree of the polynomial in a polynomial regression model. Innovative results on this problem, via the use of canonical moments, have been reported by Dette (1995) and Franke (2000), where further references are available.

To summarize, the experimental design problem for model selection has already been an active area of research and even greater activity, catering to both frequentist and Bayesian inference, is anticipated in this area in the near future. The elegant exposition given by Professors Rao and Wu in the present work will definitely act as a great stimulator for future research in this direction.

ADDITIONAL REFERENCES

Booth, K.H.V. and Cox, D.R. (1962). Some systematic supersaturated designs. *Technometrics* **4**, 489-495.

Cheng, C.S. and Mukerjee, R. (1998). Regular fractional factorial designs with minimum aberration and maximum estimation capacity. *Ann. Statist.* **26**, 2289-2300.

Cheng, C.S., Steinberg, D.M. and Sun, D.X. (1999). Minimum aberration and model robustness for two-level factorial designs. *J. Roy. Statist. Soc. B* **61**, 85-93.

Dette, H. (1995) Optimal designs for identifying the degree of a polynomial regression. *Ann. Statist.* **23**, 1248-1266.

Dey, A. and Mukerjee, R. (1999). *Fractional Factorial Plans.* John Wiley, New York.

Franke, T. (2000). *D- und D_1-optimale Versuchsplane unter Nebenbedingungen und gewichtete Maximin-Versuchsplane bei polynomialer regression.* Dissertation, Ruhr-Universitat Bochum.

Ghosh, S. (1996). Sequential assembly of fractions in factorial experiments. In *Handbook of Statistics,* **13** (S. Ghosh and C.R. Rao, eds.), 407-435. North-Holland, Amsterdam.

Gupta, B.C. (1990). A survey of search designs for factorial experiments. In *Probability, Statistics and Design of Experiments* (R.R. Bahadur, ed.), 329-345. Wiley Eastern, New Delhi.

Li, W.W. and Wu, C.F.J. (1997). Columnwise-pairwise algorithms with applications to the construction of supersaturated designs. *Technometrics* **39**, 171-179.

Lin, D.K.J. (1993). A new class of supersaturated designs. *Technometrics* **35**, 28-31.

Lin, D.K.J. (1995). Generating systematic supersaturated designs. *Technometrics* **37**, 213-225.

Shirakura, T., Takahashi, T. and Srivastava, J.N. (1996). Searching probabilities for nonzero effects in search designs for the noisy case. *Ann. Statist.* **24**, 2560-2568.

Srivastava, J.N. (1975). Designs for searching non-negligible effects. In *A Survey of Statistical Design and Linear Models* (J.N. Srivastava, ed.), 507-519. North-Holland, Amsterdam.

Srivastava, J.N. (1996). A critique of some aspects of experimental design. In *Handbook of Statistics*, **13** (S. Ghosh and C.R. Rao, eds.), 309-341. North-Holland, Amsterdam.

Wu, C.F.J. and Hamada, M. (2000). *Experiments: Planning, Analysis and Parameter Design Optimization.* John Wiley, New York.

REJOINDER

C. R. Rao and Y. Wu

The authors would like to thank S. Konishi and R. Mukerjee for their valuable comments. Konishi suggests an extension of the GIC criterion using a penalized maximum likelihood estimator of the unknown parameters. This new method may provide some robustness to the choice of a model. To what extent is the selection of the model affected by the particular choice of the prior distribution of parameters and models suggested by Konishi needs some investigation. Mukerjee raised the problem of design of experiments to provide the minimum number of observations needed for model selection ensuring some robustness. This is, indeed, a new area of research, but much depends on the accuracy of apriori information regarding the unknown parameters. For instance, in the example mentioned by Mukerjee, the number of active factors out of a large number n of factors is known to be a given number $k < n$, and the problem is that of generating a minimum number of observations to determine which subset of k factors is active. It would be interesting, perhaps more relevant in practice, to know whether supersaturated designs suggested for this purpose can also be used to select a subset of factors which are more active than the others. The problem of model selection needs more discussion in terms of objectives, the use of prior information, appropriate methodology and robustness. We hope our review with the additional material contributed by Konishi and Mukerjee will stimulate further research in statistical model selection.

Model Selection
IMS Lecture Notes - Monograph Series (2001) Volume 38

The Practical Implementation of Bayesian Model Selection

Hugh Chipman, Edward I. George and Robert E. McCulloch

The University of Waterloo, The University of Pennsylvania and *The University of Chicago*

Abstract

In principle, the Bayesian approach to model selection is straightforward. Prior probability distributions are used to describe the uncertainty surrounding all unknowns. After observing the data, the posterior distribution provides a coherent post data summary of the remaining uncertainty which is relevant for model selection. However, the practical implementation of this approach often requires carefully tailored priors and novel posterior calculation methods. In this article, we illustrate some of the fundamental practical issues that arise for two different model selection problems: the variable selection problem for the linear model and the CART model selection problem.

'Hugh Chipman is Associate Professor of Statistics, Department of Statistics and Actuarial Science, University of Waterloo, Waterloo, ON N2L 3G1, Canada; email: hachipman@uwaterloo.ca. Edward I. George is Professor of Statistics, Department of Statistics, The Wharton School of the University of Pennsylvania, 3620 Locust Walk, Philadelphia, PA 19104-6302, U.S.A; email: edgeorge@wharton.upenn.edu. Robert E. McCulloch is Professor of Statistics, Graduate School of Business, University of Chicago, 1101 East 58th Street, Chicago, IL, U.S.A; email: Robert.McCulloch@gsb.uchicago.edu. This work was supported by NSF grant DMS-98.03756 and Texas ARP grant 003658.690.

Contents

1 Introduction

The Bayesian approach to statistical problems is fundamentally probabilistic. A joint probability distribution is used to describe the relationships between all the unknowns and the data. Inference is then based on the conditional probability distribution of the unknowns given the observed data, the posterior distribution. Beyond the specification of the joint distribution, the Bayesian approach is automatic. Exploiting the internal consistency of the probability framework, the posterior distribution extracts the relevant information in the data and provides a complete and coherent summary of post data uncertainty. Using the posterior to solve specific inference and decision problems is then straightforward, at least in principle.

In this article, we describe applications of this Bayesian approach for model uncertainty problems where a large number of different models are under consideration for the data. The joint distribution is obtained by introducing prior distributions on all the unknowns, here the parameters of each model and the models themselves, and then combining them with the distributions for the data. Conditioning on the data then induces a posterior distribution of model uncertainty that can be used for model selection and other inference and decision problems. This is the essential idea and it can be very powerful. Especially appealing is its broad generality as it is based only on probabilistic considerations. However, two major challenges confront its practical implementation - the specification of the prior distributions and the calculation of the posterior. This will be our main focus.

The statistical properties of the Bayesian approach rest squarely on the specification of the prior distributions on the unknowns. But where do these prior distributions come from and what do they mean? One extreme answer to this question is the pure subjective Bayesian point of view that characterizes the prior as a wholly subjective description of initial uncertainty, rendering the posterior as a subjective post data description of uncertainty. Although logically compelling, we find this characterization to be unrealistic in complicated model selection problems where such information is typically unavailable or difficult to precisely quantify as a probability distribution. At the other extreme is the objective Bayesian point of view which seeks to find semi-automatic prior formulations or approximations when subjective information is unavailable. Such priors can serve as default inputs and make them attractive for repeated use by non-experts.

Prior specification strategies for recent Bayesian model selection implementations, including our own, have tended to fall somewhere between these two extremes. Typically, specific parametric families of proper priors are considered, thereby reducing the specification problem to that of selecting appropriate hyperparameter values. To avoid the need for subjective inputs, automatic default hyperparameter choices are often recommended. For this purpose, empirical Bayes considerations, either formal or informal, can be helpful, especially when informative choices are needed. However, subjective considerations can also be helpful, at least for roughly gauging prior location and scale and for putting small probability on implausible values. Of course, when substantial prior information is available, the Bayesian model selection implementations provide a natural environment for introducing realistic and important views.

By abandoning the pure subjective point of view, the evaluation of such Bayesian methods must ultimately involve frequentist considerations. Typically, such evaluations have taken the form of average performance over repeated simulations from hypothetical models or of cross validations on real data. Although such evaluations are necessarily limited in scope, the Bayesian procedures have consistently performed well compared to non-Bayesian alternatives. Although more work is clearly needed on this crucial aspect, there is cause for optimism, since by the complete class theorems of decision theory, we need look no further than Bayes and generalized Bayes procedures for good frequentist performance.

The second major challenge confronting the practical application of Bayesian model selection approaches is posterior calculation or perhaps more accurately, posterior exploration. Recent advances in computing technology coupled with developments in numerical and Monte Carlo methods, most notably Markov Chain Monte Carlo (MCMC), have opened up new and promising directions for addressing this challenge. The basic idea behind MCMC here is the construction of a sampler which simulates a Markov chain that is converging to the posterior distribution. Although this provides a route to calculation of the full posterior, such chains are typically run for a relatively short time and used to search for high posterior models or to estimate posterior characteristics. However, constructing effective samplers and the use of such methods can be a delicate matter involving problem specific considerations such as model structure and the prior formulations. This very active area of research continues to hold promise for future developments.

In this introduction, we have described our overall point of view to provide context for the implementations we are about to describe. In Section 2, we describe the general Bayesian approach in more detail. In Sections 3 and 4, we illustrate the practical implementation of these general ideas to Bayesian variable selection for the linear model and Bayesian CART model selection, respectively. In Section 5, we conclude with a brief discussion of related recent implementations for Bayesian model selection.

2 The General Bayesian Approach

2.1 A Probabilistic Setup for Model Uncertainty

Suppose a set of K models $\mathcal{M} = \{M_1, \ldots, M_K\}$ are under consideration for data Y, and that under M_k, Y has density $p(Y \mid \theta_k, M_k)$ where θ_k is a vector of unknown parameters that indexes the members of M_k. (Although we refer to M_k as a model, it is more precisely a model class). The Bayesian approach proceeds by assigning a prior probability distribution $p(\theta_k \mid M_k)$ to the parameters of each model, and a prior probability $p(M_k)$ to each model. Intuitively, this complete specification can be understood as a three stage hierarchical mixture model for generating the data Y; first the model M_k is generated from $p(M_1), \ldots, p(M_K)$, second the parameter vector θ_k is generated from $p(\theta_k \mid M_k)$, and third the data Y is generated from $p(Y \mid \theta_k, M_k)$.

Letting Y_f be future observations of the same process that generated Y, this prior formulation induces a joint distribution $p(Y_f, Y, \theta_k, M_k) = p(Y_f, Y \mid \theta_k, M_k)\ p(\theta_k \mid M_k)$ $p(M_k)$. Conditioning on the observed data Y, all remaining uncertainty is captured by the joint posterior distribution $p(Y_f, \theta_k, M_k \mid Y)$. Through conditioning and marginalization, this joint posterior can be used for a variety Bayesian inferences and decisions. For example, when the goal is exclusively prediction of Y_f, attention would focus on the predictive distribution $p(Y_f \mid Y)$, which is obtained by margining out both θ_k and M_k. By averaging over the unknown models, $p(Y_f \mid Y)$ properly incorporates the model uncertainty embedded in the priors. In effect, the predictive distribution sidesteps the problem of model selection, replacing it by model averaging. However, sometimes interest focuses on selecting one of the models in $\mathcal{M}$ for the data Y, the model selection problem. One may simply want to discover a useful simple model from a large speculative class of models. Such a model might, for example, provide valuable scientific insights or perhaps a less costly method for prediction than the model average. One may instead want to test a theory represented by one of a set of carefully studied models.

In terms of the three stage hierarchical mixture formulation, the model selection problem becomes that of finding the model in $\mathcal{M}$ that actually generated the data, namely the model that was generated from $p(M_1), \ldots, p(M_K)$ in the first step. The probability that M_k was in fact this model, conditionally on having observed Y, is the posterior model probability

$$p(M_k \mid Y) = \frac{p(Y \mid M_k)p(M_k)}{\sum_k p(Y \mid M_k)p(M_k)} \tag{2.1}$$

where

$$p(Y \mid M_k) = \int p(Y \mid \theta_k, M_k)p(\theta_k \mid M_k)d\theta_k \tag{2.2}$$

is the marginal or integrated likelihood of M_k. Based on these posterior probabilities, pairwise comparison of models, say M_1 and M_2, is summarized by the posterior odds

$$\frac{p(M_1 \mid Y)}{p(M_2 \mid Y)} = \frac{p(Y \mid M_1)}{p(Y \mid M_2)} \times \frac{p(M_1)}{p(M_2)}. \tag{2.3}$$

This expression reveals how the data, through the Bayes factor $\frac{p(Y \mid M_1)}{p(Y \mid M_2)}$, updates the prior odds $\frac{p(M_1)}{p(M_2)}$ to yield the posterior odds.

The model posterior distribution $p(M_1|Y), \ldots, p(M_K|Y)$ is the fundamental object of interest for model selection. Insofar as the priors $p(\theta_k \mid M_k)$ and $p(M_k)$ provide an initial representation of model uncertainty, the model posterior summarizes all the relevant information in the data Y and provides a complete post-data representation of model uncertainty. By treating $p(M_k \mid Y)$ as a measure of the "truth" of model M_k, a natural and simple strategy for model selection is to choose the most probable M_k, the one for which $p(M_k \mid Y)$ largest. Alternatively one might prefer to report a set of high posterior models along with their probabilities to convey the model uncertainty. ✓

More formally, one can motivate selection strategies based on the posterior using a decision theoretic framework where the goal is to maximize expected utility, (Gelfand, Dey and Chang 1992 and Bernardo and Smith 1994). More precisely, let α represent the action of selecting M_k, and suppose that α is evaluated by a utility function $u(\alpha, \Delta)$, where Δ is some unknown of interest, possibly Y_f. Then, the optimal selection is that α which maximizes the expected utility

$$\int u(\alpha, \Delta) p(\Delta \mid Y) d\Delta \tag{2.4}$$

where the predictive distribution of Δ given Y

$$p(\Delta \mid Y) = \sum_k p(\Delta \mid M_k, Y) p(M_k \mid Y) \tag{2.5}$$

is a posterior weighted mixture of the conditional predictive distributions.

$$p(\Delta \mid M_k, Y) = \int p(\Delta \mid \theta_k, M_k) p(\theta_k \mid M_k, Y) d\theta_k \tag{2.6}$$

$p(\Delta \mid \theta_k, M_k, Y)$

It is straightforward to show that if Δ identifies one of the M_k as the "true state of nature", and $u(\alpha, \Delta)$ is 0 or 1 according to whether a correct selection has been made, then selection of the highest posterior probability model will maximize expected utility. However, different selection strategies are motivated by other utility functions. For example, suppose α entails choosing $p(\Delta \mid M_k, Y)$ as a predictive distribution for a future observation Δ, and this selection is to be evaluated by the logarithmic score function $u(\alpha, \Delta) = \log p(\Delta \mid M_k, Y)$. Then, the best selection is that α which maximizes the posterior weighted logarithmic divergence

$$\sum_{k'} p(M_{k'} \mid Y) \int p(\Delta \mid M_{k'}, Y) \log \frac{p(\Delta \mid M_{k'}, Y)}{p(\Delta \mid M_k, Y)} \tag{2.7}$$

(San Martini and Spezzaferri 1984).

However, if the goal is strictly prediction and not model selection, then expected logarithmic utility is maximized by using the posterior weighted mixture $p(\Delta \mid Y)$ in (2.5). Under squared error loss, the best prediction of Δ is the overall posterior mean

$$E(\Delta \mid Y) = \sum_k E(\Delta \mid M_k, Y) p(M_k \mid Y). \tag{2.8}$$

Such model averaging or mixing procedures incorporate model uncertainty and have been advocated by Geisser (1993), Draper (1995), Hoeting, Madigan, Raftery and Volinsky (1999) and Clyde, Desimone and Parmigiani (1995). Note however, that if a cost of model complexity is introduced into these utilities, then model selection may dominate model averaging.

Another interesting modification of the decision theory setup is to allow for the possibility that the "true" model is not one of the M_k, a commonly held perspective in many applications. This aspect can be incorporated into a utility analysis by using the actual predictive density in place of $p(\Delta \mid Y)$. In cases where the form of the true model is completely unknown, this approach serves to motivate cross validation types of training sample approaches, (see Bernardo and Smith 1994, Berger and Pericchi 1996 and Key, Perrichi and Smith 1998).

2.2 General Considerations for Prior Selection

For a given set of models $\mathcal{M}$, the effectiveness of the Bayesian approach rests firmly on the specification of the parameter priors $p(\theta_k | M_k)$ and the model space prior $p(M_1), \ldots, p(M_K)$. Indeed, all of the utility results in the previous section are predicated on the assumption that this specification is correct. If one takes the subjective point of view that these priors represent the statistician's prior uncertainty about all the unknowns, then the posterior would be the appropriate update of this uncertainty after the data Y has been observed. However appealing, the pure subjective point of view here has practical limitations. Because of the sheer number and complexity of unknowns in most model uncertainty problems, it is probably unrealistic to assume that such uncertainty can be meaningfully described.

The most common and practical approach to prior specification in this context is to try and construct noninformative, semi-automatic formulations, using subjective and empirical Bayes considerations where needed. Roughly speaking, one would like to specify priors that allow the posterior to accumulate probability at or near the actual model that generated the data. At the very least, such a posterior can serve as a heuristic device to identify promising models for further examination.

Beginning with considerations for choosing the model space prior $p(M_1), \ldots, p(M_K)$, a simple and popular choice is the uniform prior

$$p(M_k) \equiv 1/K \tag{2.9}$$

which is noninformative in the sense of favoring all models equally. Under this prior, the model posterior is proportional to the marginal likelihood, $p(M_k | Y) \propto p(Y | M_k)$, and posterior odds comparisons in (2.3) reduce to Bayes factor comparisons. However, the apparent noninformativeness of (2.9) can be deceptive. Although uniform over models, it will typically not be uniform on model characteristics such as model size. A more subtle problem occurs in setups where many models are very similar and only a few are distinct. In such cases, (2.9) will not assign probability uniformly to model neighborhoods and may bias the posterior away from good models. As will be seen in later sections, alternative model space priors that dilute probability within model neighborhoods can be meaningfully considered when specific model structures are taken into account.

Turning to the choice of parameter priors $p(\theta_k \mid M_k)$, direct insertion of improper noninformative priors into (2.1) and (2.2) must be ruled out because their arbitrary norming constants are problematic for posterior odds comparisons. Although one can avoid some of these difficulties with constructs such as intrinsic Bayes factors (Berger and Pericchi 1996) or fractional Bayes factors (O'Hagan 1995), many Bayesian model selection implementations, including our own, have stuck with proper parameter priors, especially in large problems. Such priors guarantee the internal coherence of the Bayesian formulation, allow for meaningful hyperparameter specifications and yield proper posterior distributions which are crucial for the MCMC posterior calculation and exploration described in the next section.

Several features are typically used to narrow down the choice of proper parameter priors. To ease the computational burden, it is very useful to choose priors under which rapidly computable closed form expressions for the marginal $p(Y \mid M_k)$ in (2.2) can be obtained. For exponential family models, conjugate priors serve this purpose and so have been commonly used. When such priors are not used, as is sometimes necessary outside the exponential family, computational efficiency may be obtained with the approximations of $p(Y | M_k)$ described in Section 2.3. In any case, it is useful to parametrize $p(\theta_k \mid M_k)$ by a small number of interpretable hyperparameters. For nested model formulations, which are obtained by setting certain parameters to zero, it is often natural to center the priors of such parameters at zero, further simplifying the specification. A crucial challenge is setting the prior dispersion. It should be large enough to avoid too much prior influence, but small enough to avoid overly diffuse specifications that tend to downweight $p(Y \mid M_k)$ through (2.2), resulting in too little probability on M_k. For this purpose, we have found it useful to consider subjective inputs and empirical Bayes estimates.

2.3 Extracting Information from the Posterior

Once the priors have been chosen, all the needed information for Bayesian inference and decision is implicitly contained in the posterior. In large problems, where exact calculation of (2.1) and (2.2) is not feasible, Markov Chain Monte Carlo (MCMC) can often be used to extract such information by simulating an approximate sample from the posterior. Such samples can be used to estimate posterior characteristics or to explore the posterior, searching for models with high posterior probability.

For a model characteristic η, MCMC entails simulating a Markov chain, say $\eta^{(1)}, \eta^{(2)}, \ldots$, that is converging to its posterior distribution $p(\eta \mid Y)$. Typically, η will be an index of the models M_k or an index of the values of (θ_k, M_k). Simulation of $\eta^{(1)}, \eta^{(2)}, \ldots$ requires a starting value $\eta^{(0)}$ and proceeds by successive simulation from a probability transition kernel $p(\eta \mid \eta^{(j)})$, see Meyn and Tweedie (1993). Two of the most useful prescriptions for constructing a kernel that generates a Markov chain converging to a given $p(\eta \mid Y)$, are the Gibbs sampler (GS) (Geman and Geman 1984, Gelfand and Smith 1990) and the Metropolis-Hastings (MH) algorithms (Metropolis 1953, Hastings 1970). Introductions to these methods can be found in Casella and George (1992) and Chib and Greenberg (1995). More general treatments that detail precise convergence conditions (essentially irreducibility and aperiodicity) can found in Besag and Green (1993), Smith and Roberts (1993) and Tierney (1994).

When $\eta \in R^p$, the GS is obtained by successive simulations from the full conditional component distributions $p(\eta_i \mid \eta_{-i})$, $i = 1, \ldots, p$, where η_{-i} denotes the most recently updated component values of η other than η_i. Such GS algorithms reduce the problem of simulating from $p(\eta \mid Y)$ to a sequence of one-dimensional simulations.

MH algorithms work by successive sampling from an essentially arbitrary probability transition kernel $q(\eta \mid \eta^{(j)})$ and imposing a random rejection step at each transition. When the dimension of $\eta^{(j)}$ remains fixed, an MH algorithm is defined by:

1. Simulate a candidate η^* from the transition kernel $q(\eta \mid \eta^{(j)})$

2. Set $\eta^{(j+1)} = \eta^*$ with probability

$$\alpha(\eta^* \mid \eta^{(j)}) = \min\left\{1, \frac{q(\eta^{(j)} \mid \eta^*)}{q(\eta^* \mid \eta^{(j)})} \frac{p(\eta^* \mid Y)}{p(\eta^{(j)} \mid Y)}\right\}$$

Otherwise set $\eta^{(j+1)} = \eta^{(j)}$,

This is a special case of the more elaborate reversible jump MH algorithms (Green 1995) which can be used when dimension of η is changing. The general availability of such MH algorithms derives from the fact that $p(\eta \mid Y)$ is only needed up to the norming constant for the calculation of α above.

The are endless possibilities for constructing Markov transition kernels $p(\eta \mid \eta^{(j)})$ that guarantee convergence to $p(\eta \mid Y)$. The GS can be applied to different groupings and reorderings of the coordinates, and these can be randomly chosen. For MH algorithms, only weak conditions restrict considerations of the choice of $q(\eta \mid \eta^{(j)})$ and can also be considered componentwise. The GS and MH algorithms can be combined and used together in many ways. Recently proposed variations such as tempering, importance sampling, perfect sampling and augmentation offer a promising wealth of further possibilities for sampling the posterior. As with prior specification, the construction of effective transition kernels and how they can be exploited is meaningfully guided by problem specific considerations as will be seen in later sections. Various illustrations of the broad practical potential of MCMC are described in Gilks, Richardson, and Spieglehalter (1996).

The use of MCMC to simulate the posterior distribution of a model index η is greatly facilitated when rapidly computable closed form expressions for the marginal $p(Y \mid M_k)$ in (2.2) are available. In such cases, $p(Y \mid \eta)p(\eta) \propto p(\eta \mid Y)$ can be used to implement GS and MH algorithms. Otherwise, one can simulate an index of the values of (θ_k, M_k) (or at least M_k and the values of parameters that cannot be eliminated analytically). When the dimension of such an index is changing, MCMC implementations for this purpose typically require more delicate design, see Carlin and Chib (1995), Dellaportas, Forster and Ntzoufras (2000), Geweke (1996), Green (1995), Kuo and Mallick (1998) and Phillips and Smith (1996).

Because of the computational advantages of having closed form expressions for $p(Y \mid M_k)$, it may be preferable to use a computable approximation for $p(Y \mid M_k)$ when exact expressions are unavailable. An effective approximation for this purpose, when $h(\theta_k) \equiv \log p(Y \mid \theta_k, M_k)p(\theta_k \mid M_k)$ is sufficiently well-behaved, is obtained by Laplace's method (see Tierney and Kadane 1986) as

$$p(Y \mid M_k) \approx (2\pi)^{d_k/2} |H(\tilde{\theta}_k)|^{1/2} p(Y \mid \tilde{\theta}_k, M_k) p(\tilde{\theta}_k \mid M_k) \tag{2.10}$$

where d_k is the dimension of θ_k, $\tilde{\theta}_k$ is the maximum of $h(\theta_k)$, namely the posterior mode of $p(\tilde{\theta}_k \mid Y, M_k)$, and $H(\tilde{\theta}_k)$ is minus the inverse Hessian of h evaluated at $\tilde{\theta}_k$. This is obtained by substituting the Taylor series approximation $h(\theta_k) \approx h(\tilde{\theta}_k) - \frac{1}{2}(\theta_k - \tilde{\theta}_k)' H(\tilde{\theta}_k)(\theta_k - \tilde{\theta}_k)$ for $h(\theta_k)$ in $p(M_k \mid Y) = \int \exp\{h(\theta_k)\} d\theta_k$.

When finding $\tilde{\theta}_k$ is costly, further approximation of $p(Y \mid M)$ can be obtained by

$$p(Y \mid M_k) \approx (2\pi)^{d_k/2} |H^*(\hat{\theta}_k)|^{1/2} p(Y \mid \hat{\theta}_k, M_k) p(\hat{\theta}_k \mid M_k) \tag{2.11}$$

where $\hat{\theta}_k$ is the maximum likelihood estimate and H^* can be H, minus the inverse Hessian of the log likelihood or Fisher's information matrix. Going one step further, by ignoring the terms in (2.11) that are constant in large samples, yields the BIC approximation (Schwarz 1978)

$$\log p(Y \mid M) \approx \log p(Y \mid \hat{\theta}_k, M_k) - (d_k/2) \log n \tag{2.12}$$

where n is the sample size. This last approximation was successfully implemented for model averaging in a survival analysis problem by Raftery, Madigan and Volinsky (1996). Although it does not explicitly depend on a parameter prior, (2.12) may be considered an implicit approximation to $p(Y \mid M)$ under a "unit information prior" (Kass and Wasserman 1995) or under a "normalized" Jeffreys prior (Wasserman 2000). It should be emphasized that the asymptotic justification for these successive approximations, (2.10), (2.11), (2.12), may not be very good in small samples, see for example, McCulloch and Rossi (1991).

3 Bayesian Variable Selection for the Linear Model

Suppose Y a variable of interest, and $X_1, \ldots, X_p$ a set of potential explanatory variables or predictors, are vectors of n observations. The problem of variable selection, or subset selection as it often called, arises when one wants to model the relationship between Y and a subset of $X_1, \ldots, X_p$, but there is uncertainty about which subset to use. Such a situation is particularly of interest when p is large and $X_1, \ldots, X_p$ is thought to contain many redundant or irrelevant variables.

The variable selection problem is usually posed as a special case of the model selection problem, where each model under consideration corresponds to a distinct subset of $X_1, \ldots, X_p$. This problem is most familiar in the context of multiple regression where attention is restricted to normal linear models. Many of the fundamental developments in variable selection have occurred in the context of the linear model, in large part because its analytical tractability greatly facilitates insight and computational reduction, and because it provides a simple first order approximation to more complex relationships. Furthermore, many problems of interest can be posed as linear variable selection problems. For example, for the problem of nonparametric function estimation, the values of the unknown function are represented by Y, and a linear basis such as a wavelet basis or a spline basis are represented by $X_1, \ldots, X_p$. The problem of finding a parsimonious approximation to the function is then the linear variable selection problem. Finally, when the normality assumption is inappropriate, such as when Y is discrete, solutions for the linear model can be extended to alternatives such as general linear models (McCullagh and Nelder 1989).

We now proceed to consider Bayesian approaches to this important linear variable selection problem. Suppose the normal linear model is used to relate Y to the potential predictors $X_1, \ldots X_p$

$$Y \sim N_n(X\beta, \sigma^2 I) \tag{3.1}$$

where $X = (X_1, \ldots X_p)$, β is a $p \times 1$ vector of unknown regression coefficients, and σ^2 is an unknown positive scalar. The variable selection problem arises when there is some unknown subset of the predictors with regression coefficients so small that it would be preferable to ignore them. In Sections 3.2 and 3.4, we describe two Bayesian formulations of this problem which are distinguished by their interpretation of how small a regression coefficient must be to ignore X_i. It will be convenient throughout to index each of these 2^p possible subset choices by the vector

$$\gamma = (\gamma_1, \ldots, \gamma_p)',$$

where $\gamma_i = 0$ or 1 according to whether β_i is small or large, respectively. We use $q_\gamma \equiv \gamma' 1$ to denote the size of the γth subset. Note that here, γ plays the role of model identifier M_k described in Section 2.

We will assume throughout this section that $X_1, \ldots X_p$ contains no variable that would be included in every possible model. If additional predictors $Z_1, \ldots, Z_r$ were to be included every model, then we would assume that their linear effect had been removed by replacing Y and $X_1, \ldots X_p$ with $(I - Z(Z'Z)^{-1}Z')Y$ and $(I - Z(Z'Z)^{-1}Z')X_i$, $i = 1, \ldots, p$ where $Z = (Z_1, \ldots, Z_r)$. For example, if an intercept were to be included in every model, then we would assume that Y and $X_1, \ldots X_p$ had all been centered to have mean 0. Such reductions are simple and fast, and can be motivated from a formal Bayesian perspective by integrating out the coefficients corresponding to $Z_1, \ldots, Z_r$ with respect to an improper uniform prior.

3.1 Model Space Priors for Variable Selection

For the specification of the model space prior, most Bayesian variable selection implementations have used independence priors of the form

$$p(\gamma) = \prod w_i^{\gamma_i} (1 - w_i)^{1 - \gamma_i}, \tag{3.2}$$

which are easy to specify, substantially reduce computational requirements, and often yield sensible results, see, for example, Clyde, Desimone and Parmigiani (1996), George and McCulloch (1993, 1997), Raftery, Madigan and Hoeting (1997) and Smith and Kohn (1996). Under this prior, each X_i enters the model independently of the other coefficients, with probability $p(\gamma_i = 1) = 1 - p(\gamma_i = 0) = w_i$. Smaller w_i can be used to downweight X_i which are costly or of less interest.

A useful reduction of (3.2) has been to set $w_i \equiv w$, yielding

$$p(\gamma) = w^{q_\gamma}(1-w)^{p-q_\gamma}, \tag{3.3}$$

in which case the hyperparameter w is the a priori expected proportion of $X_i's$ in the model. In particular, setting $w = 1/2$, yields the popular uniform prior

$$p(\gamma) \equiv 1/2^p, \tag{3.4}$$

which is often used as a representation of ignorance. However, this prior puts most of its weight near models of size $q_\gamma = p/2$ because there are more of them. Increased weight on parsimonious models, for example, could instead be obtained by setting w small. Alternatively, one could put a prior on w. For example, combined with a beta prior $w \sim Beta(\alpha, \beta)$, (3.3) yields

$$p(\gamma) = \frac{B(\alpha + q_\gamma, \beta + p - q_\gamma)}{B(\alpha, \beta)} \tag{3.5}$$

where $B(\alpha, \beta)$ is the beta function. More generally, one could simply put a prior $h(q_\gamma)$ on the model dimension and let

$$p(\gamma) = \binom{p}{q_\gamma}^{-1} h(q_\gamma), \tag{3.6}$$

of which (3.5) is a special case. Under priors of the form (3.6), the components of γ are exchangeable but not independent, (except for the special case (3.3)).

Independence and exchangeable priors on γ may be less satisfactory when the models under consideration contain dependent components such as might occur with interactions, polynomials, lagged variables or indicator variables (Chipman 1996). Common practice often rules out certain models from consideration, such as a model with an X_1X_2 interaction but no X_1 or X_2 linear terms. Priors on γ can encode such preferences.

With interactions, the prior for γ can capture the dependence relation between the importance of a higher order term and those lower order terms from which it was formed. For example, suppose there are three independent main effects A, B, C and three two-factor interactions AB, AC, and BC. The importance of the interactions such as AB will often depend only on whether the main effects A and B are included in the model. This belief can be expressed by a prior for $\gamma = (\gamma_A, \ldots, \gamma_{BC})$ of the form:

$$p(\gamma) = p(\gamma_A)p(\gamma_B)p(\gamma_C)p(\gamma_{AB} \mid \gamma_A, \gamma_B)p(\gamma_{AC} \mid \gamma_A, \gamma_C)p(\gamma_{BC} \mid \gamma_B, \gamma_C). \tag{3.7}$$

The specification of terms like $p(\gamma_{AC}|\gamma_A, \gamma_C)$ in (3.7) would entail specifying four probabilities, one for each of the values of (γ_A, γ_C). Typically $p(\gamma_{AC}|0,0) < (p(\gamma_{AC}|1,0), p(\gamma_{AC} \mid 0,1)) < p(\gamma_{AC} \mid 1,1)$. Similar strategies can be considered to downweight or eliminate models with isolated high order terms in polynomial regressions or isolated high order lagged variables in ARIMA models. With indicators for a categorical predictor, it may be of interest to include either all or none of the indicators, in which case $p(\gamma) = 0$ for any γ violating this condition.

The number of possible models using interactions, polynomials, lagged variables or indicator variables grows combinatorially as the number of variables increases. In contrast to independence priors of the form (3.2), priors for dependent component models, such as (3.7), is that they concentrate mass on "plausible" models, when the number of possible models is huge. This can be crucial in applications such as screening designs, where the number of candidate predictors may exceed the number of observations (Chipman, Hamada, and Wu 1997).

Another more subtle shortcoming of independence and exchangeable priors on γ is their failure to account for similarities and differences between models due to covariate collinearity or redundancy. An interesting alternative in this regard are priors that "dilute" probability across neighborhoods of similar models, the so called dilution priors (George 1999). Consider the following simple example.

Suppose that only two uncorrelated predictors X_1 and X_2 are considered, and that they yield the following posterior probabilities:

Variables in γ	X_1	X_2	X_1, X_2
$p(\gamma \mid Y)$	0.3	0.4	0.2

Suppose now a new potential predictor X_3 is introduced, and that X_3 is very highly correlated with X_2, but not with X_1. If the model prior is elaborated in a sensible way, as is discussed below, the posterior may well look something like

Variables in γ	X_1	X_2	X_3	X_1, X_2	$\cdots$
$p(\gamma \mid Y)$	0.3	0.13	0.13	0.06	$\cdots$

The probability allocated to X_1 remains unchanged, whereas the probability allocated to X_2 and X_1, X_2 has been "diluted" across all the new models containing X_3. Such dilution seems desirable because it maintains the allocation of posterior probability across neighborhoods of similar models. The introduction of X_3 has added proxies for the models containing X_2 but not any really new models. The probability of the resulting set of equivalent models should not change, and it is dilution that prevents this from happening. Note that this dilution phenomenon would become much more pronounced when many highly correlated variables are under consideration.

The dilution phenomenon is controlled completely by the model space prior $p(\gamma)$ because $p(\gamma \mid Y) \propto p(Y \mid \gamma)p(\gamma)$ and the marginal $p(Y \mid \gamma)$ is unaffected by changes to the model space. Indeed, no dilution of neighborhood probabilities occurs under the uniform prior (3.4) where $p(\gamma|Y) \propto p(Y|\gamma)$. Instead the posterior probability of every γ is reduced while all pairwise posterior odds are maintained. For instance, when X_3 is introduced above and a uniform prior is used, the posterior probabilities become something like

Variables in γ	X_1	X_2	X_3	X_1, X_2	$\cdots$
$p(\gamma \mid Y)$	0.15	0.2	0.2	0.1	$\cdots$

If we continued to introduce more proxies for X_2, the probability of the X_1 only model could be made arbitrarily small and the overall probability of the X_2 like models could be made arbitrarily large, a disturbing feature if Y was strongly related to X_1 and unrelated to X_2. Note that any independence prior (3.2), of which (3.4) is a special case, will also fail to maintain probability allocation within neighborhoods of similar models, because the addition of a new X_j reduces all the model probabilities by w_j for models in which X_j is included, and by $(1 - w_i)$ for models in which X_j is excluded.

What are the advantages of dilution priors? Dilution priors avoid placing too little probability on good, but unique, models as a consequence of massing excess probability on large sets of bad, but similar, models. Thus dilution priors are desirable for model averaging over the entire posterior to avoid biasing averages such as (2.8) away from good models. They are also desirable for MCMC sampling of the posterior because such Markov chains gravitate towards regions of high probability. Failure to dilute the probability across clusters of many bad models would bias both model search and model averaging estimates towards those bad models. That said, it should be noted that dilution priors would not be appropriate for pairwise model comparisons because the relative strengths of two models should not depend on whether another is considered. For this purpose, Bayes factors (corresponding to selection under uniform priors) would be preferable.

3.2 Parameter Priors for Selection of Nonzero β_i

We now consider parameter prior formulations for variable selection where the goal is to ignore only those X_i for which $\beta_i = 0$ in (3.1). In effect, the problem then becomes that of selecting a submodel of (3.1) of the form

$$p(Y \mid \beta_\gamma, \sigma^2, \gamma) = N_n(X_\gamma \beta_\gamma, \sigma^2 I) \tag{3.8}$$

where X_γ is the n x q_γ matrix whose columns correspond to the γth subset of $X_1, \ldots, X_p$, β_γ is a $q_\gamma \times 1$ vector of unknown regression coefficients, and σ^2 is the unknown residual variance. Here, (β_γ, σ^2) plays the role of the model parameter θ_k described in Section 2.

Perhaps the most useful and commonly applied parameter prior form for this setup, especially in large problems, is the normal-inverse-gamma, which consists of a q_γ-dimensional normal prior on β_γ

$$p(\beta_\gamma \mid \sigma^2, \gamma) = N_{q_\gamma}(\bar{\beta}_\gamma, \sigma^2 \Sigma_\gamma), \tag{3.9}$$

coupled with an inverse gamma prior on σ^2

$$p(\sigma^2 \mid \gamma) = p(\sigma^2) = IG(\nu/2, \nu\lambda/2), \tag{3.10}$$

(which is equivalent to $\nu\lambda/\sigma^2 \sim \chi^2_\nu$). For example, see Clyde, DeSimone and Parmigiani (1996), Fernandez, Ley and Steel (2001), Garthwaite and Dickey (1992, 1996), George and McCulloch (1997), Kuo and Mallick (1998), Raftery, Madigan and Hoeting (1997) and Smith and Kohn (1996). Note that the coefficient prior (3.9), when coupled with $p(\gamma)$, implicitly assigns a point mass at zero for coefficients in (3.1) that are not contained in β_γ. As such, (3.9) may be thought of as a point-normal prior. It should also be mentioned that in one the first Bayesian variable selection treatments of the setup (3.8), Mitchell and Beauchamp (1988) proposed spike-and-slab priors. The normal-inverse-gamma prior above is obtained by simply replacing their uniform slab by a normal distribution.

A valuable feature of the prior combination (3.9) and (3.10) is analytical tractability; the conditional distribution of β_γ and σ^2 given γ is conjugate for (3.8), so that β_γ and σ^2 can be eliminated by routine integration from $p(Y, \beta_\gamma, \sigma^2 \mid \gamma) = p(Y \mid \beta_\gamma, \sigma^2, \gamma) p(\beta_\gamma \mid \sigma^2, \gamma)$ $p(\sigma^2 \mid \gamma)$ to yield

$$p(Y \mid \gamma) \propto |X'_\gamma X_\gamma + \Sigma_\gamma^{-1}|^{-1/2} |\Sigma_\gamma|^{-1/2} (\nu\lambda + S^2_\gamma)^{-(n+\nu)/2} \tag{3.11}$$

where

$$S^2_\gamma = Y'Y - Y'X_\gamma (X'_\gamma X_\gamma + \Sigma_\gamma^{-1})^{-1} X'_\gamma Y. \tag{3.12}$$

As will be seen in subsequent sections, the use of these closed form expressions can substantially speed up posterior evaluation and MCMC exploration. Note that the scale of the prior (3.9) for β_γ depends on σ^2, and this is needed to obtain conjugacy. Integrating out σ^2 with respect to (3.10), the prior for β_γ conditionally only on γ is

$$p(\beta_\gamma \mid \gamma) = T_{q_\gamma}(\nu, \bar{\beta}_\gamma, \lambda\Sigma_\gamma) \tag{3.13}$$

the q_γ-dimensional multivariate T-distribution centered at $\bar{\beta}_\gamma$ with ν degrees of freedom and scale $\lambda\Sigma_\gamma$.

The priors (3.9) and (3.10) are determined by the hyperparameters $\bar{\beta}_\gamma, \Sigma_\gamma, \lambda, \nu,$, which must be specified for implementations. Although a good deal of progress can be made through subjective elicitation of these hyperparameter values in smaller problems when substantial expert information is available, for example see Garthwaite and Dickey (1996), we focus here on the case where such information is unavailable and the goal is roughly to assign values that "minimize" prior influence.

Beginning with the choice of λ and ν, note that (3.10) corresponds to the likelihood information about σ^2 provided by ν independent observations from a $N(0, \lambda)$ distribution. Thus, λ may be thought of as a prior estimate of σ^2 and ν may be thought of as the prior sample size associated with this estimate. By using the data and treating s_Y^2, the sample variance of Y, as a rough upper bound for σ^2, a simple default strategy is to choose ν small, say around 3, and λ near s_Y^2. One might also go a bit further, by treating s_{FULL}^2, the traditional unbiased estimate of σ^2 based on a saturated model, as a rough lower bound for σ^2, and then choosing λ and ν so that (3.10) assigns substantial probability to the interval (s_{FULL}^2, s_Y^2). Similar informal approaches based on the data are considered by Clyde, Desimone and Parmigiani (1996) and Raftery, Madigan and Hoeting (1997). Alternatively, the explicit choice of λ and ν can be avoided by using $p(\sigma^2 \mid \gamma) \propto 1/\sigma^2$, the limit of (3.10) as $\nu \to 0$, a choice recommended by Smith and Kohn (1996) and Fernandez, Ley and Steel (2001). This prior corresponds to the uniform distribution on $\log \sigma^2$, and is invariant to scale changes in Y. Although improper, it yields proper marginals $p(Y \mid \gamma)$ when combined with (3.9) and so can be used formally.

Turning to (3.9), the usual default for the prior mean $\bar{\beta}_\gamma$ has been $\bar{\beta}_\gamma = 0$, a neutral choice reflecting indifference between positive and negative values. This specification is also consistent with standard Bayesian approaches to testing point null hypotheses, where under the alternative, the prior is typically centered at the point null value. For choosing the prior covariance matrix Σ_γ, the specification is substantially simplified by setting $\Sigma_\gamma = c\,V_\gamma$, where c is a scalar and V_γ is a preset form such as $V_\gamma = (X_\gamma' X_\gamma)^{-1}$ or $V_\gamma = I_{q_\gamma}$, the $q_\gamma \times q_\gamma$ identity matrix. Note that under such V_γ, the conditional priors (3.9) provide a consistent description of uncertainty in the sense that they are the conditional distributions of the nonzero components of β given γ when $\beta \sim N_p(0, c\,\sigma^2 (X'X)^{-1})$ or $\beta \sim N_p(0, c\,\sigma^2 I)$, respectively. The choice $V_\gamma = (X_\gamma' X_\gamma)^{-1}$ serves to replicate the covariance structure of the likelihood, and yields the g-prior recommended by Zellner

(1986). With $V_\gamma = I_{q_\gamma}$, the components of β_γ are conditionally independent, causing (3.9) to weaken the likelihood covariance structure. In contrast to $V_\gamma = (X_\gamma' X_\gamma)^{-1}$, the effect of $V_\gamma = I_{q_\gamma}$ on the posterior depends on the relative scaling of the predictors. In this regard, it may be reasonable to rescale the predictors in units of standard deviation to give them a common scaling, although this may be complicated by the presence of outliers or skewed distributions.

Having fixed V_γ, the goal is then to choose c large enough so that the prior is relatively flat over the region of plausible values of β_γ, thereby reducing prior influence (Edwards, Lindman and Savage 1963). At the same time, however, it is important to avoid excessively large values of c because the prior will eventually put increasing weight on the null model as $c \to \infty$, a form of the Bartlett-Lindley paradox, Bartlett (1957). For practical purposes, a rough guide is to choose c so that (3.13) assigns substantial probability to the range of all plausible values for β_γ. Raftery, Madigan and Hoeting (1997), who used a combination of $V_\gamma = I_{q_\gamma}$ and $V_\gamma = (X_\gamma' X_\gamma)^{-1}$ with standardized predictors, list various desiderata along the lines of this rough guide which lead them to the choice $c = 2.85^2$. They also note that their resulting coefficient prior is relatively flat over the actual distribution of coefficients from a variety of real data sets. Smith and Kohn (1996), who used $V_\gamma = (X_\gamma' X_\gamma)^{-1}$, recommend $c = 100$ and report that performance was insensitive to values of c between 10 and 10,000. Fernandez, Ley and Steel (2001) perform a simulation evaluation of the effect of various choices for c, with $V_\gamma = (X_\gamma' X_\gamma)^{-1}$, $p(\sigma^2 \mid \gamma) \propto 1/\sigma^2$ and $p(\gamma) = 2^{-p}$, on the posterior probability of the true model. Noting how the effect depends on the true model and noise level, they recommend $c = \max\{p^2, n\}$.

3.3 Calibration and Empirical Bayes Variable Selection

An interesting connection between Bayesian and non-Bayesian approaches to variable selection occurs when the special case of (3.9) with $\bar{\beta}_\gamma = 0$ and $V_\gamma = (X_\gamma' X_\gamma)^{-1}$, namely

$$p(\beta_\gamma \mid \sigma^2, \gamma) = N_{q_\gamma}(0, c\sigma^2 (X_\gamma' X_\gamma)^{-1}), \tag{3.14}$$

is combined with

$$p(\gamma) = w^{q_\gamma}(1-w)^{p-q_\gamma} \tag{3.15}$$

the simple independence prior in (3.3); for the moment, σ^2 is treated as known. As shown by George and Foster (2000), this prior setup can be calibrated by choices of c and w so that the same γ maximizes both the model posterior and the canonical penalized sum-of-squares criterion

$$SS_\gamma/\sigma^2 - F\, q_\gamma \tag{3.16}$$

where $SS_\gamma = \hat\beta'_\gamma X'_\gamma X_\gamma \hat\beta_\gamma$, $\hat\beta_\gamma \equiv (X'_\gamma X_\gamma)^{-1} X'_\gamma Y$ and F is a fixed penalty. This correspondence may be of interest because a wide variety of popular model selection criteria are obtained by maximizing (3.16) with particular choices of F and with σ^2 replaced by an estimate $\hat\sigma^2$. For example $F = 2$ yields C_p (Mallows 1973) and, approximately, AIC (Akaike 1973), $F = \log n$ yields BIC (Schwarz 1978) and $F = 2\log p$ yields RIC (Donoho and Johnstone 1994, Foster and George 1994). The motivation for these choices are varied; C_p is motivated as an unbiased estimate of predictive risk, AIC by an expected information distance, BIC by an asymptotic Bayes factor and RIC by minimax predictive risk inflation.

The posterior correspondence with (3.16) is obtained by reexpressing the model posterior under (3.14) and (3.15) as

$$\begin{aligned} p(\gamma \mid Y) &\propto w^{q_\gamma}(1-w)^{p-q_\gamma}(1+c)^{-q_\gamma/2} \exp\left\{-\frac{Y'Y - SS_\gamma}{2\sigma^2} - \frac{SS_\gamma}{2\sigma^2(1+c)}\right\} \\ &\propto \exp\left[\frac{c}{2(1+c)}\{SS_\gamma/\sigma^2 - F(c,w)\, q_\gamma\}\right], \end{aligned} \tag{3.17}$$

where

$$F(c,w) = \frac{1+c}{c}\left\{2\log\frac{1-w}{w} + \log(1+c)\right\}. \tag{3.18}$$

The expression (3.17) reveals that, as a function of γ for fixed Y, $p(\gamma \mid Y)$ is increasing in (3.16) when $F = F(c,w)$. Thus, both (3.16) and (3.17) are simultaneously maximized by the same γ when c and w are chosen to satisfy $F(c,w) = F$. In this case, Bayesian model selection based on $p(\gamma \mid Y)$ is equivalent to model selection based on the criterion (3.16).

This correspondence between seemingly different approaches to model selection provides additional insight and interpretability for users of either approach. In particular, when c and w are such that $F(c,w) = 2, \log n$ or $2\log p$, selecting the highest posterior model (with σ^2 set equal to $\hat\sigma^2$) will be equivalent to selecting the best AIC/C_p, BIC or RIC models, respectively. For example, $F(c,w) = 2, \log n$ and $2\log p$ are obtained when $c \simeq 3.92, n$ and p^2 respectively, all with $w = 1/2$. Similar asymptotic connections are pointed out by Fernandez, Ley and Steel (2001) when $p(\sigma^2 \mid \gamma) \propto 1/\sigma^2$ and $w = 1/2$. Because the posterior probabilities are monotone in (3.16) when $F = F(c,w)$, this correspondence also reveals that the MCMC methods discussed in Section 3.5 can also be used to search for large values of (3.16) in large problems where global maximization is not computationally feasible. Furthermore, since c and w control the expected size and proportion of the nonzero components of β, the dependence of $F(c,w)$ on c and w provides an implicit connection between the penalty F and the profile of models for which its value may be appropriate.

Ideally, the prespecified values of c and w in (3.14) and (3.15) will be consistent with the true underlying model. For example, large c will be chosen when the regression coefficients are large, and small w will be chosen when the proportion of nonzero coefficients are small. To avoid the difficulties of choosing such c and w when the true model is completely unknown, it may be preferable to treat c and w as unknown parameters, and use empirical Bayes estimates of c and w based on the data. Such estimates can be obtained, at least in principle, as the values of c and w that maximize the marginal likelihood under (3.14) and (3.15), namely

$$\begin{aligned} L(c, w \mid Y) &\propto \sum_{\gamma} p(\gamma \mid w)\, p(Y \mid \gamma, c) \\ &\propto \sum_{\gamma} w^{q_\gamma} (1-w)^{p-q_\gamma} (1+c)^{-q_\gamma/2} \exp\left\{ \frac{c\, SS_\gamma}{2\sigma^2(1+c)} \right\}. \end{aligned} \tag{3.19}$$

Although this maximization is generally impractical when p is large, the likelihood (3.19) simplifies considerably when X is orthogonal, a setup that occurs naturally in nonparametric function estimation with orthogonal bases such as wavelets. In this case, letting $t_i = b_i v_i / \sigma$ where v_i^2 is the ith diagonal element of $X'X$ and b_i is the ith component of $\hat{\beta} = (X'X)^{-1}X'Y$, (3.19) reduces to

$$L(c, w \mid Y) \propto \prod_{i=1}^{p} \left\{ (1-w) e^{-t_i^2/2} + w(1+c)^{-1/2} e^{-t_i^2/2(1+c)} \right\}. \tag{3.20}$$

Since many fewer terms are involved in the product in (3.20) than in the sum in (3.19), maximization of (3.20) is feasible by numerical methods even for moderately large p.

Replacing σ^2 by an estimate $\hat{\sigma}^2$, the estimators $\hat{c}$ and $\hat{w}$ that maximize the marginal likelihood L above can be used as prior inputs for an empirical Bayes analysis under the priors (3.14) and (3.15). In particular, (3.17) reveals that the γ maximizing the posterior $p(\gamma \mid Y, \hat{c}, \hat{w})$ can be obtained as the γ that maximizes the marginal maximum likelihood criterion

$$C_{\text{MML}} = SS_\gamma / \hat{\sigma}^2 - F(\hat{c}, \hat{w})\, q_\gamma, \tag{3.21}$$

where $F(c, w)$ is given by (3.18). Because maximizing (3.19) to obtain $\hat{c}$ and $\hat{w}$ can be computationally overwhelming when p is large and X is not orthogonal, one might instead consider a computable empirical Bayes approximation, the conditional maximum likelihood criterion

$$C_{\text{CML}} = SS_\gamma/\hat{\sigma}^2 - q_\gamma \left\{ 1 + \log_+ (SS_\gamma / \hat{\sigma}^2 q_\gamma) \right\} - 2 \left\{ \log(p - q_\gamma)^{-(p-q_\gamma)} + \log q_\gamma^{-q_\gamma} \right\} \tag{3.22}$$

where $\log_+(\cdot)$ is the positive part of $\log(\cdot)$. Selecting the γ that maximizes C_{CML} provides an approximate empirical Bayes alternative to selection based on C_{MML}.

In contrast to criteria of the form (3.16), which penalize $SS_\gamma/\hat{\sigma}^2$ by Fq_γ, with F constant, C_{MML} uses an adaptive penalty $F(\hat{c}, \hat{w})$ that is implicitly based on the estimated distribution of the regression coefficients. C_{CML} also uses an adaptive penalty, but one can be expressed by a rapidly computable closed form that can be shown to act like a combination of a modified BIC penalty $F = \log n$, which gives it same consistency property as BIC, and a modified RIC penalty $F = 2 \log p$. Insofar as maximizing C_{CML} approximates maximizing C_{MML}, these interpretations at least roughly explain the behavior of the C_{MML} penalty $F(\hat{c}, \hat{w})$ in (3.21).

George and Foster (2000) proposed the empirical Bayes criteria C_{MML} and C_{CML} and provided simulation evaluations demonstrating substantial performance advantages over the fixed penalty criteria (3.16); selection using C_{MML} delivers excellent performance over a much wider portion of the model space, and C_{CML} performs nearly as well. The superiority of empirical Bayes methods was confirmed in context of wavelet regression by Johnstone and Silverman (1998) and Clyde and George (1999). Johnstone and Silverman (1998) demonstrated the superiority of using maximum marginal likelihood estimates of c and w with posterior median selection criteria, and proposed EM algorithms for implementation. Clyde and George (1999) also proposed EM algorithms for implementation, extended the methods to include empirical Bayes estimates of σ^2 and considered both model selection and model averaging.

Finally, a fully Bayes analysis which integrates out c and w with respect to some noninformative prior $p(c, w)$ could be a promising alternative to empirical Bayes estimation of c and w. Indeed, the maximum marginal likelihood estimates $\hat{c}$ and $\hat{w}$ correspond to the posterior mode estimates under a Bayes formulation with independent uniform priors on c and w, a natural default choice. As such, the empirical Bayes methods can be considered as approximations to fully Bayes methods, but approximations which do not fully account for the uncertainty surrounding c and w. We are currently investigating the potential of such fully Bayes alternatives and plan to report on them elsewhere.

3.4 Parameter Priors for Selection Based on Practical Significance

A potential drawback of the point-normal prior (3.9) for variable selection is that with enough data, the posterior will favor the inclusion of X_i for any $\beta_i \neq 0$, no matter how small. Although this might be desirable from a purely predictive standpoint, it can also run counter to the goals of parsimony and interpretability in some problems, where it would be preferable to ignore such negligible β_i. A similar phenomenon occurs in frequentist hypothesis testing, where for large enough sample sizes, small departures from a point null become statistically significant even though they are not practically significant or meaningful.

An alternative to the point-normal prior (3.9), which avoids this potential drawback, is the normal-normal formulation used in the stochastic search variable selection (SSVS) procedure of George and McCulloch (1993, 1996, 1997). This formulation builds in the goal of excluding X_i from the model whenever $|\beta_i| < \delta_i$ for a given $\delta_i > 0$. The idea is that δ_i is a "threshold of practical significance" that is prespecified by the user. A simple choice might be $\delta_i = \Delta Y/\Delta X_i$, where ΔY is the size of an insignificant change in Y, and ΔX_i is the size of the maximum feasible change in X_i. To account for the cumulative effect of changes of other X's in the model, one might prefer the smaller conservative choice $\delta_i = \Delta Y/(p\Delta X_i)$. The practical potential of the SSVS formulation is nicely illustrated by Wakefield and Bennett (1996).

Under the normal-normal formulation of SSVS, the data always follow the saturated model (3.1) so that

$$p(Y \mid \beta, \sigma^2, \gamma) = N_n(X\beta, \sigma^2 I) \tag{3.23}$$

for all γ. In the general notation of Section 2, the model parameters here are the same for every model, $\theta_k \equiv (\beta, \sigma^2)$. The γth model is instead distinguished by a coefficient prior of the form

$$\pi(\beta \mid \sigma^2, \gamma) = \pi(\beta \mid \gamma) = N_p(0, D_\gamma R D_\gamma) \tag{3.24}$$

where R is a correlation matrix and D_γ is a diagonal matrix with diagonal elements

$$(D_\gamma)_{ii} = \begin{cases} \sqrt{v_{0i}} & \text{when} \quad \gamma_i = 0 \\ \sqrt{v_{1i}} & \text{when} \quad \gamma_i = 1 \end{cases} \tag{3.25}$$

Under the model space prior $p(\gamma)$, the marginal prior distribution of each component of β is here

$$p(\beta_i) = (1 - p(\gamma_i))N(0, v_{0i}) + p(\gamma_i)N(0, v_{1i}) \tag{3.26}$$

a scale mixture of two normal distributions.

Although β is independent of σ^2 in (3.24), the inverse Gamma prior (3.10) for σ^2 is still useful, as are the specification considerations for it discussed in Section 3.2. Furthermore, $R \propto (X'X)^{-1}$ and $R = I$ are natural choices for R in (3.24), similarly to the commonly used choices for Σ_γ in (3.9).

To use this normal-normal setup for variable selection, the hyperparameters v_{0i} and v_{1i} are set "small and large" respectively, so that $N(0, v_{0i})$ is concentrated and $N(0, v_{1i})$ is diffuse. The general idea is that when the data support $\gamma_i = 0$ over $\gamma_i = 1$, then β_i is probably small enough so that X_i will not be needed in the model. For a given threshold δ_i, higher posterior weighting of those γ values for which $|\beta_i| > \delta_i$ when $\gamma_i = 1$, can be achieved by choosing v_{0i} and v_{1i} such that $p(\beta_i \mid \gamma_i = 0) = N(0, v_{0i}) > p(\beta_i \mid \gamma_i = 1) = N(0, v_{1i})$ precisely on the interval $(-\delta_i, \delta_i)$. This property is obtained by any v_{0i} and v_{1i} satisfying

$$\log(v_{1i}/v_{0i})/(v_{0i}^{-1} - v_{1i}^{-1}) = \delta_{i\gamma}^2 \tag{3.27}$$

By choosing v_{1i} such that $N(0, v_{1i})$ is consistent with plausible values of β_i, v_{0i} can then be chosen according to (3.27). George and McCulloch (1997) report that computational problems and difficulties with v_{1i} too large will be avoided whenever $v_{1i}/v_{0i} \leq 10,000$, thus allowing for a wide variety of settings.

Under the normal-normal setup above, the joint distribution of β and σ^2 given γ is not conjugate for (3.1) because (3.24) excludes σ^2. This prevents analytical reduction of the full posterior $p(\beta, \sigma^2, \gamma \mid Y)$, which can severely increase the cost of posterior computations. To avoid this, one can instead consider the conjugate normal-normal formulation using

$$p(\beta \mid \sigma^2, \gamma) = N_p(0, \sigma^2 D_\gamma R D_\gamma), \tag{3.28}$$

which is identical to (3.24) except for the insertion of σ^2. Coupled with the inverse Gamma prior (3.10) for σ^2, the conditional distribution of β and σ^2 given γ is conjugate. This allows for the analytical margining out of β and σ^2 from $p(Y, \beta, \sigma^2 \mid \gamma) = p(Y \mid \beta, \sigma^2) p(\beta \mid \sigma^2, \gamma) p(\sigma^2 \mid \gamma)$ to yield

$$p(Y \mid \gamma) \propto |X'X + (D_\gamma R D_\gamma)^{-1}|^{-1/2} |D_\gamma R D_\gamma|^{-1/2} (\nu\lambda + S_\gamma^2)^{-(n+\nu)/2} \tag{3.29}$$

where

$$S_\gamma^2 = Y'Y - Y'X(X'X + (D_\gamma R D_\gamma)^{-1})^{-1} X'Y. \tag{3.30}$$

As will be seen in Section 3.5, this simplification confers strong advantages for posterior calculation and exploration.

Under (3.28), (3.10), and a model space prior $p(\gamma)$, the marginal distribution each component of β is

$$p(\beta_i \mid \gamma) = (1 - \gamma_i) T_1(\nu, 0, \lambda v_{0i}) + \gamma_i T_1(\nu, 0, \lambda v_{1i}), \tag{3.31}$$

a scale mixture of t-distributions, in contrast to the normal mixture (3.26). As with the nonconjugate prior, the idea is that v_{0i} and v_{1i} are to be set "small and large" respectively, so that when the data supports $\gamma_i = 0$ over $\gamma_i = 1$, then β_i is probably small enough so that X_i will not be needed in the model. However, the way in which v_{0i} and v_{1i} determine "small and large" is affected by the unknown value of σ^2, thereby making specification more difficult and less reliable than in the nonconjugate formulation. For a chosen threshold of practical significance δ_i, the pdf $p(\beta_i \mid i, \gamma_i = 0) = T(\nu, 0, \lambda v_{0i})$ is larger than the pdf $p(\beta_i \mid i, \gamma_i = 1) = T(\nu, 0, \lambda v_{1i})$ precisely on the interval $(-\delta_i, \delta_i)$, when v_{0i} and v_{1i} satisfy

$$(v_{0i}/v_{1i})^{\nu/(\nu+1)} = [v_{0i} + \delta_i^2/(\nu\lambda)]/[v_{1i} + \delta_i^2/(\nu\lambda)] \tag{3.32}$$

By choosing v_{1i} such that $T(\nu, 0, \lambda v_{1i})$ is consistent with plausible values of β_i, v_{0i} can then be chosen according to (3.32).

Another potentially valuable specification of the conjugate normal-normal formulation can be used to address the problem of outlier detection, which can be framed as a variable selection problem by including indicator variables for the observations as potential predictors. For such indicator variables, the choice $v_{0i}^* = 1$ and $v_{1i}^* = K > 0$ yields the well-known additive outlier formulation, see, for example, Petit and Smith (1985). Furthermore, when used in combination with the previous settings for ordinary predictors, the conjugate prior provides a hierarchical formulation for simultaneous variable selection and outlier detection. This has also been considered by Smith and Kohn (1996). A related treatment has been considered by Hoeting, Raftery and Madigan (1996).

3.5 Posterior Calculation and Exploration for Variable Selection

3.5.1 Closed Form Expressions for $p(Y \mid \gamma)$

A valuable feature of the previous conjugate prior formulations is that they allow for analytical margining out of β and σ^2 from $p(Y, \beta, \sigma^2 \mid \gamma)$ to yield the closed form expressions in (3.11) and (3.29) which are proportional to $p(Y \mid \gamma)$. Thus, when the model prior $p(\gamma)$ is computable, this can be used to obtain a computable, closed form expression $g(\gamma)$ satisfying

$$g(\gamma) \propto p(Y \mid \gamma) p(\gamma) \propto p(\gamma \mid Y). \tag{3.33}$$

The availability of such $g(\gamma)$ can greatly facilitate posterior calculation and estimation. Furthermore, it turns out that for certain formulations, the value of $g(\gamma)$ can be rapidly updated as γ is changed by a single component. As will be seen, such rapid updating schemes can be used to speed up algorithms for evaluating and exploring the posterior $p(\gamma \mid Y)$.

Consider first the conjugate point-normal formulation (3.9) and (3.10) for which $p(Y \mid \gamma)$ proportional to (3.11) can be obtained. When $\Sigma_\gamma = c\,(X_\gamma' X_\gamma)^{-1}$, a function $g(\gamma)$ satisfying (3.33) can be expressed as

$$g(\gamma) = (1+c)^{-q_\gamma/2} (\nu\lambda + Y'Y - (1+1/c)^{-1} W'W)^{-(n+\nu)/2} p(\gamma) \tag{3.34}$$

where $W = T'^{-1} X_\gamma' Y$ for upper triangular T such that $T'T = X_\gamma' X_\gamma$ for (obtainable by the Cholesky decomposition). As noted by Smith and Kohn (1996), the algorithm of Dongarra, Moler, Bunch and Stewart (1979) provides fast updating of T, and hence W and $g(\gamma)$, when γ is changed one component at a time. This algorithm requires $O(q_\gamma^2)$ operations per update, where γ is the changed value.

Now consider the conjugate normal-normal formulation (3.28) and (3.10) for which $p(Y \mid \gamma)$ proportional to (3.29) can be obtained. When $R = I$ holds, a function $g(\gamma)$

satisfying (3.33) can be expressed as

$$g(\gamma) = (\prod_{i=1}^{p} T_{ii}^2 [(1-\gamma_i)v_{0\gamma_{(i)}}^* + \gamma_i v_{1\gamma_{(i)}}^*])^{-1/2}(\nu\lambda + Y'Y - W'W)^{-(n+\nu)/2} p(\gamma) \quad (3.35)$$

where $W = T'^{-1}\tilde{X}'\tilde{Y}$ for upper triangular T such that $T'T = \tilde{X}'\tilde{X}$ (obtainable by the Cholesky decomposition). As noted by George and McCulloch (1997), the Chambers (1971) algorithm provides fast updating of T, and hence W and $g(\gamma)$, when γ is changed one component at a time. This algorithm requires $O(p^2)$ operations per update.

The availability of these computable, closed form expressions for $g(\gamma) \propto p(\gamma \mid Y)$ enables exhaustive calculation of $p(\gamma \mid Y)$ in moderately sized problems. In general, this simply entails calculating $g(\gamma)$ for every γ value and then summing over all γ values to obtain the normalization constant. However, when one of the above fast updating schemes can be used, this calculation can be substantially speeded up by sequential calculation of the 2^p $g(\gamma)$ values where consecutive γ differ by just one component. Such an ordering is provided by the Gray Code, George and McCulloch (1997). After computing T, W and $g(\gamma)$ for an initial γ value, subsequent values of T, W and $g(\gamma)$ can be obtained with the appropriate fast updating scheme by proceeding in the Gray Code order. Using this approach, this exhaustive calculation is feasible for p less than about 25.

3.5.2 MCMC Methods for Variable Selection

MCMC methods have become a principal tool for posterior evaluation and exploration in Bayesian variable selection problems. Such methods are used to simulate a sequence

$$\gamma^{(1)}, \gamma^{(2)}, \ldots \quad (3.36)$$

that converges (in distribution) to $p(\gamma|Y)$. In formulations where analytical simplification of $p(\beta, \sigma^2, \gamma \mid Y)$ is unavailable, (3.36) can be obtained as a subsequence of a simulated Markov chain of the form

$$\beta^{(1)}, \sigma^{(1)}, \gamma^{(1)}, \beta^{(2)}, \sigma^{(2)}, \gamma^{(2)}, \ldots \quad (3.37)$$

that converges to $p(\beta, \sigma^2, \gamma \mid Y)$. However, in conjugate formulations where β and σ^2 can be analytically eliminated form the posterior, the availability of $g(\gamma) \propto p(\gamma \mid Y)$ allows for the flexible construction of MCMC algorithms that simulate (3.36) directly as a Markov chain. Such chains are often more useful, in terms of both computational and convergence speed.

In problems where the number of potential predictors p is very small, and $g(\gamma) \propto p(\gamma \mid Y)$ is unavailable, the sequence (3.36) may be used to evaluate the entire posterior $p(\gamma \mid Y)$. Indeed, empirical frequencies and other functions of the γ values will be consistent estimates of their values under $p(\gamma \mid Y)$. In large problems where exhaustive calculation of all 2^p values of $p(\gamma|Y)$ is not feasible, the sequence (3.36) may still provide useful information. Even when the length of the sequence (3.36) is much smaller than 2^p, it may be possible to identify at least some of the high probability γ, since those γ are expected to appear more frequently. In this sense, these MCMC methods can be used to stochastically search for high probability models.

In the next two subsections, we describe various MCMC algorithms which may be useful for simulating (3.36). These algorithms are obtained as variants of the Gibbs sampler (GS) and Metropolis-Hastings (MH) algorithms described in Section 2.3.

3.5.3 Gibbs Sampling Algorithms

Under the nonconjugate normal-normal formulation (3.24) and (3.10) for SSVS, the posterior $p(\beta, \sigma^2, \gamma \mid Y)$ is p-dimensional for all γ. Thus, a simple GS that simulates the full parameter sequence (3.37) is obtained by successive simulation from the full conditionals

$$\begin{gathered} p(\beta \mid \sigma^2, \gamma, Y) \\ p(\sigma^2 \mid \beta, \gamma, Y) = p(\sigma^2 \mid \beta, Y) \\ p(\gamma_i \mid \beta, \sigma^2, \gamma_{(i)}, Y) = p(\gamma_i \mid \beta, \gamma_{(i)}),\ i = 1, \ldots, p \end{gathered} \tag{3.38}$$

where at each step, these distributions are conditioned on the most recently generated parameter values. These conditionals are standard distributions which can be simulated quickly and efficiently by routine methods.

For conjugate formulations where $g(\gamma)$ is available, a variety of MCMC algorithms for generating (3.36) directly as a Markov chain, can be conveniently obtained by applying the GS with $g(\gamma)$. The simplest such implementation is obtained by generating each γ value componentwise from the full conditionals,

$$\gamma_i \mid \gamma_{(i)}, Y \quad i = 1, 2, \ldots, p, \tag{3.39}$$

($\gamma_{(i)} = (\gamma_1, \gamma_2, \ldots, \gamma_{i-1}, \gamma_{i+1}, \ldots, \gamma_p)$) where the γ_i may be drawn in any fixed or random order. By margining out β and σ^2 in advance, the sequence (3.36) obtained by this algorithm should converge faster than the nonconjugate Gibbs approach, rendering it more effective on a per iteration basis for learning about $p(\gamma \mid Y)$, see Liu, Wong and Kong (1994).

The generation of the components in (3.39) in conjunction with $g(\gamma)$ can be obtained trivially as a sequence of Bernoulli draws. Furthermore, if $g(\gamma)$ allows for fast updating as in (3.34) or (3.35), the required sequence of Bernoulli probabilities can be computed faster and more efficiently. To see this, note that the Bernoulli probabilities are simple functions of the ratio

$$\frac{p(\gamma_i = 1, \gamma_{(i)} \mid Y)}{p(\gamma_i = 0, \gamma_{(i)} \mid Y)} = \frac{g(\gamma_i = 1, \gamma_{(i)})}{g(\gamma_i = 0, \gamma_{(i)})}. \tag{3.40}$$

At each step of the iterative simulation from (3.39), one of the values of $g(\gamma)$ in (3.40) will be available from the previous component simulation. Since γ has been varied by only a single component, the other value of $g(\gamma)$ can then be obtained by using the appropriate updating scheme. Under the point-normal prior (3.9) with $\Sigma_\gamma = c\,(X_\gamma' X_\gamma)^{-1}$, the fast updating of (3.34) requires $O(q_\gamma^2)$ operations, whereas under the conjugate normal-normal prior formulation (3.28) with $R = I$ fast updating of (3.35) requires $O(p^2)$ operations. Thus, GS algorithms in the former case can be substantially faster when $p(\gamma \mid Y)$ is concentrated on those γ for which q_γ is small, namely the parsimonious models. This advantage could be pronounced in large problems with many useless predictors.

Simple variants of the componentwise GS can be obtained by generating the components in a different fixed or random order. Note that in any such generation, it is not necessary to generate each and every component once before repeating a coordinate. Another variant of the GS can be obtained by drawing the components of γ in groups, rather than one at a time. Let $\{I_k\}$, $k = 1, 2, \ldots, m$ be a partition of $\{1, 2, \ldots, p\}$ so that, $I_k \subseteq \{1, 2, \ldots, p\}$, $\cup I_k = \{1, 2, \ldots, p\}$ and $I_{k_1} \cap I_{k_2} = \emptyset$ for $k_1 \neq k_2$. Let $\gamma_{I_k} = \{\gamma_i \,|\, i \in I_k\}$ and $\gamma_{(I_k)} = \{\gamma_i \mid i \notin I_k\}$. The grouped GS generates (3.36) by iterative simulation from

$$\gamma_{I_k} \mid \gamma_{(I_k)}, Y \quad k = 1, 2, \ldots, m. \tag{3.41}$$

Fast updating of $g(\gamma)$, when available, can also be used to speed up this simulation by computing the conditional probabilities of each γ_{I_k} in Gray Code order. The potential advantage of such a grouped GS is improved convergence of (3.36). This might be achieved by choosing the partition so that strongly correlated γ_i are contained in the same I_k, thereby reducing the dependence between draws in the simulation. Intuitively, clusters of such correlated γ_i should correspond to clusters of correlated X_i which, in practice, might be identified by clustering procedures. As before, variants of the grouped GS can be obtained by generating the γ_{I_k} in a different fixed or random order.

3.5.4 Metropolis-Hastings Algorithms

The availability of $g(\gamma) \propto p(\gamma|Y)$ also facilitates the use of MH algorithms for direct simulation of (3.36). By restricting attention to the set of γ values, a discrete space, the simple MH form described in Section 2.3 can be used. Because $g(\gamma)/g(\gamma') = p(\gamma \mid Y)/p(\gamma' \mid Y)$, such MH algorithms are here of the form:

1. Simulate a candidate γ^* from a transition kernel $q(\gamma^* \mid \gamma^{(j)})$.

2. Set $\gamma^{(j+1)} = \gamma^*$ with probability
$$\alpha(\gamma^* \mid \gamma^{(j)}) = \min\left\{\frac{q(\gamma^{(j)} \mid \gamma^*)}{q(\gamma^* \mid \gamma^{(j)})}\frac{g(\gamma^*)}{g(\gamma^{(j)})}, 1\right\}. \tag{3.42}$$
Otherwise, $\gamma^{(j+1)} = \gamma^{(j)}$.

When available, fast updating schemes for $g(\gamma)$ can be exploited. Just as for the Gibbs sampler, the MH algorthims under the point-normal formulations (3.9) with $\Sigma_\gamma = c\,(X_\gamma' X_\gamma)^{-1}$ will be the fastest scheme when $p(\gamma \mid Y)$ is concentrated on those γ for which q_γ is small.

A special class of MH algorithms, the Metropolis algorithms, are obtained from the class of transition kernels $q(\gamma^1 \mid \gamma^0)$ which are symmetric in γ^1 and γ^0. For this class, the form of (3.42) simplifies to
$$\alpha^M(\gamma^* \mid \gamma^{(j)}) = \min\left\{\frac{g(\gamma^*)}{g(\gamma^{(j)})}, 1\right\}. \tag{3.43}$$
Perhaps the simplest symmetric transition kernel is
$$q(\gamma^1 \mid \gamma^0) = 1/p \quad \text{if} \quad \sum_1^p |\gamma_i^0 - \gamma_i^1| = 1. \tag{3.44}$$
This yields the Metropolis algorithm

1. Simulate a candidate γ^* by randomly changing one component of $\gamma^{(j)}$.
2. Set $\gamma^{(j+1)} = \gamma^*$ with probability $\alpha^M(\gamma^* \mid \gamma^{(j)})$. Otherwise, $\gamma^{(j+1)} = \gamma^{(j)}$.

This algorithm was proposed in a related model selection context by Madigan and York (1995) who called it MC^3. It was used by Raftery, Madigan and Hoeting (1997) for model averaging, and was proposed for the SSVS prior formulation by Clyde and Parmigiani (1994).

The transition kernel (3.44) is a special case of the class of symmetric transition kernels of the form
$$q(\gamma^1 \mid \gamma^0) = q_d \quad \text{if} \quad \sum_1^p |\gamma_i^0 - \gamma_i^1| = d. \tag{3.45}$$
Such transition kernels yield Metropolis algorithms of the form

1. Simulate a candidate γ^* by randomly changing d components of $\gamma^{(j)}$ with probability q_d.
2. Set $\gamma^{(j+1)} = \gamma^*$ with probability $\alpha^M(\gamma^* \mid \gamma^{(j)})$. Otherwise, $\gamma^{(j+1)} = \gamma^{(j)}$.

Here q_d is the probability that γ^* will have d new components. By allocating some weight to q_d for larger d, the resulting algorithm will occasionally make big jumps to different γ values. In contrast to the algorithm obtained by (3.44) which only moves locally, such algorithms require more computation per iteration.

Finally, it may also be of interest to consider asymmetric transition kernels such as

$$q(\gamma^1 \mid \gamma^0) = q_d \text{ if } \sum_1^p (\gamma_i^0 - \gamma_i^1) = d. \tag{3.46}$$

Here q_d is the probability of generating a candidate value γ^* which corresponds to a model with d more variables $\gamma^{(j)}$. When $d < 0$, γ^* will represent a more parsimonious model than $\gamma^{(j)}$. By suitable weighting of the q_d probabilities, such Metropolis-Hastings algorithms can be made to explore the posterior in the region of more parsimonious models.

3.5.5 Extracting Information from the Output

In nonconjugate setups, where $g(\gamma)$ is unavailable, inference about posterior characteristics based on (3.36) ultimately rely on the empirical frequency estimates the visited γ values. Although such estimates of posterior characteristics will be consistent, they may be unreliable, especially if the size of the simulated sample is small in comparison to 2^p or if there is substantial dependence between draws. The use of empirical frequencies to identify high probability γ values for selection can be similarly problematic.

However, the situation changes dramatically in conjugate setups where $g(\gamma) \propto p(\gamma|Y)$ is available. To begin with, $g(\gamma)$ provides the relative probability of any two values γ^0 and γ^1 via $g(\gamma^0) \,/\, g(\gamma^1)$ and so can be used to definitively identify the higher probability γ in the sequence (3.36) of simulated values. Only minimal additional effort is required to obtain such calculations since $g(\gamma)$ must be calculated for each of the visited γ values in the execution of the MCMC algorithms described in Sections 3.5.3 and 3.5.4.

The availability of $g(\gamma)$ can also be used to obtain estimators of the normalizing constant C

$$p(\gamma \mid Y) = C\, g(\gamma) \tag{3.47}$$

based on the MCMC output (3.36) , say $\gamma^{(1)}, \ldots, \gamma^{(K)}$. Let A be a preselected subset of γ values and let $g(A) = \sum_{\gamma \in A} g(\gamma)$ so that $p(A \mid Y) = C\, g(A)$. Then, a consistent estimator of C is

$$\hat{C} = \frac{1}{g(A)K} \sum_{k=1}^{K} I_A(\gamma^{(k)}) \tag{3.48}$$

where $I_A(\,)$ is the indicator of the set A, George and McCulloch (1997). Note that if (3.36) were an uncorrelated sequence, then $\text{Var}(\hat{C}) = (C^2/K)\frac{1-p(A \mid Y)}{p(A \mid Y)}$ suggesting that the variance of (3.48) is decreasing as $p(A \mid Y)$ increases. It is also desirable to choose A such that $I_A(\gamma^{(k)})$ can be easily evaluated. George and McCulloch (1997) obtain very good results by setting A to be those γ values visited by a preliminary simulation of (3.36). Peng (1998) has extended and generalized these ideas to obtain estimators of C that improve on (3.48).

Inserting $\hat{C}$ into (3.47) yields improved estimates of the probability of individual γ values

$$\hat{p}(\gamma \mid Y) = \hat{C}\, g(\gamma), \tag{3.49}$$

as well as an estimate of the total visited probability

$$\hat{p}(B \mid Y) = \hat{C}\, g(B), \tag{3.50}$$

where B is the set of visited γ values. Such $\hat{p}(B \mid Y)$ can provide valuable information about when to stop a MCMC simulation. Since $\hat{p}(\gamma \mid Y)/p(\gamma \mid Y) \equiv \hat{C}/C$, the uniform accuracy of the probability estimates (3.49) is

$$|(\hat{C}/C) - 1|. \tag{3.51}$$

This quantity is also the total probability discrepancy since $\sum_\gamma |\hat{p}(\gamma \mid Y) - p(\gamma \mid Y)| = |\hat{C} - C| \sum_\gamma g(\gamma) = |(\hat{C}/C) - 1|$.

The simulated values (3.36) can also play an important role in model averaging. For example, suppose one wanted to predict a quantity of interest Δ by the posterior mean

$$E(\Delta \mid Y) = \sum_{\text{all } \gamma} E(\Delta \mid \gamma, Y) p(\gamma \mid Y). \tag{3.52}$$

When p is too large for exhaustive enumeration and $p(\gamma \mid Y)$ cannot be computed, (3.52) is unavailable and is typically approximated by something of the form

$$\hat{E}(\Delta \mid Y) = \sum_{\gamma \in S} E(\Delta \mid \gamma, Y) \hat{p}(\gamma \mid Y, S) \tag{3.53}$$

where S is a manageable subset of models and $\hat{p}(\gamma \mid Y, S)$ is a probability distribution over S. (In some cases, $E(\Delta \mid \gamma, Y)$ will also be approximated).

Using the Markov chain sample for S, a natural choice for (3.53) is

$$\hat{E}_f(\Delta \mid Y) = \sum_{\gamma \in S} E(\Delta \mid \gamma, Y) \hat{p}_f(\gamma \mid Y, S) \tag{3.54}$$

where $\hat{p}_f(\gamma \mid Y, S)$ is the relative frequency of γ in S, George (1999). Indeed, (3.54) will be a consistent estimator of $E(\Delta \mid Y)$. However, here too, it appears that when $g(\gamma)$ is available, one can do better by using

$$\hat{E}_g(\Delta \mid Y) = \sum_{\gamma \in S} E(\Delta \mid \gamma, Y) \hat{p}_g(\gamma \mid Y, S) \tag{3.55}$$

where

$$\hat{p}_g(\gamma \mid Y, S) = g(\gamma)/g(S) \tag{3.56}$$

is just the renormalized value of $g(\gamma)$. For example, when S is an iid sample from $p(\gamma|Y)$, (3.55) increasingly approximates the best unbiased estimator of $E(\Delta \mid Y)$ as the sample size increases. To see this, note that when S is an iid sample, $\hat{E}_f(\Delta \mid Y)$ is unbiased for $E(\Delta \mid Y)$. Since S (together with g) is sufficient, the Rao-Blackwellized estimator $E(\hat{E}_f(\Delta \mid Y) \mid S)$ is best unbiased. But as the sample size increases, $E(\hat{E}_f(\Delta \mid Y) \mid S) \to \hat{E}_g(\Delta \mid Y)$.

4 Bayesian CART Model Selection

For our second illustration of Bayesian model selection implementations, we consider the problem of selecting a classification and regression tree (CART) model for the relationship between a variable y and a vector of potential predictors $x = (x_1, \ldots, x_p)$. An alternative to linear regression, CART models provide a more flexible specification of the conditional distribution of y given x. This specification consists of a partition of the x space, and a set of distinct distributions for y within the subsets of the partition. The partition is accomplished by a binary tree T that recursively partitions the x space with internal node splitting rules of the form $\{x \in A\}$ or $\{x \notin A\}$. By moving from the root node through to the terminal nodes, each observation is assigned to a terminal node of T which then associates the observation with a distribution for y.

Although any distribution may be considered for the terminal node distributions, it is convenient to specify these as members of a single parametric family $p(y|\theta)$ and to assume all observations of y are conditionally independent given the parameter values. In this case, a CART model is identified by the tree T and the parameter values $\Theta = (\theta_1, \ldots, \theta_b)$ of the distributions at each of the b terminal nodes of T. Note that T here plays the role of M_k of model identifier as described in Section 2. The model is called a regression tree model or a classification tree model according to whether y is quantitative or qualitative, respectively. For regression trees, two simple and useful specifications for the terminal node distributions are the mean shift normal model

$$p(y \mid \theta_i) = N(\mu_i, \sigma^2), \; i = 1, \ldots, b, \tag{4.1}$$

where $\theta_i = (\mu_i, \sigma)$, and the mean-variance shift normal model

$$p(y \mid \theta_i) = N(\mu_i, \sigma_i^2), \; i = 1, \ldots, b, \tag{4.2}$$

where $\theta_i = (\mu_i, \sigma_i)$. For classification trees where y belongs to one of K categories, say

$C_1, \ldots, C_K$, a natural choice for terminal node distributions are the simple multinomials

$$p(y \mid \theta_i) = \prod_{k=1}^{K} p_{ik}^{I(y \in C_k)} \quad i = 1, \ldots, b, \tag{4.3}$$

where $\theta_i = p_i \equiv (p_{i1}, \ldots, p_{iK})$, $p_{ik} \geq 0$ and $\sum_k p_{ik} = 1$. Here $p(y \in C_k) = p_{ik}$ at the ith terminal node of T.

As illustration, Figure 1 depicts a regression tree model where $y \sim N(\theta, 2^2)$ and $x = (x_1, x_2)$. x_1 is a quantitative predictor taking values in [0,10], and x_2 is a qualitative predictor with categories $\{A, B, C, D\}$. The binary tree has 9 nodes of which $b = 5$ are terminal nodes that partition the x space into 5 subsets. The splitting rules are displayed at each internal node. For example, the leftmost terminal node corresponds to $x_1 \leq 3.0$ and $x_2 \in \{C, D\}$. The θ_i value identifying the mean of y given x is displayed at each terminal node. Note that in contrast to a linear model, θ_i decreases in x_1 when $x_2 \in \{A, B\}$, but increases in x_1 when $x_2 \in \{C, D\}$.

The two basic components of the Bayesian approach to CART model selection are prior specification and posterior exploration. Prior specification over CART models entails putting a prior on the tree space and priors on the parameters of the terminal node distributions. The CART model likelihoods are then used to update the prior to yield a posterior distribution that can be used for model selection. Although straightforward in principle, practical implementations require subtle and delicate attention to details. Prior formulation must be interpretable and computationally manageable. Hyperparameter specification can be usefully guided by overall location and scale measures of the data. A feature of this approach is that the prior specification can be used to downweight undesirable model characteristics such as tree complexity or to express a preference for certain predictor variables. Although the entire posterior cannot be computed in nontrivial problems, posterior guided MH algorithms can still be used to search for good tree models. However, the algorithms require repeated restarting or other modifications because of the multimodal nature of the posterior. As the search proceeds, selection based on marginal likelihood rather than posterior probability is preferable because of the dilution properties of the prior. Alternatively, a posterior weighted average of the visited models can be easily obtained.

CART modelling was popularized in the statistical community by the seminal book of Breiman, Friedman, Olshen and Stone (1984). Earlier work by Kass (1980) and Hawkins and Kass (1982) developed tree models in a statistical framework. There has also been substantial research on trees in the field of machine learning, for example the C4.5 algorithm and its predecessor, ID3 (Quinlan 1986, 1993). Here, we focus on the method of Breiman et al. (1984), which proposed a nonparametric approach for tree selection based on a greedy algorithm named CART. A concise description of this approach, which

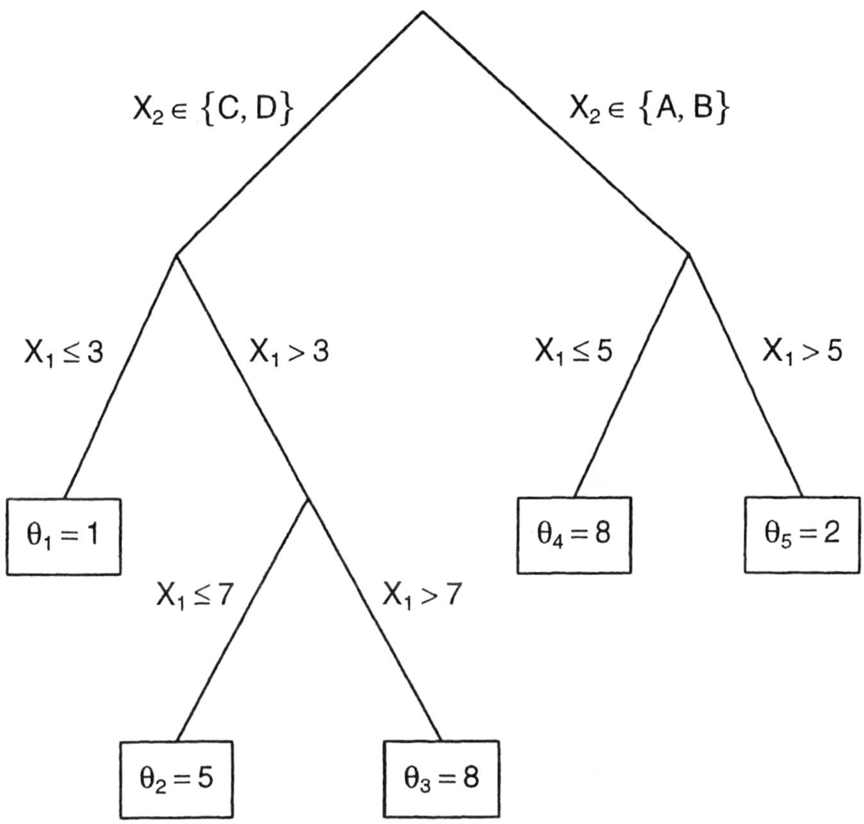

Figure 1: *A regression tree where* $y \sim N(\theta, 2^2)$ *and* $x = (x_1, x_2)$.

seeks to partition the x space into regions where the distribution of y is 'homogeneous', and its implementation in S appears in Clark and Pregibon (1992). Bayesian approaches to CART are enabled by elaborating the CART tree formulation to include parametric terminal node distributions, effectively turning it into a statistical model and providing a likelihood. Conventional greedy search algorithms are also replaced by the MCMC algorithms that provide a broader search over the tree model space.

The Bayesian CART model selection implementations described here were proposed by Chipman, George and McCulloch (1998) and Denison, Mallick and Smith (1998a), hereafter referred to as CGM and DMS, respectively. An earlier Bayesian approach to classification tree modelling was proposed by Buntine (1992) which, compared to CGM and DMS, uses similar priors for terminal node distributions, but different priors on the space of trees, and deterministic, rather than stochastic, algorithms for model search. Priors for tree models based on Minimum Encoding ideas were proposed by Quinlan and Rivest (1989) and Wallace and Patrick (1993). Oliver and Hand (1995) discuss and provide an empirical comparison of a variety of pruning and Bayesian model averaging approaches based on CART. Paass and Kindermann (1997) applied a simpler version of the CGM approach and obtained results which uniformly dominated a wide variety of competing methods. Other alternatives to greedy search methods include Sutton (1991) and Lutsko and Kuijpers (1994) who use simulated annealing, Jordan and Jacobs (1994) who use the EM algorithm, Breiman (1996), who averages trees based on bootstrap samples, and Tibshirani and Knight (1999) who select trees based on bootstrap samples.

4.1 Prior Formulations for Bayesian CART

Since a CART model is identified by (Θ, T), a Bayesian analysis of the problem proceeds by specifying priors on the parameters of the terminal node distributions of each tree $p(\Theta \mid T)$ and a prior distribution $p(T)$ over the set of trees. Because the prior for T does not depend on the form of the terminal node distributions, $p(T)$ can be generally considered for both regression trees and classification trees.

4.1.1 Tree Prior Specification

A tree T partitions the x space and consists of both the binary tree structure and the set of splitting rules associated with the internal nodes. A general formulation approach for $p(T)$ proposed by CGM, is to specify $p(T)$ implicitly by the following tree-generating stochastic process which "grows" trees from a single root tree by randomly "splitting" terminal nodes:

1. Begin by setting T to be the trivial tree consisting of a single root (and terminal)

node denoted η.

2. Split the terminal node η with probability $p_\eta = \alpha(1 + d_\eta)^{-\beta}$ where d_η is the depth of the node η, and $\alpha \in (0, 1)$ and $\beta \geq 0$ are prechosen control parameters.

3. If the node splits, randomly assign it a splitting rule as follows: First choose x_i uniformly from the set of available predictors. If x_i is quantitative, assign a splitting rule of the form $\{x_i \leq s\}$ vs $\{x_i > s\}$ where s is chosen uniformly from the available observed values of x_i. If x_i is qualitative, assign a splitting rule of the form $\{x_i \in C\}$ vs $\{x_i \notin C\}$ where C is chosen uniformly from the set of subsets of available categories of x_i. Next assign left and right children nodes to the split node, and apply steps 2 and 3 to the newly created tree with η equal to the new left and the right children (if nontrivial splitting rules are available).

By available in step 3, we mean those predictors, split values and category subsets that would not lead to empty terminal nodes. For example, if a binary predictor was used in a splitting rule, it would no longer be available for splitting rules at nodes below it. Each realization of such a process can simply be considered as a random draw from $p(T)$. Furthermore, this specification allows for straightforward evaluation of $p(T)$ for any T, and can be effectively coupled with the MH search algorithms described in Section 4.2.1.

Although other useful forms can easily be considered for the splitting probability in step 2 above, the choice of $p_\eta = \alpha(1+d_\eta)^{-\beta}$ is simple, interpretable, easy to compute and dependent only on the depth d_η of the node η. The parameters α and β control the size and shape of the binary tree produced by the process. To see how, consider the simple specification, $p_\eta \equiv \alpha < 1$ when $\beta = 0$. In this case the probability of any particular binary tree with b terminal nodes (ignoring the constraints of splitting rule assignments in step 3) is just $\alpha^{b-1}(1 - \alpha)^b$, a natural generalization of the geometric distribution. (A binary tree with b terminal nodes will always have exactly $(b - 1)$ internal nodes). Setting α small will tend to yield smaller trees and is a simple convenient way to control the size of trees generated by growing process.

The choice of $\beta = 0$ above assigns equal probability to all binary trees with b terminal nodes regardless of their shape. Indeed any prior that is only a function of b will do this; for example, DMS recommend this with a truncated Poisson distribution on b. However, for increasing $\beta > 0$, p_η is a decreasing function of d_η making deeper nodes less likely to split. The resulting prior $p(T)$ puts higher probability on "bushy" trees, those whose terminal nodes do not vary too much in depth. Choosing α and β in practice can guided by looking at the implicit marginal distributions of characteristics such as b. Such marginals can be easily simulated and graphed.

Turning to the splitting rule assignments, step 3 of the tree growing process represents the prior information that at each node, available predictors are equally likely to be effective, and that for each predictor, available split values or category subsets are equally likely to be effective. This specification is invariant to monotone transformations of the quantitative predictors, and is uniform on the observed quantiles of a quantitative predictor with no repeated values. However, it is not uniform over all possible splitting rules because it assigns lower probability to splitting rules based on predictors with more potential split values or category subsets. This feature is necessary to maintain equal probability on predictor choices, and essentially yields the dilution property discussed in Sections 2.2 and 3.1. Predictors with more potential split values will give rise to more trees. By downweighting the splitting rules of such predictors, $p(T)$ serves to dilute probability within neighborhoods of similar trees.

Although the uniform choices for $p(T)$ above seem to be reasonable defaults, non-uniform choices may also be of interest. For example, it may be preferable to place higher probability on predictors that are thought to be more important. A preference for models with fewer variables could be expressed by putting greater mass on variables already assigned to ancestral nodes. For the choice of split value, tapered distribution at the extremes would increase the tendency to split more towards the interior range of a variable. One might also consider the distribution of split values to be uniform on the available range of the predictor and so could weight the available observed values accordingly. For the choice of category subset, one might put extra weight on subsets thought to be more important.

As a practical matter, note that all of the choices above consider only the observed predictor values as possible split points. This induces a discrete distribution on the set of splitting rules, and hence the support of $p(T)$ will be a finite set of trees in any application. This is not really a restriction since it allows for all possible partitions of any given data set. The alternative of putting a continuous distribution on the range of the predictors would needlessly increase the computational requirements of posterior search while providing no gain in generality. Finally, we note that the dependence of $p(T)$ on the observed x values is typical of default prior formulations, as was the case for some of the coefficient prior covariance choices discussed in Sections 3.2 and 3.4.

4.1.2 Parameter Prior Specifications

As discussed in Section 2.3, the computational burden of posterior calculation and exploration is substantially reduced when the marginal likelihood, here $p(Y \mid T)$, can be obtained in closed form. Because of the large size of the space of CART models, this computational consideration is key in choosing the prior $p(\Theta \mid T)$ for the parameters of the terminal node distributions. For this purpose, we recommend the conjugate prior forms below for the parameters of the models (4.1)-(4.3). For each of these priors, Θ can be analytically margined out via (2.2), namely

$$p(Y \mid T) = \int p(Y \mid \Theta, T) p(\Theta \mid T) d\Theta, \tag{4.4}$$

where Y here denotes the observed values of y.

For regression trees with the mean-shift normal model (4.1), perhaps the simplest prior specification for $p(\Theta|T)$ is the standard normal-inverse-gamma form where $\mu_1, \dots, \mu_b$ are iid given σ and T with

$$p(\mu_i \mid \sigma, T) = N(\bar{\mu}, \sigma^2/a) \tag{4.5}$$

and

$$p(\sigma^2 \mid T) = p(\sigma^2) = IG(\nu/2, \nu\lambda/2). \tag{4.6}$$

Under this prior, standard analytical simplification yields

$$p(Y \mid T) \propto \frac{c\, a^{b/2}}{\prod_{i=1}^{b} (n_i + a)^{1/2}} \left(\sum_{i=1}^{b} (s_i + t_i) + \nu\lambda \right)^{-(n+\nu)/2} \tag{4.7}$$

where c is a constant which does not depend on T, s_i is $(n_i - 1)$ times the sample variance of the ith terminal node Y values, $t_i = \frac{n_i a}{n_i + a} (\bar{y}_i - \bar{\mu})^2$, and $\bar{y}_i$ is the sample mean of the ith terminal node Y values.

In practice, the observed Y can be used to guide the choice of hyperparameter values for $(\nu, \lambda, \bar{\mu}, a)$. Considerations similar to those discussed for Bayesian variable selection in Section 3.2 are also useful here. To begin with, because the mean-shift model attempts to explain the variation of Y, it is reasonable to expect that σ^2 will be smaller than s_Y^2, the sample variance of Y. Similarly, it is reasonable to expect that σ^2 will be larger than a pooled variance estimate obtained from a deliberate overfitting of the data by a greedy algorithm, say s_G^2. Using these values as guides, ν and λ would then be chosen so that the prior for σ^2 assigns substantial probability to the interval (s_G^2, s_Y^2). Once ν and λ have been chosen, $\bar{\mu}$ and a would be selected so that the prior for μ is spread out over the range of Y values.

For the more flexible mean-variance shift model (4.2) where σ_i can also vary across the terminal nodes, the normal-inverse-gamma form is easily extended to

$$p(\mu_i \mid \sigma_i, T) = N(\bar{\mu}, \sigma_i^2/a) \tag{4.8}$$

and

$$p(\sigma_i^2 \mid T) = p(\sigma_i^2) = IG(\nu/2, \nu\lambda/2), \tag{4.9}$$

with the pairs $(\mu_1, \sigma_1), \ldots, (\mu_b, \sigma_b)$ independently distributed given T. Under this prior, analytical simplification is still straightforward, and yields

$$p(Y \mid T) \propto \prod_{i=1}^{b} \pi^{-n_i/2}(\lambda\nu)^{\nu/2} \frac{\sqrt{a}}{\sqrt{n_i + a}} \frac{\Gamma((n_i + \nu)/2)}{\Gamma(\nu/2)} (s_i + t_i + \nu\lambda)^{-(n_i+\nu)/2} \tag{4.10}$$

where s_i and t_i are as above. Interestingly, the MCMC computations discussed in the next section are facilitated by the factorization of this marginal likelihood across nodes, in contrast to the marginal likelihood (4.7) for the equal variance model.

Here too, the observed Y can be used to guide the choice of hyperparameter values for $(\nu, \lambda, \bar{\mu}, a)$. The same ideas above may be used with an additional consideration. In some cases, the mean-variance shift model may explain variance shifts much more so than mean shifts. To handle this possibility, it may be better to choose ν and λ so that σ_Y^2 is more toward the center rather than the right tail of the prior for σ^2. We might also tighten up our prior for μ about the average y value. In any case, it can be useful to explore the consequences of several different prior choices.

For classification trees with the simple multinomial model (4.3), a useful conjugate prior specification for $\Theta = (p_1, \ldots, p_b)$ is the standard Dirichlet distribution of dimension $K - 1$ with parameter $\alpha = (\alpha_1, \ldots, \alpha_K)$, $\alpha_k > 0$, where $p_1, \ldots, p_b$ are iid given T with

$$p(p_i \mid T) = \text{Dirichlet}(p_i \mid \alpha) \propto p_{i1}^{\alpha_1 - 1} \cdots p_{iK}^{\alpha_K - 1}. \tag{4.11}$$

When $K = 2$ this reduces to the familiar Beta prior. Under this prior, standard analytical simplification yields

$$p(Y \mid T) \propto \left(\frac{\Gamma\left(\sum_k \alpha_k\right)}{\prod_k \Gamma(\alpha_k)}\right)^b \prod_{i=1}^{b} \frac{\prod_k \Gamma(n_{ik} + \alpha_k)}{\Gamma\left(n_i + \sum_k \alpha_k\right)} \tag{4.12}$$

where n_{ik} is the number of ith terminal node Y values in category C_k, $n_i = \sum_k n_{ik}$ and $k = 1, \ldots, K$ over the sums and products above. For a given tree, $p(Y \mid T)$ will be larger when the Y values within each node are more homogeneous. To see this, note that assignments for which the Y values at the same node are similar will lead to more disparate values of $n_{i1}, \ldots, n_{iK}$, which in turn will lead to larger values of $p(Y \mid T)$.

The natural default choice for α is the vector $(1, \ldots, 1)$ for which the Dirichlet prior (4.11) is the uniform. However, by setting certain α_k to be larger for certain categories, $p(Y \mid T)$ will become more sensitive to misclassification at those categories. This would be desirable when classification into those categories is most important.

One detail of analytical simplifications yielding integrated likelihoods (4.7), (4.10) or (4.12) merits attention. Independence of parameters across terminal nodes means that integration can be carried out separately for each node. Normalizing constants in integrals for each node that would usually be discarded (for example $a^{b/2}$ in (4.7)) need to be kept, since the number of terminal nodes, b, varies across trees. This means that these normalizing constants will be exponentiated to a different power for trees of different size.

All the prior specifications above assume that given the tree T, the parameters in the terminal nodes are independent. When terminal nodes share many common parents, it may be desirable to introduce dependence between their θ_i values. Chipman, George, and McCulloch (2000) introduce such a dependence for the regression tree model, resulting in a Bayesian analogue of the tree shrinkage methods considered by Hastie and Pregibon (1990) and Leblanc and Tibshirani (1998).

4.2 Stochastic Search of the CART Model Posterior

Combining any of the closed form expressions (4.7), (4.10) or (4.12) for $p(Y \mid T)$ with $p(T)$ yields a closed form expression $g(T)$ satisfying

$$g(T) \propto p(Y \mid T)p(T) \propto p(T \mid Y). \tag{4.13}$$

Analogous to benefits of the availability $g(\gamma)$ in (3.33) for Bayesian variable selection, the availability of $g(T)$ confers great advantages for posterior computation and exploration in Bayesian CART model selection.

Exhaustive evaluation of $g(T)$ over all T will not be feasible, except in trivially small problems, because of the sheer number of possible trees. This not only prevents exact calculation of the norming constant, but also makes it nearly impossible to determine exactly which trees have largest posterior probability. In spite of these limitations, MH algorithms can still be used to explore the posterior. Such algorithms simulate a Markov chain sequence of trees

$$T^0, T^1, T^2, \ldots \tag{4.14}$$

which are converging in distribution to the posterior $p(T \mid Y)$. Because such a simulated sequence will tend to gravitate towards regions of higher posterior probability, the simulation can be used to stochastically search for high posterior probability trees. We now proceed to describe the details of such algorithms and their effective implementation.

4.2.1 Metropolis-Hastings Search Algorithms

By restricting attention to a finite set of trees, as discussed in the last paragraph of Section 4.1.1, the simple MH form described in Section 2.3 can be used for direct simulation of the Markov chain (4.14). Because $g(T)/g(T') = p(T \mid Y)/p(T' \mid Y)$, such MH algorithms are obtained as follows. Starting with an initial tree T^0, iteratively simulate the transitions from T^j to T^{j+1} by the two steps:

1. Simulate a candidate T^* from the transition kernel $q(T \mid T^j)$.

2. Set $T^{j+1} = T^*$ with probability

$$\alpha(T^* \mid T^j) = \min\left\{\frac{q(T^j \mid T^*)\, g(T^*)}{q(T^* \mid T^j)\, g(T^j)}, 1\right\}. \tag{4.15}$$

   Otherwise, set $T^{j+1} = T^j$.

The key to making this an effective MH algorithm is the choice of transition kernel $q(T \mid T^j)$. A useful strategy in this regard is to construct $q(T \mid T^j)$ as a mixture of simple local moves from one tree to another - moves that have a chance of increasing posterior probability. In particular, CGM use the following $q(T \mid T^j)$, which generates T from T^j by randomly choosing among four steps:

- GROW: Randomly pick a terminal node. Split it into two new ones by randomly assigning it a splitting rule using the same random splitting rule assignment used to determine $p(T)$.

- PRUNE: Randomly pick a parent of two terminal nodes and turn it into a terminal node by collapsing the nodes below it.

- CHANGE: Randomly pick an internal node, and randomly reassign it using the same random splitting rule assignment used to determine $p(T)$.

- SWAP: Randomly pick a parent-child pair which are both internal nodes. Swap their splitting rules unless the other child has the identical rule. In that case, swap the splitting rule of the parent with that of both children.

In executing the GROW, CHANGE and SWAP steps, attention is restricted to splitting rule assignments that do not force the tree have an empty terminal node. CGM also recommend further restricting attention to splitting rule assignments which yield trees with at least a small number (such as five) observations at every terminal node. A similar $q(T \mid T^j)$, without the SWAP step, was proposed by DMS. An interesting general approach for constructing such moves was proposed by Knight, Kustra and Tibshirani (1998).

The transition kernel $q(T \mid T^j)$ above has some appealing features. To begin with, every step from T to T^* has a counterpart that moves from T^* to T. Indeed, the GROW and PRUNE steps are counterparts of one another, and the CHANGE and

SWAP steps are their own counterparts. This feature guarantees the irreducibility of the algorithm, which is needed for convergence. It also makes it easy to calculate the ratio $q(T^j \mid T^*)/q(T^* \mid T^j)$ in (4.15). Note that other reversible moves may be much more difficult to implement because their counterparts are impractical to construct. For example, pruning off more than a pair of terminal nodes would require a complicated and computationally expensive reverse step. Another computational feature occurs in the GROW and PRUNE steps, where there is substantial cancellation between g and q in the calculation of (4.15) because the splitting rule assignment for the prior is used.

4.2.2 Running the MH Algorithm for Stochastic Search

The MH algorithm described in the previous section can be used to search for desirable trees. To perform an effective search it is necessary to understand its behavior as it moves through the space of trees. By virtue of the fact that its limiting distribution is $p(T|Y)$, it will spend more time visiting tree regions where $p(T|Y)$ is large. However, our experience in assorted problems (see the examples in CGM) has been that the algorithm quickly gravitates towards such regions and then stabilizes, moving locally in that region for a long time. Evidently, this is a consequence of a transition kernel that makes local moves over a sharply peaked multimodal posterior. Once a tree has reasonable fit, the chain is unlikely to move away from a sharp local mode by small steps. Because the algorithm is convergent, we know it will eventually move from mode to mode and traverse the entire space of trees. However, the long waiting times between such moves and the large size of the space of trees make it impractical to search effectively with long runs of the algorithm. Although different move types might be implemented, we believe that any MH algorithm for CART models will have difficulty moving between local modes.

To avoid wasting time waiting for mode to mode moves, our search strategy has been to repeatedly restart the algorithm. At each restart, the algorithm tends to move quickly in a direction of higher posterior probability and eventually stabilize around a local mode. At that point the algorithm ceases to provide new information, and so we intervene in order to find another local mode more quickly. Although the algorithm can be restarted from any particular tree, we have found it very productive to repeatedly restart at the trivial single node tree. Such restarts have led to a wide variety of different trees, apparently due to large initial variation of the algorithm. However, we have also found it productive to restart the algorithm at other trees such as previously visited intermediate trees or trees found by other heuristic methods. For example, CGM demonstrate that restarting the algorithm at trees found by bootstrap bumping (Tibshirani and Knight 1999) leads to further improvements over the start points.

A practical implication of restarting the chain is that the number of restarts must be traded off against the length of the chains. Longer chains may more thoroughly explore a local region of the model space, while more restarts could cover the space of models more completely. In our experience, a preliminary run with a small number of restarts can aid in deciding these two parameters of the run. If the marginal likelihood stops increasing before the end of each run, lengthening runs may be less profitable than increasing the number of restarts.

It may also be useful to consider the slower "burn in" modification of the algorithm proposed by DMS. Rather than let their MH algorithm move quickly to a mode, DMS intervene, forcing the algorithm to move around small trees with around 6 or fewer nodes, before letting it move on. This interesting strategy can take advantage of the fact that the problems caused by the sharply peaked multimodal posterior are less acute when small trees are constructed. Indeed, when trees remain small, the change or swap steps are more likely to be permissible (since there are fewer children to be incompatible with), and help move around the model space. Although this "burn in" strategy will slow down the algorithm, it may be a worthwhile tradeoff if it sufficiently increases the probability of finding better models.

4.2.3 Selecting the "Best" Trees

As many trees are visited by each run of the algorithm, a method is needed to identify those trees which are of most interest. Because $g(T) \propto p(T \mid Y)$ is available for each visited tree, one might consider selecting those trees with largest posterior probability. However, this can be problematic because of the dilution property of $p(T)$ discussed in Section 4.1.1. Consider the following simple example. Suppose we were considering all possible trees with two terminal nodes and a single rule. Suppose further that we had only two possible predictors, a binary variable with a single available splitting rule, and a multilevel variable with 100 possible splits. If the marginal likelihood $p(Y \mid T)$ was the same for all 101 rules, then the posterior would have a sharp mode on the binary variable because the prior assigns small probability to each of the 100 candidate splits for the multilevel predictor, and much larger probability to the single rule on the binary predictor. Selection via posterior probabilities is problematic because the relative sizes of posterior modes does not capture the fact that the total posterior probability allocated to the 100 trees splitting on the multilevel variable is the same as that allocated to the single binary tree.

It should be emphasized that the dilution property is not a failure of the prior. By using it, the posterior properly allocates high probability to tree neighborhoods which are collectively supported by the data. This serves to guide the algorithm towards such regions. The difficulty is that relative sizes of posterior modes do not capture the relative allocation of probability to such regions, and so can lead to misleading comparisons of single trees. Note also that dilution is not a limitation for model averaging. Indeed, one could approximate the overall posterior mean by the average of the visited trees using weights proportional to $p(Y \mid T)p(T)$. Such model averages provide a Bayesian alternative to the tree model averaging proposed by see Breiman (1996) and Oliver and Hand (1995).

A natural criterion for tree model selection, which avoids the difficulties described above, is to use the marginal likelihood $p(Y \mid T)$. As illustrated in CGM, a useful tool in this regard is a plot of the largest observed values of $p(Y \mid T)$ against the number of terminal nodes of T, an analogue of the C_p plot (Mallows 1973). This allows the user to directly gauge the value of adding additional nodes while removing the influence of $p(T)$. In the same spirit, we have also found it useful to consider other commonly used tree selection criteria such as residual sums of squares for regression trees and misclassification rates for classification trees.

After choosing a selection criterion, a remaining issue is what to do when many different models are found, all of which fit the data well. Indeed, our experience with stochastic search in applications has been to find a large number of good tree models, distinguished only by small differences in marginal likelihood. To deal with such output, in Chipman, George and McCulloch (1998b, 2001a), we have proposed clustering methods for organizing multiple models. We found such clustering to reveal a few distinct neighborhoods of similar models. In such cases, it may be better to select a few representative models rather than a single "best" model.

5 Much More to Come

Because of its broad generality, the formulation for Bayesian model uncertainty can be applied to a wide variety of problems. The two examples that we have discussed at length, Bayesian variable selection for the linear model and Bayesian CART model selection, illustrate some of the main ideas that have been used to obtain effective practical implementations. However, there have been many other recent examples. To get an idea of the extent of recent activity, consider the following partial list of some of the highlights just within the regression framework.

To begin with, the Bayesian variable selection formulation for the linear model has been extended to the multivariate regression setting by Brown, Vannucci and Fearn (1998). It has been applied and extended to nonparametric spline regression by Denison, Mallick and Smith (1998bc), Gustafson (2000), Holmes and Mallick (2001), Liang, Truong and Wong (2000), Smith and Kohn (1996, 1997), Smith, Wong and Kohn

(1998); and to nonparametric wavelet regression by Abramovich, Sapatinas and Silverman (1998), Chipman, Kolaczyk and McCulloch (1997), Clyde and George (1999,2000), Clyde, Parmigiani and Vidakovic (1998), Holmes and Mallick (2000) and Johnstone and Silverman (1998). Related Bayesian approaches for generalized linear models and time series models have been put forward by Chen, Ibrahim and Yiannoutsos (1999), Clyde (1999), George, McCulloch and Tsay (1995), Ibrahim and Chen (2000), Mallick and Gelfand (1994), Raftery (1996), Raftery, Madigan and Volinsky (1996), Raftery and Richardson (1996), Shively, Kohn and Wood (1999), Troughton and Godsill (1997) and Wood and Kohn (1998); for loglinear models by Dellaportas and Foster (1999) and Albert (1996); and to graphical model selection by Giuduci and Green (1999) and Madigan and York (1995). Bayesian CART has been extended to Bayesian treed modelling by Chipman, George and McCulloch (2001); an related Bayesian partitioning approach has been proposed by Holmes, Denison and Mallick (2000). Alternative recent Bayesian methods for the regression setup include the predictive criteria of Laud and Ibrahim (1995), the information distance approach of Goutis and Robert (1998) and the utility approach of Brown, Fearn and Vannucci (1999) based on the early work of Lindley (1968). An excellent summary of many of the above articles and additional contributions to Bayesian model selection can be found in Ntzoufras (1999).

Spurred on by applications to new model classes, refinements in prior formulations and advances in computing methods and technology, implementations of Bayesian approaches to model uncertainty are widening in scope and becoming increasingly prevalent. With the involvement of a growing number of researchers, the cross-fertilization of ideas is further accelerating developments. As we see it, there is much more to come.

REFERENCES

Abramovich, F., Sapatinas, T., and Silverman, B.W. (1998). Wavelet thresholding via a Bayesian approach. *J. Roy. Statist. Soc. Ser. B* **60**, 725-749.

Akaike, H. (1973). Information theory and an extension of the maximum likelihood principle. In *2nd Internat. Symp. Inform. Theory* (B.N. Petrov and F. Csaki, eds.) 267-81, Akademia Kiado, Budapest.

Albert, J.H. (1996). The Bayesian Selection of Log-linear Models. *Canad. J. Statist.* **24**, 327-347.

Bartlett, M. (1957). A comment on D. V. Lindley's Statistical Paradox. *Biometrika* **44**, 533-534.

Bernardo, J. M., and Smith, A.F.M. (1994). *Bayesian Theory*, Wiley, New York.

Berger, J.O. and Pericchi, L.R. (1996). The intrinsic Bayes factor for model selection and prediction. *J. Amer. Statist. Asso.* **91**, 109-122.

Besag, J and Green, P.J. (1993). Spatial statistics and Bayesian computation (with discussion). *J. Roy. Statist. Soc. Ser. B* **55** 25-37.

Breiman, L (1996). Bagging predictors. *Machine Learning* **24**, 123-140.

Breiman, L., Friedman, J. Olshen, R. and Stone, C. (1984). *Classification and Regression Trees.* Wadsworth.

Brown, P.J., Vannucci,M., and Fearn, T. (1998). Multivariate Bayesian variable selection and prediction. *J. Roy. Statist. Soc. Ser. B* **60**, 627-642.

Brown, P.J., Fearn, T. and Vannucci, M. (1999). The choice of variables in multivariate regression: a non-conjugate Bayesian decision theory approach. *Biometrika* **86**, 635-648.

Buntine, W. (1992). Learning Classification Trees. *Statist. Comput.* **2**, 63-73.

Carlin, B.P. and Chib, S. (1995). Bayesian model choice via Markov Chain Monte Carlo. *J. Roy. Statist. Soc. Ser. B* **77**, 473-484.

Casella, G. and George, E.I. (1992). Explaining the Gibbs sampler. *The American Statistician*, **46**, 167-174.

Chambers, J.M. (1971). Regression updating. *J. Amer. Statist. Asso.* **66**, 744-748.

Chen, M.H., Ibrahim, J.G. and Yiannoutsos, C. (1999). Prior elicitation, variable selection and Bayesian computation for logistic regression models. *J. Roy. Statist. Soc. Ser. B* **61**, 223-243.

Chib, S. and Greenberg, E. (1995). Understanding the Metropolis-Hastings algorithm. *The American Statistician*, **49**, 327-335.

Chipman, H. (1996). Bayesian variable selection with related predictors. *Canad. J. Statist.* **24**, 17-36.

Chipman, H. A., George, E. I., and McCulloch, R. E. (1998a). Bayesian CART model search (with discussion). *J. Amer. Statist. Asso.* **93**, 935-960.

Chipman, H. A., George, E. I. and McCulloch, R. E. (1998b). Making sense of a forest of trees. *Computing Science and Statistics, Proc. 30th Symp. Interface* (S. Weisberg, ed.) 84-92, Interface Foundation of North America, Fairfax, VA.

Chipman, H., George, E. I., and McCulloch, R. E. (2000). Hierarchical priors for Bayesian CART shrinkage. *Statist. Comput.* **10**, 17-24.

Chipman, H., George, E.I. and McCulloch, R.E. (2001a). Managing multiple models. *Artificial Intelligence and Statistics 2001*, (Tommi Jaakkola and Thomas Richardson, eds.) 11-18, Morgan Kaufmann, San Francisco, CA.

Chipman, H., George, E. I., and McCulloch, R. E. (2001b). Bayesian treed models. *Machine Learning.* To appear.

Chipman, H., Hamada, M. and Wu, C. F. J., (1997). A Bayesian variable selection approach for analyzing designed experiments with complex aliasing. *Technometrics*, 39, 372-381.

Chipman, H., Kolaczyk, E., and McCulloch, R. (1997). Adaptive Bayesian wavelet shrinkage. *J. Amer. Statist. Asso.* **92**, 1413-1421.

Clark, L. and Pregibon, D. (1992). Tree-Based Models. In *Statistical Models in S* (J. Chambers and T. Hastie, eds.) 377-419, Wadsworth.

Clyde, M.A. (1999). Bayesian model averaging and model search strategies (with discussion). In *Bayesian Statistics 6* (J.M. Bernardo, J.O. Berger, A.P. Dawid and A.F.M. Smith, eds.) 157-185, Oxford Univ. Press.

Clyde, M.A., DeSimone, H., and Parmigiani, G. (1996). Prediction via orthogonalized model mixing. *J. Amer. Statist. Asso.* **91**, 1197-1208.

Clyde, M. and George, E.I. (1999) Empirical Bayes estimation in wavelet nonparametric regression. In *Bayesian Inference in Wavelet Based Models* (P. Muller and B. Vidakovic, eds.) 309-22, Springer-Verlag, New York.

Clyde, M. and George, E.I. (2000). Flexible empirical Bayes estimation for wavelets. *J. Roy. Statist. Soc. Ser. B* **62**, 681-698.

Clyde, M.A. and Parmigiani, G. (1994). Bayesian variable selection and prediction with mixtures, *J. Biopharmaceutical Statist.*

Clyde, M., Parmigiani, G., Vidakovic, B. (1998). Multiple shrinkage and subset selection in wavelets. *Biometrika* **85**, 391-402.

Denison, D.G.T., Mallick, B.K. and Smith, A.F.M. (1998a). A Bayesian CART algorithm. *Biometrika* **85**, 363-377.

Denison, D.G.T., Mallick, B.K. and Smith, A.F.M. (1998b). Automatic Bayesian curve fitting. *J. Roy. Statist. Soc. Ser. B* **60**, 333-350.

Denison, D.G.T., Mallick, B.K. and Smith, A.F.M. (1998c). Bayesian MARS. *Statist. Comput.* **8**, 337-346.

Dellaportas, P. and Foster, J.J. (1999). Markov Chain Monte Carlo Model determination for hierarchical and graphical log-linear Models. *Biometrika* **86**, 615-634.

Dellaportas, P. Forster, J.J. and Ntzoufras, I. (2000). On Bayesian model and variable selection using MCMC. *Statist. Comput.* To appear.

Dongarra, J.J., Moler C.B., Bunch, J.R. and Stewart, G.W. (1979). Linpack Users' Guide. SIAM, Philadelphia.

Donoho, D.L. and Johnstone, I.M. (1994). Ideal spatial adaptation by wavelet shrinkage. *Biometrika* **81**, 425-56.

Draper, D. (1995). Assessment and propagation of model uncertainty (with discusssion). *J. Roy. Statist. Soc. Ser. B* **57**, 45-98.

Edwards, W Lindman H. and Savage, L.J. (1963). Bayesian statistical inference for psychological research. *Psychological Review* **70** 193-242.

Fernandez, C., Ley, E. and Steel, M.F.J. (2001). Benchmark priors for Bayesian model averaging. *J. Econometrics* **100**, 381-427.

Foster, D.P. and George, E.I. (1994). The risk inflation criterion for multiple regression. *Ann. Statist.* **22**, 1947-75.

Garthwaite,P. H. and Dickey, J.M. (1992). Elicitation of prior distributions for variable-selection problems in regression. *Ann. Statist.* **20**, 1697-1719.

Garthwaite, P. H. and Dickey, J.M. (1996). Quantifying and using expert opinion for variable-selection problems in regression (with discussion). *Chemomet. Intel. Lab. Syst.* **35**, 1-34.

Gelfand, A. E., Dey, D. K., and Chang, H. (1992). Model determination using predictive distributions With implementations via sampling-based methods. In *Bayesian Statistics 4* (J. M. Bernardo, J. O. Berger, A. P. Dawid and A. F. M. Smith, eds.) 147–167, Oxford Univ. Press.

Gelfand, A., and Smith, A. F. M. (1990). Sampling-based approaches to calculating marginal densities, *J. Amer. Statist. Asso.* **85**, 398-409.

George, E.I. (1998). Bayesian model selection. *Encyclopedia of Statistical Sciences. Update Volume 3* (S. Kotz, C. Read and D. Banks, eds.) 39-46, Wiley, New York.

George, E.I. (1999). Discussion of "Bayesian model averaging and model search strategies" by M.A. Clyde. *Bayesian Statistics 6* (J.M. Bernardo, J.O. Berger, A.P. Dawid and A.F.M. Smith, eds.) 175-177, Oxford Univ. Press.

George, E.I. and Foster, D.P. (2000) Calibration and empirical Bayes variable selection. *Biometrika* **87**, 731-748.

George, E.I. and McCulloch, R.E. (1993). Variable selection via Gibbs sampling. *J. Amer. Statist. Asso.* **88**, 881-889.

George, E.I. and McCulloch, R.E. (1996). Stochastic search variable selection. In *Markov Chain Monte Carlo in Practice* (W.R. Gilks, S. Richardson and D.J. Spiegelhalter, eds.) 203-214, Chapman and Hall, London.

George, E.I. and McCulloch, R.E. (1997). Approaches for Bayesian variable selection. *Statist. Sinica* **7**, 339-73.

George, E.I., McCulloch, R.E. and Tsay, R. (1995) Two approaches to Bayesian Model selection with applications. In *Bayesian Statistics and Econometrics: Essays in Honor of Arnold Zellner* (D. Berry, K. Chaloner, and J. Geweke, eds.) 339-348, Wiley, New York.

Geisser, S. (1993). *Predictive Inference: An Introduction.* Chapman and Hall, London.

Geman, S. and Geman, D. (1984). Stochastic relaxation, Gibbs distributions and the Bayesian restoration of images. *IEEE Trans. Pattn. Anal. Mach. Intell.* **6**, 721-741.

Geweke, J. (1996). Variable selection and model comparison in regression. In *Bayesian Statistics 5* (J. M. Bernardo, J. O. Berger, A. P. Dawid and A. F. M. Smith, eds.) 609-620, Oxford Univ. Press.

Gilks, W.R., Richardson S. and Spiegelhalter, D.J. (1996) *Markov Chain Monte Carlo in Practice.* Chapman and Hall, London.

Giuduci, P. and Green, P.J. (1999). Decomposable graphical Gausssian model determination. *Biometrika* **86**, 785-802.

Goutis, C. and Robert, C.P. (1998). Model choice in generalized linear models: A Bayesian approach via Kullback-Leibler projections. *Biometrika* **82**, 711-732.

Green, P. (1995). Reversible Jump Markov chain Monte Carlo computation and Bayesian model determination. *Biometrika* **82**, 711-732.

Gustafson, P. (2000). Bayesian regression modelling with interactions and smooth effects. *J. Amer. Statist. Asso.* **95**, 795-806.

Hastie, T., and Pregibon, D. (1990). Shrinking trees. AT&T Bell Laboratories Technical Report.

Hastings, W.K. (1970). Monte Carlo sampling methods using Markov chain and their applications. *Biometrika* **57**, 97-109.

Hawkins, D. M. and Kass, G. V. (1982). Automatic interaction detection. In *Topic in Applied Multivariate Analysis* (D. M. Hawkins, ed.) 267-302, Cambridge Univ. Press.

Hoeting, J. A., Raftery, A.E. and Madigan, D. (1996). A method for simultaneous variable selection and outlier identification in linear regression. *Computational Statistics and Data Analysis* **22**, 251-270.

Hoeting, J. A., Madigan, D., Raftery, A.E., and Volinsky, C.T. (1999). Bayesian Model averaging: A tutorial (with discussion). *Statist. Sci.* **14**:4, 382-417. (Corrected version available at http://www.stat.washington.edu/www/research/online/hoeting1999.pdf).

Holmes C.C., Denison, D.G.T. and Mallick, B.K. (2000). Bayesian prediction via partitioning. Technical Report, Dept. of Mathematics, Imperial College, London.

Holmes, C.C. and Mallick, B.K. (2000). Bayesian wavelet networks for nonparametric regression. *IEEE Trans. Neur. Netwrks.* **10**, 1217-1233.

Holmes, C.C. and Mallick, B.K. (2001). Bayesian regression with multivariate linear splines. *J. Roy. Statist. Soc. Ser. B* **63**, 3-18.

Ibrahim, J.G. and Chen, M.-H. (2000). Prior Elicitation and Variable Selection for Generalized Linear Mixed Models. In *Generalized Linear Models: A Bayesian Perspective*, (Dey, D.K., Ghosh, S.K. and Mallick, B.K. eds.) 41-53, Marcel Dekker, New York.

Johnstone, I.M. and Silverman, B.W. (1998). Empirical Bayes approaches to mixture problems and wavelet regression. Technical Report, Univ. of Bristol.

Jordan, M.I. and Jacobs, R.A. (1994). Mixtures of experts and the EM algorithm. *Neural Computation* **6**, 181-214.

Kass, G. V. (1980). An exploratory technique for investigating large quantities of categorical data. *Appl. Statist.* **29**, 119-127.

Kass, R. E. and Wasserman, L. (1995). A reference Bayesian test for nested hypotheses and its relationship to the Schwarz criterion. *J. Amer. Statist. Asso.* **90**, 928-34.

Key, J.T., Pericchi, L.R. and Smith, A.F.M. (1998). Choosing among models when none of them are true. In *Proc. Workshop Model Selection, Special Issue of Rassegna di Metodi Statistici ed Applicazioni* (W. Racugno, ed.) 333-361, Pitagora Editrice, Bologna.

Knight, K., Kustra, R. and Tibshirani, R. (1998). Discussion of "Bayesian CART model search" by Chipman, H. A., George, E. I., and McCulloch, R. E. *J. Amer. Statist. Asso.* **93**, 950-957.

Kuo, L. and Mallick, B. (1998). Variable selection for regression models. *Sankhyā Ser. B* **60**, 65-81.

Laud, P.W. and Ibrahim, J.G. (1995). Predictive model selection. *J. Roy. Statist. Soc. Ser. B* **57**, 247-262.

Leblanc, M. and Tibshirani, R. (1998). Monotone shrinkage of trees. *J. Comput. Graph. Statist.* **7**, 417-433.

Liang F., Truong, Y.K. and Wong, W.H. (2000). Automatic Bayesian model averaging for linear regression and applications in Bayesian curve fitting. Technical Report, Dept. of Statistics, Univ. of California, Los Angeles.

Lindley, D.V. (1968). The choice of variables in regression. *J. Roy. Statist. Soc. Ser. B* **30**, 31-66.

Liu, J.S., Wong, W.H., and Kong, A. (1994). Covariance structure and convergence rate of the Gibbs sampler with applications to the comparisons of estimators and augmentation schemes. *Biometrika* **81**, 27-40.

Lutsko, J. F. and Kuijpers, B. (1994). Simulated annealing in the construction of near-optimal decision trees. In *Selecting Models from Data: AI and Statistics IV* (P. Cheeseman and R. W. Oldford, eds.) 453-462.

Madigan, D. and York, J. (1995). Bayesian graphical models for discrete data. *Internat. Statist. Rev.* **63**, 215-232.

Mallick, B.K. and Gelfand, A.E. (1994). Generalized linear models with unknown number of components. *Biometrika* **81**, 237-245.

Mallows, C. L. (1973). Some Comments on C_p. *Technometrics* **15**, 661-676.

McCullagh, P. and Nelder, J.A. (1989). *Generalized Liner Models, 2nd Ed.* Chapman and Hall, New York.

McCulloch, R.E. and Rossi P. (1991). A Bayesian approach to testing the arbitrage pricing theory. *J. Econometrics* **49**, 141-168.

Metropolis, N., Rosenbluth, A.W., Rosenbluth, M.N., Teller, A.H. and Teller,E. (1953). Equations of state calculations by fast computing machines. *J. Chem. Phys.* **21**, 1087-1091.

Meyn, S.P. and Tweedie, R.L. (1993). *Markov Chains and Stochastic Stability.* Springer-Verlag, New York.

Mitchell, T.J. and Beauchamp, J.J. (1988). Bayesian variable selection in linear regression (with discussion). *J. Amer. Statist. Asso.* **83**, 1023-1036.

Ntzoufras, I. (1999). Aspects of Bayesian model and variable selection using MCMC. Ph.D. dissertation, Dept. of Statistics, Athens Univ. of Economics and Business, Athens, Greece.

O'Hagan, A. (1995). Fractional Bayes factors for model comparison (with discussion). *Jour. of the Roy. Statist. Soc. Ser. B* **57**, 99-138.

Oliver, J.J. and Hand, D.J. (1995). On pruning and averaging decision trees. *Proc. Internat. Machine Learning Conf.* 430-437.

Paass, G. and Kindermann, J. (1997). Describing the uncertainty of Bayesian predictions by using ensembles of models and its application. *1997 Real World Comput. Symp.* 118-125, Real World Computing Partnership, Tsukuba Research Center, Tsukuba, Japan.

Petit, L.I., and Smith, A. F. M. (1985). Outliers and influential observations in linear models. In *Bayesian Statistics 2*, (J.M. Bernardo, M.H. DeGroot, D.V. Lindley and A.F.M. Smith, eds.) 473-494, North-Holland, Amsterdam.

Peng, L. (1998). Normalizing constant estimation for discrete distribution simulation.

Ph.D. dissertation, Dept. MSIS, Univ. of Texas, Austin.

Phillips, D. B., and Smith, A. F. M. (1995). Bayesian model comparison via jump diffusions, In *Practical Markov Chain Monte Carlo in Practice* (W.R. Gilks, S. Richardson and D.J. Spiegelhalter, eds.) 215-239, Chapman and Hall, London.

Quinlan, J. R. (1986). Induction of decision trees. *Machine Learning* **1**, 81-106.

Quinlan, J. R. (1993). *C4.5: Tools for Machine Learning*, Morgan Kauffman, San Mateo, CA.

Quinlan, J.R. and Rivest, R.L. (1989). Inferring decision trees using the minimum description length principle. *Information and Computation* **80**, 227-248.

Raftery, A.E. (1996). Approximate Bayes factors and accounting for model uncertainty in generalized linear models. *Biometrika* **83**, 251-266.

Raftery, A. E., Madigan, D. and Hoeting, J. A. (1997). Bayesian model averaging for linear regression models. *J. Amer. Statist. Asso.* **92**, 179-191.

Raftery, A.E., Madigan, D. M., and Volinsky C.T. (1995). Accounting for model uncertainty in survival analysis improves predictive performance (with discussion). In *Bayesian Statistics 5* (J. M. Bernardo, J. O. Berger, A. P. Dawid and A. F. M. Smith, eds.) 323-350, Oxford Univ. Press.

Raftery, A.E. and Richardson, S. (1996). Model selection for generalized linear models via GLIB: application to nutrition and breast cancer. *Bayesian Biostatistics* (D.A. Berry and D.K. Strangl, eds.) 321-353, Marcel Dekker, New York.

San Martini, A. and Spezzaferri, F. (1984). A predictive model selection criterion *J. Roy. Statist. Soc. Ser. B* **46**, 296-303.

Schwarz, G. (1978). Estimating the dimension of a model. *Ann. Statist.* **6**, 461-4.

Shively, T.S., Kohn, R. and Wood, S. (1999). Variable selection and function estimation in additive nonparametric regression using a data-based prior (with discussion). *J. Amer. Statist. Asso.* **94**, 777-806.

Smith, A.F.M and Roberts, G.O. (1993). Bayesian computation via the Gibbs sampler and related Markov chain Monte Carlo methods. *J. Roy. Statist. Soc. Ser. B* **55**, 3-24.

Smith, M. and Kohn, R.(1996). Nonparametric regression using Bayesian variable selection. *Journal of Econometrics*, 75, 317-344.

Smith, M. and Kohn, R. (1997). A Bayesian approach to nonparametric bivariate regression. *J. Amer. Statist. Asso.* **92**, 1522-1535.

Smith, M., Wong, C.M. and Kohn, R. (1998). Additive nonparametric regression with autocorrelated errors. *J. Roy. Statist. Soc. Ser. B* **60**, 311-331.

Sutton, C. (1991). Improving classification trees with simulated annealing. *Computing Science and Statistics, Proc. 23rd Symp. Interface* (E. Keramidas, ed.) 396-402, Interface Foundation of North America.

Tibshirani, R. and Knight, K. (1999). Model search by bootstrap "bumping". *J. Comput. Graph. Statist.* **8**, 671-686.

Tierney, L. (1994). Markov chains for exploring posterior distributions (with discussion). *Ann. Statist.* **22**, 1674-1762.

Tierney, L. and Kadane, J.B. (1986). Accurate approximations for posterior moments and marginal densities. *J. Amer. Statist. Asso.* **81**, 82-86.

Troughton, P.T. and Godsill, S.J. (1997). Bayesian model selection for time series using Markov Chain Monte Carlo. *Proc. IEEE Internat. Conf. Acoustics, Speech and Signal Processing*, 3733-3736.

Wakefield, J.C. and Bennett, J.E. (1996). The Bayesian modelling of covariates for population pharmacokinetic models. *J. Amer. Statist. Asso.* **91**, 917-928.

Wallace, C.C. and Patrick, J.D. (1993). Coding decision trees. *Machine Learning* **11**, 7-22.

Wasserman, L. (2000). Bayesian model selection and model averaging. *J. Math. Psychology* **44**, 92-107.

Wood, S. and Kohn, R. (1998). A Bayesian approach to robust binary nonparametric regression. *J. Amer. Statist. Asso.* **93**, 203-213.

Zellner, A. (1986). On assessing prior distributions and Bayesian regression analysis with g-prior distributions. In *Bayesian Inference and Decision Techniques: Essays in Honor of Bruno de Finietti* (P.K. Goel and A. Zellner, eds.) 233-43, North-Holland, Amsterdam.

DISCUSSION

M. Clyde

Duke University

I would like to begin by thanking the authors for their many contributions to Bayesian model selection and for providing an excellent summary of the growing body of literature on Bayesian model selection and model uncertainty. The development of computational tools such as the Gibbs sampler and Markov chain Monte Carlo approaches, has led to an explosion in Bayesian approaches for model selection. On the surface, Bayesian model averaging (BMA) and model selection is straightforward to implement: one specifies the distribution of the data, and the prior probabilities of models and model specific parameters; Bayes theorem provides the rest. As the authors point out the two major challenges confronting the practical implementation of Bayesian model selection are choosing prior distributions and calculating posterior distributions. In my experience, I have found that this is especially true in high dimensional problems, such as wavelet regression or non-parametric models, where subjective elicitation of prior distributions is practically infeasible and enumeration of the number of potential models is intractable (Clyde et al. 1998; Clyde and George 2000).

Choice of Prior Distributions

The specification of prior distributions is often broken down into two parts: (1) elicitation of distributions for parameters specific to each model, such as the distribution for regression coefficients in linear models, $p(\beta|\gamma, \sigma^2)$, and (2) selection of a prior distribution over models $p(\gamma)$. For high dimensional problems one cannot specify the prior probability of each γ directly, and practical implementations of Bayesian selection have usually made prior assumptions that the presence or absence of a variable is independent of the presence or absence of other variables. As a special case of this, the uniform prior distribution over models is appealing in that posterior probabilities of models depend only on the likelihood. However, this prior distribution may not be sensible for model averaging when there are groups of similar variables and does not provide the proper "dilution" of posterior mass over similar models (see Clyde 1999; Hoeting et al. 1999), and discussion therein (George 1999; 1999a). In this regard, uniform and independent prior distributions must be used carefully with highly correlated explanatory variables and special consideration should be given to constructing the model space. Even with

Merlise Clyde is Associate Professor, Institute of Statistics and Decision Sciences, Duke University, Durham NC 27708-0251, U.S.A. email: clyde@stat.Duke.EDU.

uniform priors on models, the posterior distribution over models naturally penalizes adding redundant variables, however, this may not be enough to lead to the proper rate of dilution with nearly redundant variables. One approach to constructing dilution prior distributions is to start with a uniform prior over models and use imaginary training data to construct a posterior distribution for γ based on the training data; this posterior would become the prior distribution for γ for the observed data. Selection of the training data and hyperparameters are non-trivial issues, however, this approach would likely provide better dilution properties than starting with an independent prior. Clearly, construction of prior distributions for models that capture similarities among variables in a sensible way is a difficult task and one that needs more exploration.

For (1), by far the most common choice is a normal prior distribution for β, such as in the conjugate setup for point prior selection models (section 3.2 CGM), where $\beta_\gamma \sim N(0, \sigma^2 \Sigma_\gamma)$. Again, as one cannot typically specify a separate prior distribution for β under each model, any practical implementation for Bayesian model uncertainty usually resorts to structured families of prior distributions. Another important consideration is whether prior specifications for β are "compatible" across models (Dawid and Lauritzen 2000) . For example, suppose that Model 1 contains variables 1 and 2, Model 2 contains variables 2 and 3, and Model 3 includes only variable 2. With apologies for possible abuse of notation, let β_2 denote the coefficient for variable 2 in each model. With completely arbitrary choices for Σ_γ, under Model 1 the variance for β_2 given that $\beta_1 = 0$ may not be the same as the variance for β_2 given that $\beta_3 = 0$ derived under Model 2, and both may differ from the variance for β_2 under the prior distribution given Model 3.

To avoid this incompatibility, choices for Σ_γ are often obtained from conditional specifications (i.e. conditional on $\beta_i = 0$ for $\gamma_i = 0$) derived from a prior distribution for β under the full model. For example, Zellner's g-prior (Zellner 1986) is commonly used, which leads to $\Sigma = c(X'X)^{-1}$ for the full model and $\Sigma_\gamma = c(X'_\gamma X_\gamma)^{-1}$ for model γ.

While in many cases conditioning leads to sensible choices, the result may depend on the choice of parameterization, which can lead to a Borel paradox (Kass and Raftery 1995; Dawid and Lauritzen 2000). Other structured families may be induced by marginalization or projections, leading to possibly different prior distributions. While structured prior distributions may reduce the number of hyperparameters that must be selected, i.e. to one parameter c, how to specify c is still an issue.

The choice of c requires careful consideration, as it appears in the marginal likelihoods of the data and hence the posterior model probabilities, and in my experience, can be influential. In the situation where little prior information is available, being too "non-informative" about β (taking c large) can have the un-intended consequence of favoring the null model *a posteriori* (Kass and Raftery 1995). While "default" prior distributions (both proper and improper) can be calibrated based on information criteria such as AIC (Akaike Information Criterion - Akaike 1973), BIC (Bayes Information Criterion – Schwarz 1978), or RIC (Risk Inflation Criterion – Foster and George 1994), there remains the question of which one to use (Clyde 2000; George and Foster 2000; Fernandez et al. 1998); such decisions may relate more to utilities for model selection rather than prior beliefs (although it may be impossible to separate the two issues). Based on simulation studies, Fernandez et al.(1998) recommend RIC-like prior distributions when $n < p^2$ and BIC-like prior distributions otherwise. In wavelet regression, where $p = n$, there are cases where priors calibrated based on BIC have better predictive performance than prior distributions calibrated using RIC, and vice versa. From simulation studies and asymptotic arguments, it is clear that there is no one default choice for c that will "perform" well for all contingencies (Fernandez et al. 1998; Hansen and Yu 1999).

Empirical Bayes approaches provide an adaptive choice for c. One class of problems where BMA has had outstanding success is in non-parametric regression using wavelet bases. In nonparametric wavelet regression where subjective information is typically not available, empirical Bayes methods for estimating the hyperparameters have (empirically) led to improved predictive performance over a number of fixed hyperparameter specifications as well as default choices such as AIC, BIC, and RIC (Clyde and George 1999, 2000; George and Foster 2000; Hansen and Yu 1999) for both model selection and BMA. Because of the orthogonality in the design matrix under discrete wavelet transformations, EB estimates can be easily obtained using EM algorithms (Clyde and George 1999, 2000; Johnstone and Silverman 1998) and allow for fast analytic expressions for Bayesian model averaging and model selection despite the high dimension of the parameter space ($p = n$) and model space (2^n), bypassing MCMC sampling altogether.

George and Foster (2000) have explored EB approaches to hyperparameter selection for linear models with correlated predictors. EM algorithms for obtaining estimates of c and ω, as in Clyde and George (2000), can be adapted to the non-orthogonal case with unknown σ^2 using the conjugate point prior formulation and independent model space priors (equation 3.2 in CGM). For the EM algorithm both model indicators γ and σ^2 are treated as latent data and imputed using current estimates of c and $\omega = (\omega_1, \ldots, \omega_p)$, where ω_j is the prior probability that variable j is included under the independence prior. At iteration i, this leads to

$$\hat{\gamma}^{(i)} = \frac{p(Y|\gamma, c^{(i)})p(\gamma|\omega^{(i)})}{\sum_{\gamma'} p(Y|\gamma, c^{(i)})p(\gamma|\omega^{(i)})} \tag{1}$$

$$S_{\gamma}^{2(i)} = Y'Y - \frac{c^{(i)}}{1+c^{(i)}}\mathrm{SSR}(\gamma) \tag{2}$$

where $SSR(\gamma)$ is the usual regression sum of squares and $S_{\gamma}^{2(i)}$ is a Bayesian version of residual sum of squares using the current estimate $c^{(i)}$. Values of c and ω that maximize the posterior distribution for c and ω given the observed data and current values of the latent data are

$$\hat{\omega}_j^{(i+1)} = \sum_{\gamma \text{ such that } \gamma_j=1} \hat{\gamma}^{(i)} \tag{3}$$

$$\hat{c}^{(i+1)} = \max\left\{0, \left(\sum_{\gamma} \hat{\gamma}^{(i)} \frac{\mathrm{SSR}(\gamma)/(p \sum_j \hat{\omega}_j^{(i+1)})}{(\lambda\nu + S_\gamma^{2(i)})/(n+\nu)}\right) - 1\right\} \tag{4}$$

where ν and λ are hyperparameters in the inverse gamma prior distribution for σ^2 (CGM equation 3.10). These steps are iterated until estimates converge. EB estimates based on a common ω for all variables are also easy to obtain. For ridge-regression independent priors with $\Sigma_\gamma = cI$ or other prior distributions for β, estimates for ω_j have the same form, but estimates for c are slightly more complicated and require numerical optimization.

The ratio in the expression for $\hat{c}$ has the form of a generalized F-ratio, which is weighted by estimates of posterior model probabilities. The EM algorithm highlights an immediate difficulty with a common c for all models, as one or two highly significant coefficients may influence the EB estimate of c. For example, the intercept may be centered far from 0, and may have an absolute t-ratio much larger than the t-ratios of other coefficients. As the intercept is in all models, it contributes to all of the $SSR(\gamma)$ terms, which has the effect of increasing c as the absolute t-ratio increases. Since the same c appears in the prior variance of all other coefficients, if c becomes too large in order to account for the size of the intercept, we risk having the null model being favored (Bartlett's paradox (Kass and Raftery 1995)). While one could use a different prior distribution for the intercept (even a non-informative prior distribution, which would correspond to centering all variables), the problem may still arise among the other variables if there are many moderate to large coefficients, and a few that have extreme standardized values. Implicit in the formulation based on a common c is that the non-zero standardized coefficients follow a normal distribution with a common variance. As such, this model cannot accommodate one or a few extremely large standardized coefficients without increasing the odds that the remaining coefficients are zero. Using a heavy-tailed prior distribution for β may result in more robust EB estimates of c (Clyde and George 2000). Other possible solutions including adding additional structure into the prior that would allow for different groups of coefficients with a different c in each group. In the context of wavelet regression, coefficients are grouped based on the multi-resolution wavelet decomposition; in other problems there may not be any natural a priori groupings. Related to EB methods is the minimum description length (MDL) approach to model selection, which effectively uses a different c_γ estimated from the data for each model (Hansen and Yu 1999). While EB methods have led to improvements in performance, part of the success depends on careful construction of the model/prior. Some of the problems discussed above highlight possible restrictions of the normal prior.

Unlike the EM estimates for orthogonal regression, the EB approach with correlated predictors involves a summation over all models, which is clearly impractical in large problems. As in the inference problem, one can base EB estimation on a sample of models. This approach has worked well in moderate sized problems, where leaps and bounds (Furnival and Wilson 1974) was used to select a subset of models; these were then used to construct the EB prior distribution, and then estimates under BMA with the estimated prior. For larger problems, leaps and bounds may not be practical, feasible, or suitable (such as CART models), and models must be sampled using MCMC or other methods. How to scale the EB/EM approach up for larger problems where models must be sampled is an interesting problem.

In situations where there is uncertainty regarding a parameter, the Bayesian approach is to represent that prior uncertainty via a distribution. In other words, why not add another level to the hierarchy and specify a prior distribution on c rather than using a fixed value? While clearly feasible using Gibbs sampling and MCMC methods, analytic calculation of marginal likelihoods is no longer an option. Empirical Bayes (EB) estimation of c often provides a practical compromise between the fully hierarchical Bayes model and Bayes procedures where c is fixed in advance. The EB approach plugs in the modal c into $g(\gamma)$ which ignores uncertainty regarding c, while a fully Bayes approach would integrate over c to obtain the marginal likelihood. As the latter does not exist in closed form, Monte Carlo frequencies of models provide consistent estimates of posterior model probabilities. However, in large dimensional problems where frequencies of most models may be only 0 or 1, it is not clear that Monte Carlo frequencies of models $p_f(\gamma|Y, S)$ from implementing MCMC for the fully Bayesian approach are superior to using renormalized marginal likelihoods evaluated at the EB estimate of c. When the EB estimate of c corresponds to the posterior mode for c, renormalized marginal likelihoods $g(\gamma)$ evaluated at the EB estimate of c are closely related to Laplace approximations (Tierney and Kadane 1986) for integrating the posterior with respect to c (the Laplace approximation would involve a term with the determinant of the negative Hessian of the log posterior). A hybrid approach where MCMC samplers are used to identify/sample models from the fully hierarchical Bayes model, but one evaluates posterior model probabilities for the unique models using Laplace approximations may provide better estimates that account for uncertainty in c.

Implementing Sampling of Models

In the variable selection problem for linear regression, marginal likelihoods are available in closed form (at least for nice conjugate prior distributions); for generalized linear models and many other models, Laplace's method of integration can provide accurate approximations to marginal distributions. The next major problem is that the model space is often too large to allow enumeration of all models, and beyond 20-25 variables, estimation of posterior model probabilities, model selection, and BMA must be based on a sample of models.

Deterministic search for models using branch and bounds or leaps and bounds algorithms (Furnival and Wilson 1974) is efficient for problems with typically fewer than 30 variables. For larger problems, such as in non-parametric models or generalized additive models, these methods are too expensive computationally or do not explore a large enough region of the model space, producing poor fits (Hanson and Kooperberg 1999). While Gibbs and MCMC sampling have worked well in high dimensional orthogonal problems, Wong et al. (1997) found that in high dimensional problems such as non-parametric regression using non-orthogonal basis functions that Gibbs samplers were unsuitable, from both a computational efficiency standpoint as well as for numerical reasons, as the sampler tended to get stuck in local modes. Their proposed focused sampler "focuses" on variables that are more "active" at each iteration, and in simulation studies provided better MSE performance than other classical non-parametric approaches or Bayesian approaches using Gibbs or reversible jump MCMC sampling.

Recently, Holmes and Mallick (1998) adapted perfect sampling (Propp and Wilson 1996) to the context of orthogonal regression. While more computationally intensive per iteration, this may prove to be more efficient for estimation than Gibbs sampling or MH algorithms in problems where the method is applicable. Whether perfect sampling can be used with non-orthogonal designs is still open.

With the exception of deterministic search, most methods for sampling models rely on algorithms that sample models with replacement. In cases where $g(\gamma)$ is known, model probabilities may be estimated using renormalized model probabilities (Clyde et al. 1996). As no additional information is provided under resampling models, algorithms based on sampling models without replacement may be more efficient. Under random sampling without replacement (with equal probabilities), the estimates of model probabilities (CGM equation 3.56) are ratios of Horvitz-Thompson estimators (Horvitz and Thompson 1952) and are simulation consistent. Current work (joint with M. Littman) involves designing adaptive algorithms for sampling without replacement where sampling probabilities are sequentially updated. This appears to be a promising direction for implementation of Bayesian model selection and model averaging.

Summary

CGM have provided practical implementations for Bayesian model selection in linear and generalized linear models, non-parametric regression, and CART models, as well as spurred advances in research so that it is feasible to account for model uncertainty in a wide variety of problems. Demand for methods for larger problems seems to outpace growth in computing resources, and there is a growing need for Bayesian model selection methods that "scale up" as the dimension of the problem increases. Guidance in prior selection is also critical, as in many cases prior information is limited. For example, current experiments using gene-array technology result in high dimensional design matrices, $p > 7000$, however, the sample size may only be on the order of 10-100 (Spang et al. 2000). Identifying which genes (corresponding to columns of X) are associated with outcomes (response to treatment, disease status, etc) is a challenging problem for Bayesian model selection, from both a computational standpoint, as well as the choice of prior distributions.

ADDITIONAL REFERENCES

Clyde, M. (2000). Model uncertainty and health effect studies for particulate matter. To appear in *Environmentrics.*

Dawid, A. and Lauritzen, S. (2000). Compatible prior distributions. Technical Report.

Fernandez, C., Ley, E., and Steel, M.F. (1998). Benchmark priors for Bayesian model averaging. Technical report, Dept. of Econometrics, Tillburg Univ., Netherlands.

Furnival, G.M. and Wilson, Robert W.J. (1974). Regression by Leaps and Bounds. *Technometrics,* **16**, 499-511.

George, E.I. (1999a). Discussion of "Bayesian Model Averaging: A Tutorial" by Hoeting, J.A., Madigan, D., Raftery, A.E., and Volinsky, C.T. *Statist. Sci.* **14**, 401-404.

Hansen, M. and Yu, B. (1999). Model selection and the principle of minimum description. Technical Report *http://cm.bell-labs.com/who/cocteau/papers.*

Hanson, M. and Kooperberg, C. (1999). Spline adaptation in extended linear models. Technical Report *http://cm.bell-labs.com/who/cocteau/papers.*

Hoeting, H.A., Madigan, D., Raftery, A.E., and Volinsky, C.T. (1999). Bayesian model averaging: A tutorial (with discussion). *Statist. Sci.* **14**, 382-417.

Holmes, C. and Mallick, B.K. (1998). Perfect simulation for orthogonal model mixing. Technical Report, Dept. of Math., Imperial College, London.

Horvitz, D. and Thompson, D. (1952). A generalization of sampling without replacement from a finite universe. *J. Amer. Statist. Asso.* **47**, 663-685.

Kass, R.E. and Raftery, A.E. (1995). Bayes factors. *J. Amer. Statist. Asso.* **90**, 773-795.

Propp, J. and Wilson, D. (1996). Exact sampling with coupled Markov chains and applications to statistical mechanics. *Random Structures and Algorithms,* **9**, 223-252.

Spang, R., Zuzan, H., West, M., Nevins, J, Blanchette, C., and Marks, J.R. (2000). Prediction and uncertainty in the analysis of gene expression profiles. Discussion paper, ISDS, Duke Univ.

Wong, F., Hansen,M.H., Kohn, R., and Smith, M. (1997). Focused Sampling and its Application to Nonparametric and Robust Regression. Bell Labs Technical Report. Technical Report *http://cm.bell-labs.com/who/cocteau/papers.*

Dean P. Foster and Robert A. Stine

University of Pennsylvania

We want to draw attention to three ideas in the paper of Chipman, George and McCulloch (henceforth CGM). The first is the importance of an adaptive variable selection criterion. The second is the development of priors for interaction terms. Our perspective is information theoretic rather than Bayesian, so we briefly review this alternative perspective. Finally, we want to call attention to the practical importance of having a fully automatic procedure. To convey the need for automatic procedures, we discuss the role of variable selection in developing a model for credit risk from the information in a large database.

Adaptive variable selection

A method for variable selection should be *adaptive.* By this, we mean that the prior, particularly $p(\gamma)$, should adapt to the complexity of the model that matches the data rather than impose an external presumption of the number of variables in the model. One may argue that in reasonable problems the modeler should have a good idea how many predictors are going to be useful. It can appear that a well-informed modeler does not need an adaptive prior and can use simpler, more rigid alternatives that reflect knowledge of the substantive context. While domain knowledge is truly useful, it does

Dean P. Foster and Robert P. Stine are Associate Professors, Department of Statistics, The Wharton School of the University of Pennsylvania, Philadelphia, PA 19104-6302, U.S.A; emails: foster@diskworld.wharton.upenn.edu and stine@wharton.upenn.edu.

not follow that such knowledge conveys how many predictors belong in a model. The problem is made most transparent in the following admittedly artificial setting.

A small error in the choice of the basis in an orthogonal regression can lead to a proliferation in the number of required predictors. Suppose that we wish to predict future values of a highly periodic sequence, one dominated by a single sinusoid with frequency ω. If we approach this as a problem in variable selection and use the common Fourier basis to define the collection of predictors, the number of predictors is influenced by how close the frequency of the dominant cycle comes to a Fourier frequency. Fourier frequencies are of the form $\omega_j = 2\pi j/n$, indicating sinusoids that complete precisely j cycles during our n observations. If it so happens that $\omega = \omega_k$, then our model will likely need but one sinusoid to model the response. If ω is not of this form, however, our model will require many sinusoids from the Fourier basis to fit the data well. For example, with $n = 256$ and $\omega = 2\pi 5.5/n$, it takes 8 sinusoids at Fourier frequencies to capture 90% of the variation in this signal. The optimal basis would need but one sinusoid. Adaptive thresholding — the empirical Bayes approach — is forgiving of such errors, whereas dogmatic methods that anticipate, say, a single sinusoid are not.

Information theory and the choice of priors

A difficult choice in the use of Bayesian models for variable selection is the choice of a prior, particularly a prior for the subspace identifying the predictors. We have found coding ideas drawn from information theory useful in this regard, particularly the ideas related to Rissanen's minimum description length (*MDL*). The concreteness of coding offers appealing metaphors for picking among priors that produce surprisingly different selection criteria. In the Bayesian setting, calibration also offers a framework for contrasting the range of variable selection criteria.

The problem we consider from information theory is compression. This problem is simple to state. An *encoder* observes a sequence of n random variables $Y = (Y_1, \ldots, Y_n)$, and his objective is to send a *message* conveying these observations to a *decoder* using as few bits as possible. In this context, a *model* is a completely specified probability distribution, a distribution that is shared by the encoder and decoder. Given that both encoder and decoder share a model $P(Y)$ for the data, the optimal message length (here, the so-called "idealized length" since we ignore fractional bits and the infinite precision of real numbers) is

$$\ell(Y) = \log_2 \frac{1}{P(Y)} \text{ bits.}$$

If the model is a good representation for the data, then $P(Y)$ is large and the resulting message length is small. Since the encoder and decoder share the model $P(Y)$ they can use a technique known as arithmetic coding to realize this procedure. But what model should they use?

Common statistical models like the linear model are parametric models P_{θ_q}, indexed by a q-dimensional parameter θ_q. For example, suppose that the data Y are generated by the Gaussian linear model

$$Y = \theta_1 X_1 + \theta_2 X_2 + \cdots + \theta_q X_q + \epsilon, \quad \epsilon \sim N(0, \sigma^2) .$$

To keep the analysis straightforward, we will assume σ^2 is known (see Barron, Rissanen and Yu 1998, for the general case). Given this model, the shortest code for the data is obtained by maximizing the probability of Y, namely using maximum likelihood (i.e., least squares) to estimate θ_q and obtain a message with length

$$\ell_{\hat{\theta}_q}(Y) = \frac{\log_2 e}{2\sigma^2} RSS(\hat{\theta}_q) + \frac{n}{2} \log_2(2\pi\sigma^2) ,$$

where $RSS(\hat{\theta}_q)$ is the residual sum of squares. This code length is not realizable, however, since $P_{\hat{\theta}_q}$ is *not* a model in our sense. The normal density for Y with parameters $\hat{\theta}_q = \theta_q(Y)$ integrates to more than one, $C_{n,q} = \int_Y P_{\theta_q(Y)}(Y)dY > 1$.

Once normalized with the help of some benign constraints that make the integral finite but do not interfere with variable selection (see, e.g., Rissanen 1986), the code length associated with the model $P_{\hat{\theta}_q}/C_{n,q}$ is

$$\ell_q(Y) = \log_2 C_{n,q} + \ell_{\hat{\theta}_q}(Y) . \tag{1}$$

The need for such normalization reminds us that coding does not allow improper priors; improper priors generate codes of infinite length. We can think of the first summand in (1) as the length of a code for the parameters $\hat{\theta}_q$ (thus defining a prior for θ_q) and the second as a code for the compressed data. This perspective reveals how coding guards against over-fitting: adding parameters to the model will increase $C_{n,q}$ while reducing the length for the data.

So far, so good, but we have not addressed the problem of variable selection. Suppose that both the encoder and decoder have available a collection of p possible predictors to use in this q-variable regression. Which predictors should form the code? In this expanded context, our code at this point is incomplete since it includes $\hat{\theta}_q$, but does not identify the q predictors. It is easy to find a remedy: simply prefix the message with the p bits in γ. Since codes imply probabilities, the use of p bits to encode γ implies a prior, p_1 say, for these indicators. This prior is the iid Bernoulli model with probability $\Pr(\gamma_i = 1) = \frac{1}{2}$ for which the optimal code length for γ is indeed p,

$$\log_2 1/p_1(\gamma) = \log_2 2^p = p .$$

Since adding a predictor does not affect the length of this prefix – it's always p bits – we add the predictor X_{q+1} if the gain in data compression (represented by the reduction in RSS) compensates for the increase in the normalizing constant $C_{n,q}$. Using a so-called two-part code to approximate the code length (1), we have shown (Foster and Stine 1996) that this approach leads to a thresholding rule. For orthogonal predictors, this criterion amounts to choosing those predictors whose z-statistic $z_j = \hat{\theta}_j/\text{SE}(\hat{\theta}_j)$ exceeds a threshold near 2. Such a procedure resembles the frequentist selection procedure *AIC*, which uses a threshold of $\sqrt{2}$ in this context.

Now suppose that p is rather large. Using the p bits to represent γ seems foolish if we believe but one or two predictors are likely to be useful. If indeed few predictors are useful, we obtain a shorter message by instead forming a prefix from the *indices* of the those $\gamma_j = 1$. Each index now costs us about $\log_2 p$ bits and implies a different prior for γ. This prior, p_2 say, is again iid Bernoulli, but with small probability $\Pr(\gamma_i = 1) = 1/p$; the optimal code length for γ under p_2 is

$$\log_2 1/p_2(\gamma) = q \log p - (p-q)\log(1-q/p) \approx q \log p \,,$$

for $q = \sum_j \gamma_j \ll p$. Notice how this code assigns a higher cost to adding a predictor to the model. Adding a predictor does not affect the length of the prefix given by $p_1(\gamma)$. With $p_2(\gamma)$ as the prior, however, adding a predictor adds both a coefficient as well as its index to the message. The prefix grows by an additional $\log_2 p$ bits. For orthogonal regression and two-part codes, this approach implies a threshold for z_j near $\sqrt{2\log p}$ which also happens to correspond roughly to another frequentist procedure. This is the well-known Bonferroni method which retains predictors whose p-value is less than α/p for some $0 \leq \alpha \leq 1$.

Both of these codes have some appeal and correspond to frequentist methods as well as Bayesian priors, but neither is adaptive. The prior for the first code with a fixed p-bit prefix expects half of the predictors to be useful. The second has a prior that expects only one predictor to enter the model. As codes, both are flawed. The gold standard in coding is to compress the data down to the limit implied by the entropy of the underlying process, whatever that process may be. Both $p_1(\gamma)$ and $p_2(\gamma)$ only approach that limit when they happen to be right (e.g., when in fact only one predictor is needed in the model). Alternatively, so-called universal codes exist for representing binary sequences, and these compress γ almost as well as if the underlying probability were known. Assuming that the elements γ_j are iid (one can drop this condition as well), a universal code represents γ_q using about $pH(q/p)$ bits, where H denotes the Boolean entropy function

$$H(u) = u \log_2 \frac{1}{u} + (1-u)\log_2 \frac{1}{1-u}, \quad 0 \leq u \leq 1 \,.$$

Universal codes are *adaptive* in that they perform well for all values of q/p, doing almost as well as either of the previous codes when they happen to be right, but much better in other cases. Returning to the setting of an orthogonal regression, a universal code also implies a threshold for adding a predictor. The threshold in this case now depends on how many predictors are in the model. One adds the predictor X_j to a model that already has q predictors if its absolute z-statistic $|\hat{\theta}_j/\mathrm{SE}(\hat{\theta}_j)| > \sqrt{2\log p/q}$. This is essentially the empirical Bayes selection rule discussed by CGM in Section 3.3. The threshold decreases as the model grows, adapting to the evidence in favor of a larger collection of predictors. Again, this procedure is analogous to a frequentist method, namely step-up testing as described, for example, in Benjamini and Hochberg (1995).

Coding also suggests novel priors for other situations when the elements of γ are not so "anonymous". For example, consider the treatment of interaction terms. In the application we discuss in the next section, we violate the principle of marginality and treat interactions in a non-hierarchical fashion. That is, we treat them just like any other coefficient. Since we start with 350 predictors, the addition of interactions raises the number of possible variables to about $p = 67,000$. Since they heavily outnumber the linear terms, interactions dominate the predictors selected for our model. Coding the model differently leads to a different prior. For example, consider a variation on the second method for encoding γ by giving the index of the predictor. One could modify this code to handle interactions by appending a single bit to all indices for interactions. This one bit would indicate whether the model included the underlying linear terms as well. In this way, the indices for X_j, X_k and $X_j * X_k$ could be coded in $1 + \log_2 p$ bits rather than $3\log_2 p$ bits, making it much easier for the selection criterion to add the linear terms.

An application of automatic, adaptive selection

Methods for automatic variable selection matter most in problems that confront the statistician with many possibly relevant predictors. If the available data set holds, say, 1000 observations but only 10 predictors, then variable selection is not going to be very important. The fitted model with all 10 of these predictors is going to do about as well as anything. As the number of predictors increases, however, there comes a point where an automatic method is necessary. What constitutes a large number of possible predictors? Probably several thousand or more.

Such problems are not simply imaginary scenarios and are the common fodder for "data mining". Here is one example of such a problem, one that we discuss in detail in Foster and Stine (2000). The objective is to anticipate the onset of bankruptcy for credit card holders. The available data set holds records of credit card usage for a collection of some 250,000 accounts. For each account, we know a variety of demographic characteristics, such as place and type of residence of the card holder. When combined with several months of past credit history and indicators of missing data, we have more than 350 possible predictors. The presence of missing data adds further features, and indeed we have found the absence of certain data to be predictive of credit risk. Though impressive at first, the challenge of choosing from among 350 features is nonetheless small by data mining standards. For predicting bankruptcy, we have found interactions between pairs or even triples to be very useful. Considering pairwise interactions swells the number of predictors to over 67,000. It would be interesting to learn how to apply a Gibbs sampler to such problems with so many possible features.

Though challenging for any methodology, problems of this size make it clear that we must have automated methods for setting prior distributions. To handle 67,000 predictors, we use adaptive thresholding and stepwise regression. Beginning from the model with no predictors, we identify the first predictor X_{j_1} that by itself explains the most variation in the response. We add this predictor to the model if its t-statistic $t_{11} = \hat{\beta}_{j_1,1}/se(\hat{\beta}_{j_1,1})$ (in absolute value) exceeds the threshold $\sqrt{2\log p}$. If $\hat{\beta}_{j_1,1}$ meets this criterion, we continue and find the second predictor X_{j_2} that when combined with X_{j_1} explains the most variation in Y. Rather than compare the associated t-statistic $t_{j_2,2}$ to the initial threshold, we reduce the threshold to $\sqrt{2\log p/2}$, making it now easier for X_{j_2} to enter the model. This process continues, greedily adding predictors so long as the t-statistic for each exceeds the declining threshold,

$$\text{Step } q\text{: Add predictor } X_{j_q} \quad \Longleftrightarrow \quad |t_{j_q,q}| > \sqrt{2\log p/q}$$

Benjamini and Hochberg (1995) use essentially the same procedure in multiple testing where it is known as step-up testing. This methodology works in this credit modelling in that it finds structure without over-fitting. Figure 1 shows a plot of the residual sum of squares as a function of model size; as usual, RSS decreases with p. The plot also shows the cross-validation sum of squares (CVSS) computed by predicting an independent sample. The validation sample for these calculations has about 2,400,000 observations; we scaled the CVSS to roughly match the scale of the RSS. Our search identified 39 significant predictors, and each of these — with the exception of the small "bump" — improves the out-of-sample performance of the model. Although the CVSS curve is flat near $p = 39$, it does not show the rapid increase typically associated with over-fitting.

Gibbs sampling and the practical Bayesian methods discussed by CGM offer an interesting alternative to our search and selection procedure. They have established the foundations, and the next challenge would seem to be the investigation of how their search procedure performs in the setting of real applications such as this. Greedy

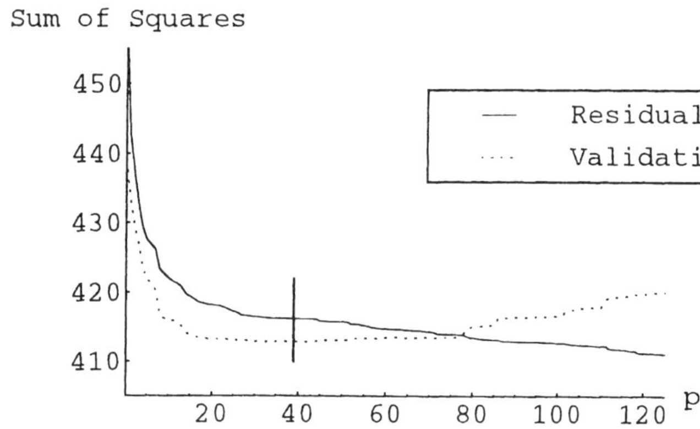

Figure 2: *Residual and cross-validation sums of squares for predicting bankruptcy.*

selection methods such as stepwise regression have been well-studied and probably do not find the best set of predictors. Once X_{j_1} becomes the first predictor, it must be in the model. Such a strategy is clearly optimal for orthogonal predictors, but can be 'tricked' by collinearity. Nonetheless, stepwise regression is fast and comparisons have shown it to be competitive with all possible subsets regression (e.g., see the discussion in Miller 1990). Is greedy good enough, or should one pursue other ways of exploring the space of models via Gibbs sampling?

ADDITIONAL REFERENCES

Barron, A., Rissanen, J. and Yu, B. (1998). The minimum description length principle in coding and modelling. *IEEE Trans. Info. Theory* **44**, 2743-2760.

Benjamini, Y. and Hoohberg, Y. (1995). Controlling the false discovery rate: a practical and powerful approach to multiple testing. *J. Roy. Statist. Soc. Ser. B* **57**, 289-300.

Foster, D.P. and Stine, R.A. (1996). Variable selection via information theory. Technical Report, Northwestern Univ., Chicago.

Foster, D.P. and Stine, R.A. (2000). Variable selection in data mining: Building a predictive model for bankruptcy. Unpublished Manuscript.

Miller, A.J. (1990). *Subset Selection in Regression.* Chapman and Hall, London.

Rissanen, J. (1986). Stochastic complexity and modelling. *Ann. Statist.* **14**, 1080-1100.

REJOINDER

Hugh Chipman, Edward I. George and Robert E. McCulloch

First of all, we would like to thank the discussants, Merlise Clyde, Dean Foster and Robert Stine, for their generous discussions. They have each made profound contributions to model selection and this comes through in their insightful remarks. Although there is some overlap in the underlying issues they raise, they have done so from different vantage points. For this reason, we have organized our responses around each of their discussions separately.

Clyde

Clyde raises key issues surrounding prior selection. For choosing model space priors for the linear model with redundant variables, she confirms the need to move away from uniform and independence priors towards dilution priors. This is especially true in high dimensional problems where independence priors will allocate most of their probability to neighborhoods of redundant models. Clyde's suggestion to use an imaginary training data to construct a dilution prior is a very interesting idea. Along similar lines, we have considered dilution priors for the linear model where $p(\gamma)$ is defined as the probability that $Y^* \sim N_n(0, I)$ is "closer" to the span of X_γ than the span of any other $X_{\gamma'}$. Here Y^* can be thought of as imaginary training data reflecting ignorance about the direction of Y. Further investigation of the construction of effective dilution priors for linear models is certainly needed.

Clyde comments on the typical choices of Σ_γ for the coefficient prior $p(\beta_\gamma \mid \sigma^2, \gamma) = N_{q_\gamma}(\bar{\beta}_\gamma, \sigma^2 \Sigma_\gamma)$ in (3.9), namely $\Sigma_\gamma = c\,(X_\gamma' X_\gamma)^{-1}$ and $\Sigma_\gamma = c\, I_{q_\gamma}$. In a sense, these choices are the two extremes, $c\,(X_\gamma' X_\gamma)^{-1}$ serves to reinforce the likelihood covariance while $c\, I_{q_\gamma}$ serves to break it apart. As we point out, the coefficient priors under such Σ_γ, are the natural conditional distributions of the nonzero components of β given γ when $\beta \sim N_p(0, c\,\sigma^2 (X'X)^{-1})$ and $\beta \sim N_p(0, c\,\sigma^2 I)$, respectively. The joint prior $p(\beta_\gamma, \gamma \mid \sigma^2)$ then corresponds to a reweighting of the conditional distributions according to the chosen model space prior $p(\gamma)$. With respect to such joint priors, the conditional distributions are indeed compatible in the sense of Dawid and Lauritzen (2000). Although not strictly necessary, we find such compatible specifications to provide an appealingly coherent description of prior information.

We agree with Clyde that the choice of c can be absolutely crucial. As the calibration result in (3.17) shows, different values of c, and hence different selection criteria such as AIC, BIC and RIC, are appropriate for different states of nature. For larger models with many small nonzero coefficients, smaller values of c are more appropriate, whereas for parsimonious models with a few large coefficients, larger values of c are better. Of course, when such information about the actual model is unavailable, as is typically the case, the adaptive empirical Bayes methods serve to insure against poor choices. It is especially appealing that by avoiding the need to specify hyperparameter values, empirical Bayes methods are automatic, a valuable feature for large complicated problems. Similar features are also offered by fully Bayes methods that margin out the hyperparameters with respect to hyperpriors. The challenge for the implementation of effective fully Bayes methods is the selection of hyperpriors that offer strong performance across the model space while avoiding the computational difficulties described by Clyde.

Clyde points out an important limitation of using empirical Bayes methods with conditional priors of the form $p(\beta_\gamma \mid \sigma^2, \gamma) = N_{q_\gamma}(\bar{\beta}_\gamma, \sigma^2 c V_\gamma)$. When the actual model has many moderate sized coefficients and but a few very large coefficients, the few large coefficients will tend to inflate the implicit estimate of c, causing the moderate sized coefficients to be ignored as noise. In addition to the heavy-tailed and the grouped prior formulations she describes for mitigating such situations, one might also consider elaborating the priors by adding a shape hyperparameter.

Finally, Clyde discusses the growing need for fast Bayesian computational methods that "scale up" for very large high dimensional problems. In this regard, it may be useful to combine heuristic strategies with Bayesian methods. For example, George and McCulloch (1997) combined globally greedy strategies with local MCMC search in applying Bayesian variable selection to build tracking portfolios. In our response to Foster and Stine below, we further elaborate on the potential of greedy algorithms for such purposes.

Foster and Stine

Foster and Stine begin by emphasizing the need for adaptive procedures. We completely agree. The adaptive empirical Bayes methods described in Section 3.3 offer improved performance across the model space while automatically avoiding the need for hyperparameter specification. For more complicated settings, adaptivity can be obtained by informal empirical Bayes approaches that use the data to gauge hyperparameter values, such as those we described for the inverse gamma distribution in Sections 3.2 and 4.1.2. In the sinusoid modelling example of Foster and Stine, a simple adaptive resolution is obtained by a Bayesian treatment with a prior on ω_k. This nicely illustrates the fundamental adaptive nature of Bayesian analysis. By using priors rather than fixed arbitrary values to describe the uncertainty surrounding the unknown characteristics in a statistical problem, Bayesian methods are automatically adaptive. We attribute the adaptivity of empirical Bayes methods to their implicit approximation of a fully Bayes approach.

Foster and Stine go on to discuss some revealing analogies between strategies for minimum length coding and formulations for Bayesian model selection. The key idea is that the probability model for the data, namely the complete Bayesian formulation, also serves to generate the coding strategy. Choosing the probability model that best predicts the data is tantamount to choosing the optimal coding strategy. Foster and Stine note that improper priors are unacceptable because they generate infinite codes. This is consistent with our strong preference for proper priors for model selection. They point out the potential inefficiencies of Bernoulli model prior codes for variable selection, and use them to motivate a universal code that adapts to the appropriate model size. This is directly analogous to our observation in Section 3.3 that different hyperparameter choices for the Bernoulli model prior (3.15) correspond to different model sizes, and that an empirical Bayes hyperparameter estimate adapts to the appropriate model size. It should be the case that the universal prior corresponds to a fully Bayes prior that is approximated by the empirical Bayes procedure. Finally, their coding scheme for interactions is interesting and clearly effective for parsimonious models. Such a coding scheme would seem to correspond to a hierarchical prior that puts a Bernoulli $1/p$ prior on each potential triple - two linear terms and their interaction - and a conditionally uniform prior on the elements of the triple.

The credit risk example given by Foster and Stine raises several interesting issues. It illustrates that with this large dataset, an automatic stepwise search algorithm can achieve promising results. Figure 1 shows how their adaptive threshold criterion guards against overfitting, although the cross validation results seem also to suggest that a smaller number of terms, around 20, is adequate for prediction. Another automatic adaptive alternative to consider here would be a stepwise search based on the empirical Bayes criterion C_{CML} in (3.22). It would also be interesting to investigate the potential of one of the Bayesian variable selection approaches using the hierarchical priors described in Section 3.1 to account for potential relationships between linear and interaction terms. As opposed to treating all potential predictors independently, such priors tend to concentrate prior mass in a smaller, more manageable region of the model space.

For example, Chipman, Hamada and Wu (1997) considered an 18 run designed experiment with 8 predictors used in a blood-glucose experiment. The non-orthogonal design made it possible to consider a total of 113 terms, including quadratic terms and interactions. They found that independence priors of the form (3.2) led to such a diffuse posterior that, in 10,000 steps of a Gibbs sampling run, the most frequently visited model was visited only 3 times. On the other hand, hierarchical priors like (3.7) raised posterior mass on the most probable model to around 15%. In the same problem stepwise methods were unable to find all the different models identified by stochastic search. In effect, priors that account for interactions (or other structure, such as correlations between predictors which can lead to the dilution problem discussed in Section 3.1) can narrow the posterior to models which are considered more "plausible". We note, however, that the credit risk example is much larger than this example, and because the number of observations there is much larger than the number of predictors, such a hierarchical prior may have only a minor effect.

The credit risk example is also a clear illustration of the everlasting potential of greedy search algorithms on very large problems. At the very least, greedy algorithms can provide a "baseline" against which MCMC stochastic search results can be compared and then thrown out if an improvement is not found. Furthermore, greedy algorithms can provide a fast way to get rough estimates of hyperparameter values, and can be used directly for posterior search. Greedy algorithms also offer interesting possibilities for enhancement of stochastic search. At the most basic level, the models identified by greedy algorithms can be used as starting points for stochastic searches. Stochastic search algorithms can also be made more greedy, for example, by exponentiating the probabilities in the accept/reject step of the MH algorithms. The use of a wide variety of search algorithms, including MCMC stochastic search, can only increase the chances of finding better models.

Model Selection
IMS Lecture Notes - Monograph Series (2001) Volume 38

Objective Bayesian Methods for Model Selection: Introduction and Comparison

James O. Berger and Luis R. Pericchi

Duke University and University of Puerto Rico

Abstract

The basics of the Bayesian approach to model selection are first presented, as well as the motivations for the Bayesian approach. We then review four methods of developing default Bayesian procedures that have undergone considerable recent development, the Conventional Prior approach, the Bayes Information Criterion, the Intrinsic Bayes Factor, and the Fractional Bayes Factor. As part of the review, these methods are illustrated on examples involving the normal linear model. The later part of the chapter focuses on comparison of the four approaches, and includes an extensive discussion of criteria for judging model selection procedures.

James O. Berger is the Arts and Sciences Professor of Statistics, Institute of Statistics and Decision Sciences, Duke University, Durham, NC 27708-0251, U.S.A; email: jberger@stat.duke.edu. Luis R. Pericchi is Professor, Department of Mathematics and Computer Science, University of Puerto Rico, Rio Piedras Campus, P.O. Box 23355, San Juan, PR 00931-3355, U.S.A; email: pericchi@goliath.cnnet.clu.edu. This research was supported by the National Science Foundation (U.S.A.), Grants DMS-9303556 and DMS-9802261, and by CONICIT-Venezuela G-97000592. The second author held a Guggenheim Fellowship during part of his research. An earlier version of this manuscript was presented at the workshop *Bayesian Model Selection*, held in Cagliari in June, 1997.

Contents

1 Introduction

1.1 Bayes Factors and Posterior Model Probabilities

Suppose that we are comparing q models for the data $\mathbf{x}$,

$$M_i\colon \mathbf{X} \text{ has density } f_i(\mathbf{x}|\boldsymbol{\theta}_i), \quad i = 1, \ldots, q,$$

where the $\boldsymbol{\theta}_i$ are unknown model parameters. Suppose that we have available prior distributions, $\pi_i(\boldsymbol{\theta}_i)$, $i = 1, \ldots, q$, for the unknown parameters. Define the marginal or predictive densities of $\mathbf{X}$,

$$m_i(\mathbf{x}) = \int f_i(\mathbf{x}|\boldsymbol{\theta}_i)\pi_i(\boldsymbol{\theta}_i)\, d\boldsymbol{\theta}_i.$$

The *Bayes factor* of M_j to M_i is given by

$$B_{ji} = \frac{m_j(\mathbf{x})}{m_i(\mathbf{x})} = \frac{\int f_j(\mathbf{x}|\boldsymbol{\theta}_j)\pi_j(\boldsymbol{\theta}_j)\, d\boldsymbol{\theta}_j}{\int f_i(\mathbf{x}|\boldsymbol{\theta}_i)\pi_i(\boldsymbol{\theta}_i)\, d\boldsymbol{\theta}_i}. \tag{1.1}$$

The Bayes factor is often interpreted as the "odds provided by the data for M_j versus M_i." Thus $B_{ji} = 10$ would suggest that the data favor M_j over M_i at odds of ten to one. Alternatively, B_{ji} is sometimes called the "weighted likelihood ratio of M_j to M_i," with the priors being the "weighting functions." These interpretations are particularly appropriate when, as here, we focus on conventional or default choices of the priors.

If prior probabilities $P(M_j)$, $j = 1, \ldots, q$, of the models are available, then one can compute the posterior probabilities of the models from the Bayes factors. Indeed, it is easy to see that *posterior probability* of M_i, given the data $\mathbf{x}$, is

$$P(M_i \,|\, \mathbf{x}) = \frac{P(M_i)m_i(\mathbf{x})}{\sum_{j=1}^q P(M_j)m_j(\mathbf{x})} = \left[\sum_{j=1}^q \frac{P(M_j)}{P(M_i)} B_{ji}\right]^{-1}. \tag{1.2}$$

A particularly common choice of the prior model probabilities is $P(M_j) = 1/q$, so that each model has the same initial probability. The posterior model probabilities are then the same as the renormalized marginal probabilities, given by

$$\bar{m}_i(\mathbf{x}) = \frac{m_i(\mathbf{x})}{\sum_{j=1}^q m_j(\mathbf{x})}. \tag{1.3}$$

Indeed, in scientific reporting, it is common to provide the $\bar{m}_i(\mathbf{x})$, rather than the $P(M_j \,|\, \mathbf{x})$, since the prior probabilities of models can be a contentious matter and anyone can use the $\bar{m}_i(\mathbf{x})$ to determine their personal posterior probabilities via (1.2), noting that $B_{ji} = \bar{m}_j(\mathbf{x})/\bar{m}_i(\mathbf{x})$. In most of this chapter we thus focus on determination of default $\bar{m}_i(\mathbf{x})$ or, equivalently, default Bayes factors.

1.2 Motivation for the Bayesian Approach to Model Selection

Reason 1: *Bayes factors and posterior model probabilities are easy to understand.* The interpretation of Bayes factors as odds and the direct probability interpretation of posterior model probabilities are readily understandable by even non-statisticians. For instance, in Example 5 (section 4.5) the three models under consideration will be seen to have (for the preferred default Bayesian analysis) posterior probabilities given by $\bar{m}_1(\mathbf{x}) = 0.18$, $\bar{m}_2(\mathbf{x}) = 0.48$ and $\bar{m}_3(\mathbf{x}) = 0.34$. In contrast, alternative model selection schemes are not based on readily interpretable quantities. For instance, many schemes are based directly or indirectly on p-values corresponding to various models, but the extreme difficulty of properly interpreting p-values is well known (see, e.g., Edwards, Lindeman, and Savage 1963, Berger and Sellke 1987, Berger and Delampady 1987, and Sellke, Bayarri and Berger 2000).

Reason 2: *Bayesian model selection is consistent.* This means that, if one of the entertained models is actually the true model, then Bayesian model selection will (under very mild conditions) guarantee selection of the true model if enough data is observed. Rather surprisingly, use of most classical model selection tools, such as p-values, C_p, and AIC, does not guarantee consistency.

It is sometimes argued that consistency is not a highly relevant concept because none of the models being considered are likely to be exactly true. While the true model might indeed typically be outside the candidate set of models, use of a procedure that fails to be consistent in the 'nice' case is disturbing. Furthermore, even when the true model is not among those being considered, results in Berk (1966) and Dmochowski (1996) show that (asymptotically and under mild conditions) Bayesian model selection will choose that model among the candidates that is closest to the true model in terms of Kullback-Leibler divergence. While this is a very nice property, the situation is complicated by the fact that approximate Bayesian model selection procedures may not share the optimality properties of real Bayesian procedures. For instance, Shibata (1981) shows that BIC, a popular approximate Bayesian procedure (see section 2.4), is not optimal for certain situations in which the true model is not in the candidate set. Further developments along these lines can be found in Berger, Ghosh, and Mukhopadhyay (1999), and suggest caution in use of approximate Bayesian model selection procedures.

Reason 3. *Bayesian model selection procedures are automatic Ockham's razors,* favoring simpler models over more complex models when the data provides roughly comparable fits for the models. Overfitting is a continual problem in model selection, since more complex models will always provide a somewhat better fit to the data than will simpler models. In classical statistics, overfitting is addressed by introduction of a penalty term (as in AIC), which increases as the complexity (i.e., the number of unknown parameters) of the model increases. There is a huge literature discussing the (unanswerable) question of which penalty term is best (see, e.g., Shao 1997). In contrast, Bayesian procedures naturally penalize model complexity, and need no introduction of a penalty term. For an interesting historical example and general discussion and references, see Jefferys and Berger (1992).

Reason 4. *The Bayesian approach to model selection is conceptually the same, regardless of the number of models under consideration.* In contrast, there is a significant distinction in the classical approach between consideration of two models and consideration of more than two models; the former case is approached with the tools of hypothesis testing, while the latter is approached with the often quite different tools of model selection. Besides requiring the learning of different statistical technologies for testing and model selection, having a distinction between the two cases is philosophically unappealing.

Reason 5. *The Bayesian approach does not require nested models, standard distributions, or regular asymptotics.* Essentially all of classical model selection is based on at least one of these assumptions. Example 3 (see section 4.3) is an example of model selection with irregular models.

Reason 6. *The Bayesian approach can account for model uncertainty.* Selecting an hypothesis or model on the basis of data, and then using the same data to estimate model parameters or make predictions based upon the model, is well known to yield (often severely) overoptimistic estimates of accuracy. In the classical approach it is often thus recommended to use part of the data to select a model and the remaining part of the data for estimation and prediction. When only limited data is available, this can be difficult. Furthermore, this approach still ignores the fact that the selected model might very well be wrong, so that predictions based on assuming the model is true could be overly optimistic.

The Bayesian approach takes a different tack: ideally, all models are left in the analysis with, say, prediction being done using a weighted average of the predictive distributions from each model, the weights being determined from the posterior probabilities of each model. This is known as 'Bayesian model averaging,' and is widely used today as the basic methodology of accounting for model uncertainty. See Geisser (1993), Draper (1995), Raftery, Madigan and Hoeting (1997), and Clyde (1999) for discussion and references.

Although keeping all models in the analysis is an ideal, this can be cumbersome for communication and descriptive purposes. If only one or two models receive substantial posterior probability, it would not be an egregious sin to eliminate the other models from consideration. Even if one must report only one model, the fact mentioned above - that Bayesian model selection acts as a strong Ockham's razor - means that at least the selected model will not be an overly complex model, and so estimates and predictions based on this model will not be quite so overly optimistic.

Reason 7. *The Bayesian approach can yield optimal conditional frequentist procedures.* As mentioned earlier, model selection when only two models are under consideration can be viewed as hypothesis testing. The standard frequentist testing procedure, Neyman-Pearson testing, has the disadvantage of requiring the report of a fixed error probability α, no matter what the data. The common data-adaptive versions of classical testing, namely p-values, are not true frequentist procedures and also suffer from the rather severe interpretational problems discussed earlier. Thus, until recently, frequentists did not have satisfactory data-adaptive testing procedures.

In Berger, Brown, and Wolpert (1994), Berger, Boukai, and Wang (1997), Dass and Berger (1998), and Sellke, Bayarri and Berger (2001), it is shown that tests based on Bayes factors can be constructed such that the posterior probabilities of the hypotheses have direct interpretations as conditional frequentist error probabilities. The reported error probabilities thus vary with the data, and yet have valid frequentist interpretations.

The necessary technical detail to make this work is the defining of suitable conditioning sets upon which to compute the conditional error probabilities. These sets necessarily include data in both the acceptance and the rejection regions, and can roughly be described as the sets which include data points providing equivalent strength of evidence (as measured by p-values) for each of the hypotheses.

There are more surprises arising from this equivalence of Bayesian and conditional frequentist testing. One is that, in sequential testing using these tests, the stopping rule is largely irrelevant to the stated error probabilities. Thus there is no need to consider spending α or any of the difficult computations involved with unconditional sequential Neyman-Pearson testing. See Berger, Boukai and Wang (1999) for an illustration.

1.3 Utility Functions and Prediction

As with any statistical problem, one should, in principle, approach model selection from the perspective of decision analysis. See Bernardo and Smith (1994) for discussion of a variety of decision and utility-based approaches to model selection. Recent articles advocating particular such approaches include Dupuis and Robert (1998), Gelfand and Ghosh (1998), Goutis and Robert (1998), and Key, Pericchi and Smith (1999). It should also be noted that posterior probabilities do not necessarily arise as components of such analyses. Frequently, however, the statistician's goal is not to perform a formal decision analysis, but to summarize information from a study in such a way that others can perform decision analyses (perhaps informally) based on this information. Also, models are typically used for a wide variety of subsequent purposes, making advance selection through problem-specific utility considerations difficult. (Note, however, that some of the works in this direction, including the above-mentioned articles, propose generic utility functions that are argued to be broadly reflective of the purposes of model selection.) Even though these issues are by no means settled, it is likely that posterior probabilities (and Bayes factors) will always remain important tools of Bayesian model selection.

The most common use of models is for prediction of future observables, and there is considerable model selection methodology that is specifically oriented towards this goal. One such methodology is Bayesian model averaging (mentioned above), that explicitly bases predictions on the posterior weighted average of all the model predictions. Indeed, this can be shown to yield optimal Bayesian predictions for a variety of loss functions. Since the posterior model probabilities are an integral part of Bayesian model averaging, the discussion in this chapter is of direct relevance to that approach.

Often, one is constrained to select a single model that will be used for subsequent prediction and, somewhat surprisingly, it is not always optimal to select the model with the largest posterior probability. The largest posterior probability model is optimal under very general conditions if only two models are being entertained (see Berger 1999) and is often optimal for variable selection in linear models having orthogonal design matrices (cf. Clyde and George. 2000). For other cases, such as in nested linear models, the optimal single model for prediction is the 'median probability model,' defined and illustrated in Barbieri and Berger (2001). Again, however, this model is found through analysis of the posterior model probabilities, so that the developments in this chapter are of direct relevance.

1.4 Motivation for Objective Bayesian Model Selection

There has been a long debate in the Bayesian community as to the roles of subjective and objective Bayesian analysis. Few would disagree that subjective Bayesian analysis is an attractive ideal, but the objective Bayesians argue that it is frequently not a realistic possibility, either because of outside constraints (e.g., the appearance of objectivity is needed), or because it is simply not feasible to obtain the extensive needed elicitations from subject experts.

In model selection, this last argument is particularly compelling, because one often initially entertains a wide variety of models, and careful subjective specification of prior distributions for all the parameters of all the models is essentially impossible. Indeed, this would typically be an egregious waste of the (always limited) time for which subject experts are available; one would typically want to use this available expert time for model formulation and, possibly, prior elicitation for the model that is ultimately selected. There is actually little debate over this issue in model selection; virtually all the analyses one sees involve default methods.

1.5 Difficulties in Objective Bayesian Model Selection

There are four main difficulties with the development of default Bayesian methods of model selection.

Difficulty 1. *Computation can be difficult.* Calculation of the Bayes factor in (1.1) can be challenging when the parameter spaces are high dimensional. Also, the total number of models under consideration, for which computations need to be done, can be enormous, especially in model selection problems such as variable selection. We do not address computational issues here; some recent papers on the subject are Carlin and Chib (1995), Green (1995), Kass and Raftery (1995), Verdinelli and Wasserman (1995), Raftery, Madigan and Hoeting (1997), Clyde (1999), Chib and Jeliazkov (2001), Dellaportas, Forster and Ntzoufras (2001), Godsill (2001), and Han and Carlin (2001).

Difficulty 2. *When the models have parameter spaces of differing dimensions, use of improper noninformative priors yields indeterminate answers.* To see this, suppose that improper noninformative priors π_i^N and π_j^N are entertained for models M_i and M_j, respectively. The 'formal' Bayes factor using these priors, B_{ji}, would then be given by (1.1). But, because the priors are improper, one could have just as well used, as the noninformative priors, $c_i\pi_i^N$ and $c_j\pi_j^N$, in which case the Bayes factor would be $(c_j/c_i)B_{ji}$. Since the choice of c_j/c_i is arbitrary, the Bayes factor is clearly indeterminate.

When the parameters $\boldsymbol{\theta}_i$ and $\boldsymbol{\theta}_j$ can be thought of as essentially similar, choosing $c_j = c_i$ is reasonable. Situations in which this can be justified (through group invariance arguments) are given in Berger, Pericchi and Varshavsky (1998). If the parameter spaces of M_i and M_j are the same, it is common practice to also choose $c_j = c_i$, although we know of no formal way of justifying the practice. When (as is typically the case) the parameter spaces are of differing dimensions, choosing $c_j = c_i$ can be a bad idea, although it may not be egregiously bad if the dimensions are close. For special situations, there have been efforts to assign reasonable values to the c_j based on an extrinsic argument; an example is the approach of Spiegelhalter and Smith (1982). Ghosh and Samanta (2001) present an interesting generalization.

Difficulty 3. *Use of 'vague proper priors' usually gives bad answers in Bayesian model selection.* It is virtually never the case in Bayesian analysis that use of vague proper priors is superior to use of improper noninformative priors, and this is especially so in model selection. To see the danger, consider the following example.

Example. Suppose we observe $\mathbf{X} = (X_1, \ldots, X_n)$, where the X_i are iid $\mathcal{N}(0,1)$ under M_1 and $\mathcal{N}(\theta, 1)$ under M_2. Suppose θ under M_2 is given the $\mathcal{N}(0, K)$ prior, with variance K large; this is the usual vague proper prior for a normal mean. An easy calculation using (1.1) yields

$$B_{21} = (nK+1)^{-1/2} \exp\left(\frac{Kn^2}{2(1+Kn)}\bar{x}^2\right). \tag{1.4}$$

For large K (or large n), this is roughly $(nK)^{-1/2}\exp(z^2/2)$, where $z = \sqrt{n}\bar{x}$. So B_{21} depends very strongly on the *arbitrarily chosen* 'large' value of K. Note that even popular hierarchical priors, with vague proper priors on the hyperparameters, can run afoul of this difficulty.

In contrast, the usual noninformative prior for θ in this situation is $\pi_2^N(\theta) = 1$. The resulting Bayes factor is $B_{21} = \sqrt{n/2\pi}\exp(z^2/2)$, which is a reasonable value. The short story here is thus *never use 'arbitrary' vague proper priors for model selection, but improper noninformative priors may give reasonable results.*

Difficulty 4. *Even 'common parameters' can change meaning from one model to another, so that prior distributions must change in a corresponding fashion.* Here is an example.

Example. We wish to predict automotive fuel consumption, Y, from the weight, X_1, and engine size, X_2, of a vehicle. Two models are entertained:

$$M_1 : Y = X_1\beta_1 + \varepsilon_1, \quad \varepsilon_1 \sim \mathcal{N}(0, \sigma_1^2)$$
$$M_2 : Y = X_1\beta_1 + X_2\beta_2 + \varepsilon_2, \quad \varepsilon_2 \sim \mathcal{N}(0, \sigma_2^2).$$

Thinking, first, about M_2, suppose the elicited prior density is of the form $\pi_2(\beta_1, \beta_2, \sigma_2) = \pi_{21}(\beta_1)\pi_{22}(\beta_2)\pi_{23}(\sigma_2)$. Since β_1 is 'common' to the two models, one frequently sees the same prior, $\pi_{21}(\beta_1)$, also used for this parameter in M_1. This is not reasonable, since β_1 has a very different meaning (and value) under M_1 than under M_2. Indeed, regressing fuel consumption on weight alone will clearly yield a larger coefficient than regressing on both weight and engine size, because of the considerable positive correlation between weight and engine size. Similarly, one often sees the variances σ_1^2 and σ_2^2 being equated and assigned the same prior, even though it is clear that σ_1^2 will typically be larger than σ_2^2.

It should be noted that this problem affects subjective, as well as default Bayesian model selection. Here, for instance, an automotive expert might center a subjective prior for β_1 under M_1 at 0.8, but might center the prior under M_2 at 0.5. Obtaining such properly compatible assessments is far from easy.

1.6 Preview

Four methods of developing default Bayesian model selection procedures have undergone considerable recent development. These methods are the Conventional Prior (CP) approach, the Bayes Information Criterion (BIC), the Intrinsic Bayes Factor (IBF) approach, and the Fractional Bayes Factor (FBF) approach. We will mention other more recent approaches, as appropriate, but our attention will focus on these four methods.

The most obvious default Bayesian method that can be employed for model selection is simply to choose 'conventional' proper prior distributions, priors that seem likely to be reasonable for typical problems. This was the approach espoused by Jeffreys (1961), who recommended specific proper priors for certain standard testing problems. We shall call this approach the *Conventional Prior Approach*, and discuss it in section 2.1, with application to selection from among linear models.

While this approach is arguably about the best that can be done from a default perspective, it is difficult to implement in general, requiring careful selection of a default prior for each specific situation. This difficulty has lead to use, in practice, of rather crude approximations to Bayes factors, typified by the *Bayesian Information Criterion (BIC)*. This is defined in section 2.2 and illustrated on the linear model.

Concerns over the accuracy and applicability of BIC have resulted in the recent development of alternative default methods of Bayesian model selection. The two most prominent of these methods are the *Fractional Bayes Factor* approach of O'Hagan (1995) and the *Intrinsic Bayes Factor* approach of Berger and Pericchi (1996a, 1996b), and Berger, Pericchi and Varshavsky (1998). These are introduced in sections 2.3 and 2.4, respectively, and applied to the linear model.

The main purpose of this chapter is to review and compare these four approaches to default model selection. To this end, it is necessary to first discuss - in section 3 - our views as to how model selection methodologies should be evaluated. This is a mixture of application and theory. 'Testing' the various methodologies in specific applications is clearly important, as is evaluating their ease of use in application. On the theoretical side, we will argue that the most enlightening approach is to investigate correspondence of the procedures with actual Bayesian procedures.

Section 4 presents the evaluations of the four studied methodologies. The section is oriented around six important or challenging examples: a situation involving group invariance, the situation of improper likelihoods, irregular models, one-sided testing, an example in which the dimension grows with the sample size, and an example in which none of the entertained models is true. Section 5 presents a summary of the comparisons.

Section 6 gives some final recommendations. In brief, our personal view of the situation is that all four discussed methods have value and should be utilized under appropriate circumstances. The challenge is thus to outline the circumstances under which each should be used or, perhaps more importantly, when each should not be used. This chapter is thus a summary of our experience in this regard. Clearly, evaluations such as this will be continually evolving as practical experience with the procedures increases.

2 Objective Bayesian Model Selection Methods, with Illustrations in the Linear Model

In this section, we discuss the four default Bayesian approaches to model selection that will be considered in this chapter, the Conventional Prior (CP) approach, the Bayes Information Criterion (BIC), the Intrinsic Bayes Factor (IBF) approach, and the Fractional Bayes Factor (FBF) approach. These approaches will be illustrated through application to model selection in the linear model.

2.1 Conventional Prior Approach

Jeffreys (1961, Chapter 5) dealt with the issue of indeterminacy of noninformative priors by (i) using noninformative priors only for common (orthogonal) parameters in the models, so that the arbitrary multiplicative constant for the priors would cancel in all Bayes factors, and (ii) using default *proper* priors (but not vague proper priors) for parameters that would occur in one model but not the other. He presented arguments justifying certain default proper priors, but mostly on a case-by-case basis. This line of development has been successfully followed by many others (for instance, by Zellner and Siow 1980; see Berger and Pericchi 1996a, for other references.)

Illustration 1: Normal Mean, Jeffreys' Conventional Prior

Suppose the data is $\mathbf{X} = (X_1, \ldots, X_n)$, where the X_i are iid $\mathcal{N}(\mu, \sigma_2^2)$ under M_2. Under M_1, the X_i are $\mathcal{N}(0, \sigma_1^2)$. Note that, because of Difficulty 4 in section 1.5, we differentiate between σ_1^2 and σ_2^2. However, in this situation the mean and variance can be shown to be orthogonal parameters (i.e., the expected Fisher information matrix is diagonal), in which case Jeffreys argues that σ_1^2 and σ_2^2 do have the same meaning across models and can be identified as $\sigma_1^2 = \sigma_2^2 = \sigma^2$. Because of this identification, Jeffreys suggests that the variances can be assigned the same (improper) noninformative prior

$\pi^J(\sigma) = 1/\sigma$, since the indeterminate multiplicative constant for the prior would cancel in the Bayes factor. (See Sansó, Pericchi and Moreno 1996, for a formal justification.)

As the unknown mean μ occurs in only M_2, it needs to be assigned a proper prior. Through a series of ingenious arguments, Jeffreys obtains the following desiderata that this proper prior should satisfy: i) it should be centered at zero (i.e., centered at M_1); ii) have scale σ; iii) be symmetric around zero; and iv) have no moments. He argues that the simplest distribution that satisfies these conditions is the Cauchy$(0, \sigma^2)$. In summary, Jeffreys's conventional prior for this problem is:

$$\pi_1^J(\sigma_1) = \frac{1}{\sigma_1}, \quad \pi_2^J(\mu, \sigma_2) = \frac{1}{\sigma_2} \cdot \frac{1}{\pi\sigma_2(1 + \mu^2/\sigma_2^2)}.$$

Although this solution appears to be rather ad hoc, it is quite reasonable; choosing the scale of the prior for μ to be σ_2 (the only available non-subjective 'scaling' in the problem) and centering it at M_1 are natural choices, and Cauchy priors are known to be robust in various ways. Although it is easy to object to having such choices imposed on the analysis, it is crucial to keep in mind that there is no real default Bayesian alternative here. Alternative objective methods either thenselves correspond to imposition of some (proper) default prior or, worse, end up not corresponding to *any* actual Bayesian analysis. (See, in this regard, the evaluation *Principle* in section 3.)

Illustration 2: Linear Model, Zellner and Siow Conventional Priors

In Zellner and Siow (1980), a generalization of the above conventional Jeffreys prior is suggested for comparing two nested models within the normal linear model. Let $\mathbf{X} = [\mathbf{1} : \mathbf{Z}_1 : \mathbf{Z}_2]$ be the design matrix for the 'full' linear model under consideration, where $\mathbf{1}$ is the vector of 1's, and (without loss of generality) it is assumed that the regressors are measured in terms of deviations from their sample means, so that $\mathbf{1}^t\mathbf{Z}_j = 0$, $j = 1, 2$. It is also assumed that the model has been parameterized in an orthogonal fashion, so that $\mathbf{Z}_1^t\mathbf{Z}_2 = 0$. (This can also be achieved without essential loss of generality.) The corresponding normal linear model, M_2, for n observations $\mathbf{y} = (y_1, \ldots, y_n)^t$ is

$$\mathbf{y} = \alpha\mathbf{1} + \mathbf{Z}_1\boldsymbol{\beta}_1 + \mathbf{Z}_2\boldsymbol{\beta}_2 + \varepsilon,$$

where ε is $\mathcal{N}_n(\mathbf{0}, \sigma^2\mathbf{I}_n)$, the n-variate normal distribution with mean vector $\mathbf{0}$ and covariance matrix σ^2 times the identity. (Because the regression parameters are orthogonal to the variance in linear models, one can again justify using a common noninformative prior for the variance; we will thus 'cheat' and not differentiate between the σ^2 in different models.) Here, the dimensions of $\boldsymbol{\beta}_1$ and $\boldsymbol{\beta}_2$ are $k_1 - 1$ and p, respectively, the odd notation chosen for compatibility with subsequent developments.

For comparison of M_2 with the model $M_1 : \boldsymbol{\beta}_2 = 0$, Zellner and Siow (1980) propose the following default conventional priors:

$$\pi_1^{ZS}(\alpha, \boldsymbol{\beta}_1, \sigma) = 1/\sigma,$$

$$\pi_2^{ZS}(\alpha, \boldsymbol{\beta}_1, \sigma, \boldsymbol{\beta}_2) = h(\boldsymbol{\beta}_2|\sigma)/\sigma,$$

where $h(\boldsymbol{\beta}_2|\sigma)$ is the Cauchy$_p(\mathbf{0}, \mathbf{Z}_2^t\mathbf{Z}_2/(n\sigma^2))$ density

$$h(\boldsymbol{\beta}_2|\sigma) = c\frac{|\mathbf{Z}_2^t\mathbf{Z}_2|^{1/2}}{(n\sigma^2)^{p/2}}\left(1 + \frac{\boldsymbol{\beta}_2^t\mathbf{Z}_2^t\mathbf{Z}_2\boldsymbol{\beta}_2}{n\sigma^2}\right)^{-(p+1)/2},$$

with $c = \Gamma[(p+1)/2]/\pi^{(p+1)/2)}$. Thus the improper priors of the "common" $(\alpha, \boldsymbol{\beta}_1, \sigma)$ are assumed to be the same for the two models (again justifiable by orthogonality), while the conditional prior of the (unique to M_2) parameter $\boldsymbol{\beta}_2$, given σ, is assumed to be the (proper) p-dimensional Cauchy distribution, with location at $\mathbf{0}$ (so that it is 'centered' at M_1) and covariance matrix $\mathbf{Z}_2^t\mathbf{Z}_2/(n\sigma^2)$, "...a matrix suggested by the form of the information matrix," to quote Zellner and Siow (1980).

Computation for these prior distributions cannot be done in closed form. However, using the fact that a Cauchy distribution can be written as a scale mixture of normal distributions, it is possible to compute the needed marginal distributions, $m_i(\mathbf{y})$, with one-dimensional numerical integration.

When there are more than two models, or the models are non-nested, there are various possible extensions of the above strategy. Zellner and Siow (1984) utilize what is often called the 'encompassing' approach (first introduced in Cox 1961), wherein one compares each submodel, M_i, to the *encompassing* model, M_0, that contains all possible covariates from the submodels. One then obtains, using the above priors, the pairwise Bayes factors B_{0i}, $i = 1, \ldots, q$. The Bayes factor of M_j to M_i is then *defined* to be

$$B_{ji} = B_{0i}/B_{0j}. \tag{2.1}$$

Illustration 2 (continued): Linear Model, Conjugate *g*-priors

Another common choice of prior for the normal linear model is the conjugate prior, called a g-prior in Zellner (1986). For a linear model (adopting a more compact notation)

$$M : \mathbf{y} = \mathbf{X}\boldsymbol{\beta} + \boldsymbol{\varepsilon}, \quad \boldsymbol{\varepsilon} \sim \mathcal{N}_n(\mathbf{0}, \sigma^2\mathbf{I}_n),$$

where σ^2 and $\boldsymbol{\beta} = (\beta_1, \ldots, \beta_k)^t$ are unknown and $\mathbf{X}$ is an $(n \times k)$ given design matrix of rank $k < n$, the *g-prior* density is defined by

$$\pi(\sigma) = \frac{1}{\sigma}, \quad \pi(\boldsymbol{\beta}\,|\,\sigma) \text{ is } \mathcal{N}_k(\mathbf{0}, g\sigma^2(\mathbf{X}^t\mathbf{X})^{-1}).$$

Sometimes $g = n$ is chosen (see, also, Shively, Kohn and Wood 1999), while sometimes g is estimated by empirical Bayes methods (see, e.g., George and Foster 2000, and Clyde and George 2000).

The key advantage of g-priors is that the marginal density, $m(\mathbf{y})$, is available in closed form and, indeed, is given by

$$m(\mathbf{y}) = \frac{\Gamma(n/2)}{2\pi^{n/2}(1+g)^{k/2}} \left(\mathbf{y}^t\mathbf{y} - \frac{g}{(1+g)}\mathbf{y}^t\mathbf{X}(\mathbf{X}^t\mathbf{X})^{-1}\mathbf{X}^t\mathbf{y}\right)^{-n/2}.$$

Thus the Bayes factors and posterior model probabilities for comparing any two linear models are available in closed form.

Unfortunately, g-priors have some some undesirable properties when used for model selection. For instance, suppose one is interested in comparing the linear model above with $M^* : \boldsymbol{\beta} = \mathbf{0}$. It can be shown that, as the least squares estimate $\hat{\boldsymbol{\beta}}$ goes to infinity, so that one becomes *certain* that M^* is wrong, the Bayes factor of M^* to M goes to the nonzero constant $(1+g)^{(k-n)/2}$. It was essentially this undesirable property that caused Jeffreys (1961) to reject g-priors for model selection, in favor of the priors discussed above (for which the Bayes factor will go to zero when the evidence is overwhelmingly against M^*).

We find the conventional prior approach to be appealing in principle, but the above discussion reveals that it is difficult to implement, even in a 'simple' situation such as the linear model. Of perhaps more concern is that there seems to be no general method for determining such conventional priors. (This is, in contrast, to the situation for, say, estimation problems, where general techniques for deriving default priors do exist; see, e.g., Berger and Bernardo 1992.) The remaining three model comparison methods that are discussed in this section have the advantage of applying automatically to quite general situations.

2.2 Intrinsic Bayes Factor (IBF) Approach

For the q models $M_1, \ldots, M_q$, suppose that (ordinary, usually improper) noninformative priors $\pi_i^N(\boldsymbol{\theta}_i)$, $i = 1, \ldots, q$, are available. In general, we recommend that these be chosen to be 'reference priors' (see Berger and Bernardo 1992), but other choices will also typically give excellent results. Define the corresponding marginal or predictive densities of $\mathbf{X}$,

$$m_i^N(\mathbf{x}) = \int f_i(\mathbf{x}|\boldsymbol{\theta}_i)\pi_i^N(\boldsymbol{\theta}_i)\, d\boldsymbol{\theta}_i.$$

The general strategy for defining IBF's starts with the definition of a proper and minimal 'training sample,' which is simply to be viewed as some subset of the entire data $\mathbf{x}$. Because we will consider a variety of training samples, we index them by l.

Definition. A training sample, $\mathbf{x}(l)$, is called *proper* if $0 < m_i^N(\mathbf{x}(l)) < \infty$ for all M_i, and *minimal* if it is proper and no subset is proper.

The "standard" use of a training sample to define Bayes factors is to use $\mathbf{x}(l)$ to "convert" the improper $\pi_i^N(\theta_i)$ to proper posteriors, $\pi_i^N(\boldsymbol{\theta}_i|\mathbf{x}(l))$, and then use the latter to define Bayes factors for the remaining data. The result, for comparing M_j to M_i, can easily be seen (under most circumstances) to be

$$B_{ji}(l) = B_{ji}^N(\mathbf{x}) \cdot B_{ij}^N(\mathbf{x}(l)),$$

where

$$B_{ji}^N = B_{ji}^N(\mathbf{x}) = \frac{m_j^N(\mathbf{x})}{m_i^N(\mathbf{x})} \quad \text{and} \quad B_{ij}^N(l) = B_{ij}^N(\mathbf{x}(l)) = \frac{m_i^N(\mathbf{x}(l))}{m_j^N(\mathbf{x}(l))} \tag{2.2}$$

are the Bayes factors that would be obtained for the full data $\mathbf{x}$ and training sample $\mathbf{x}(l)$, respectively, if one were to blindly use π_i^N and π_j^N.

While $B_{ji}(l)$ no longer depends on the scales of π_j^N and π_i^N, it does depend on the arbitrary choice of the (minimal) training sample $\mathbf{x}(l)$. To eliminate this dependence and to increase stability, we "average" the $B_{ji}(l)$ over all possible training samples $\mathbf{x}(l)$, $l = 1, \ldots, L$. A variety of different averages are possible; here we consider only the *arithmetic IBF* (AIBF) and the *median IBF* (MIBF) defined, respectively, as

$$B_{ji}^{AI} = B_{ji}^N \cdot \frac{1}{L}\sum_{l=1}^{L} B_{ij}^N, \quad B_{ji}^{MI} = B_{ji}^N \cdot \text{Med}[B_{ij}^N(l)], \tag{2.3}$$

where "Med" denotes median. For the AIBF, it is typically necessary to place the more "complex" model in the numerator, i.e., to let M_j be the more complex model, and then *define* B_{ij}^{AI} by $B_{ij}^{AI} = 1/B_{ji}^{AI}$. The IBFs defined in (1.2) are *resampling* summaries of the evidence of the data for the comparison of models, since in the averages there is sample re-use. These are the only resampling methods systematically studied in the chapter.

These IBFs were defined in Berger and Pericchi (1996a) along with alternate versions, such as the *encompassing* IBF and the *expected* IBF, which we recommended for certain scenarios. We will refer to these other IBFs in some of the illustrations and examples. Originally, our focus was on finding the 'optimal' IBF for given scenarios. The MIBF was not optimal for any of the scenarios we considered, so that we did not give it much emphasis. We subsequently found, however, that the MIBF is the most robust and widely applicable IBF (see Berger and Pericchi 1998) so that, for those who desire *one* simple default model selection tool, the MIBF is what we would recommend.

One additional aspect of the IBF approach should be mentioned. As part of the general evaluation strategy discussed in section 3, we propose investigation of so-called *intrinsic priors* corresponding to a model selection method. A strong argument can be made that, in nested models, intrinsic priors corresponding to the AIBF are very reasonable as *conventional priors* for model selection. Hence the IBF approach can also be thought of as the long sought device for generation of good conventional priors for model selection in nested scenarios.

Illustration 1 (continued): Normal Mean, AIBF and MIBF

We start with the noninformative priors $\pi_1^N(\sigma_1) = 1/\sigma_1$ and $\pi_2^N(\mu, \sigma_2) = 1/\sigma_2^2$. Note that π_2^N is not the reference prior that we recommend one use to begin computation of the IBF; but π_2^N yields simpler expressions for illustrative purposes. It turns out that minimal training samples consist of any two distinct observations $\mathbf{x}(l) = (x_i, x_j)$, and calculation shows that

$$m_1^N(\mathbf{x}(l)) = \frac{1}{2\pi(x_i^2 + x_j^2)}, \quad m_2^N(\mathbf{x}(l)) = \frac{1}{\sqrt{\pi}(x_i - x_j)^2}.$$

Computation yields the following (unscaled) Bayes factor for data $\mathbf{x}$, when using π_1^N and π_2^N directly as the priors:

$$B_{21}^N = \sqrt{\frac{2\pi}{n}} \cdot \left(1 + \frac{n\overline{x}^2}{s^2}\right)^{n/2}, \tag{2.4}$$

where $s^2 = \sum_{i=1}^n (x_i - \overline{x})^2$. Using (2.3), the AIBF is then clearly equal to

$$B_{21}^{AI} = B_{21}^N \cdot \frac{1}{L} \sum_{l=1}^{L} \frac{(x_1(l) - x_2(l))^2}{2\sqrt{\pi}[x_1^2(l) + x_2^2(l)]}, \tag{2.5}$$

while the MIBF is given by

$$B_{21}^{MI} = B_{21}^N \cdot \text{Med}_{l=1,\ldots,L}\left[\frac{(x_1(l) -_2 (l))^2}{2\sqrt{\pi}[x_1^2(l) + x_2^2(l)]}\right].$$

Illustration 2 (continued): Linear Models, AIBF and MIBF

The IBF for linear and related models are studied in Berger and Pericchi (1996a, 1996b and 1997). Suppose, for $j = 1, \ldots, q$, that model M_j for $\mathbf{Y}$ $(n \times 1)$ is the linear model

$$M_j : \mathbf{y} = \mathbf{X}_j \boldsymbol{\beta}_j + \varepsilon, \quad \varepsilon_j \sim \mathcal{N}_n(\mathbf{0}, \sigma_j^2 \mathbf{I}_n),$$

where σ_j^2 and $\boldsymbol{\beta}_j = (\beta_{j1}, \beta_{j2}, \ldots, \beta_{jk_j})^t$ are unknown, and $\mathbf{X}_j$ is an $(n \times k_j)$ given design matrix of rank $k_j < n$. We will consider priors of the form

$$\pi_j^N(\boldsymbol{\beta}_j, \sigma_j) = \sigma_j^{-(1+q_j)}, \quad q_j > -1.$$

Common choices of q_j are $q_j = 0$ (the reference prior, Berger and Bernardo 1992), or $q_j = k_j$ (Jeffreys rule prior). When comparing model M_i nested in M_j, Berger and Pericchi (1996a) also consider a *modified Jeffreys* prior, having $q_i = 0$ and $q_j = k_j - k_i$. This is intermediate between reference and Jeffreys priors.

For these priors, a minimal training sample $\mathbf{y}(l)$, with corresponding design matrix $\mathbf{X}(l)$ (under M_j), is a sample of size $m = \max\{k_j\} + 1$ such that all $(\mathbf{X}_j^t \mathbf{X}_j)$ are nonsingular. Computation then yields

$$B_{ji}^N = \frac{\pi^{(k_j-k_i)/2}}{2^{(q_i-q_j)/2}} \cdot \frac{\Gamma((n-k_j+q_j)/2)}{\Gamma((n-k_i+q_i)/2)} \cdot \frac{|\mathbf{X}_i^t \mathbf{X}_i|^{1/2}}{|\mathbf{X}_j^t \mathbf{X}_j|^{1/2}} \cdot \frac{R_i^{(n-k_i+q_i)/2}}{R_j^{(n-k_j+q_j)/2}}, \tag{2.6}$$

where R_i and R_j are the residual sums of squares under models M_i and M_j, respectively. Similarly, $B_{ij}^N(l)$ is given by the inverse of this expression with $n, \mathbf{X}_i, \mathbf{X}_j, R_i$ and R_j replaced by $m, \mathbf{X}_i(l), \mathbf{X}_j(l), R_i(l)$ and $R_j(l)$, respectively; here $R_i(l)$ and $R_j(l)$ are the residual sums of squares corresponding to the training sample $\mathbf{y}(l)$.

Inserting these expressions in (2.3) results in the Arithmetic and Median IBFs for the three default priors being considered. For instance, using the modified Jeffreys prior and defining $p = k_j - k_i > 0$, the AIBF is

$$B_{ji}^{AI} = \frac{|\mathbf{X}_i^t \mathbf{X}_i|^{1/2}}{|\mathbf{X}_j^t \mathbf{X}_j|^{1/2}} \cdot \left(\frac{R_i}{R_j}\right)^{(n-k_i)/2} \cdot \frac{1}{L} \sum_{l=1}^{L} \frac{|\mathbf{X}_j^t(l) \mathbf{X}_j(l)|^{1/2}}{|\mathbf{X}_i^t(l) \mathbf{X}_i(l)|^{1/2}} \cdot \left(\frac{R_j(l)}{R_i(l)}\right)^{(p+1)/2}. \tag{2.7}$$

To obtain the MIBF, simply replace the arithmetic average by the median. Note that the MIBF does not require M_i to be nested in M_j, as does the AIBF. note

When multiple linear models are being compared, IBFs can have the unappealing feature of violating the basic Bayesian coherency condition $B_{jk} = B_{ji}B_{ik}$. To avoid this, one can utilize the *encompassing* approach, described in the paragraph preceding (2.1). This leads to what is called the Encompassing IBF. See Lingham and Sivaganesan (1997, 1999) and Kim and Sun (2000) for different applications.

2.3 The Fractional Bayes Factor (FBF) Approach

The fractional Bayes factor (developed in O'Hagan 1995) is based on a similar intuition to that behind the IBF but, instead of using part of the *data* to turn noninformative priors into proper priors, it uses a fraction, b, of each *likelihood function*, $L_i(\boldsymbol{\theta}_i) = f_i(\mathbf{x}|\boldsymbol{\theta}_i)$, with the remaining $1 - b$ fraction of the likelihood used for model discrimination. It is easy to show that the fractional Bayes factor of model M_j to model M_i is then given by

$$B_{ji}^F = B_{ji}^N(\mathbf{x}) \frac{\int L^b(\boldsymbol{\theta}_i)\pi_i^N(\boldsymbol{\theta}_i)d\boldsymbol{\theta}_i}{\int L^b(\boldsymbol{\theta}_j)\pi_j^N(\boldsymbol{\theta}_j)d\boldsymbol{\theta}_j} = B_{ji}^N(\mathbf{x}) \frac{m_i^b(\mathbf{x})}{m_j^b(\mathbf{x})}. \tag{2.8}$$

One common choice of b (see the examples in O'Hagan 1995, and the discussion by Berger and Mortera of O'Hagan 1995) is $b = m/n$, where m is the "minimal training sample size," i.e. the number of observations contained in a minimal training sample (as defined in section 2.2), assuming that this number is uniquely defined. O'Hagan (1995, 1997) also discuss other possible choices.

Illustration 1 (continued). Normal Mean

Assume, as in section 2.2, that $\pi_1^N(\sigma_1) = 1/\sigma_1$ and $\pi_2^N(\mu, \sigma_2) = 1/\sigma_2^2$. Consider $b = r/n$, where r is to be specified. Then the correction factor to B_{ji}^N in (2.8) can be computed to be

$$\mathrm{CF}_{21}^{F(b)} = \frac{m_1^b(\mathbf{x})}{m_2^b(\mathbf{x})} = \left(\frac{r}{2\pi}\right)^{1/2}\left(1 + \frac{n\bar{x}^2}{s^2}\right)^{-r/2}, \tag{2.9}$$

and thus

$$B_{21}^{F(b)} = B_{21}^N \mathrm{CF}_{21}^{F(b)} = \left(\frac{r}{n}\right)^{1/2}\left(1 + \frac{n\bar{x}^2}{s^2}\right)^{(n-r)/2}, \tag{2.10}$$

where B_{21}^N and s^2 are as in subsection 2.2. Since minimal training samples are of size two, the usual choice of r, as mentioned above, is $r = 2$. This choice will thus be utilized in subsequent comparisons.

Illustration 2 (continued). Linear Model

Using the notation of subsection 2.2, the correction factor is

$$\mathrm{CF}_{ij}^{F(b)} = \frac{(b/2)^{(q_j-q_i)/2}}{\pi^{(k_j-k_i)/2}} \cdot \frac{\Gamma[(r-k_i+q_i)/2]}{\Gamma[(r-k_j+q_j)/2]} \cdot \frac{|\mathbf{X}_j^t\mathbf{X}_j|^{1/2}}{|\mathbf{X}_i^t\mathbf{X}_i|^{1/2}} \cdot \frac{R_j^{(r-k_j+q_j)/2}}{R_i^{(r-k_i+q_i)/2}}. \tag{2.11}$$

The FBF, found by multiplying B_{ji}^N by this correction factor, is thus

$$B_{ji}^{F(b)} = b^{(q_j-q_i)/2} \cdot \frac{\Gamma[(n-k_j+q_j)/2]\Gamma[(r-k_i+q_i)/2]}{\Gamma[(n-k_i+q_i)/2]\Gamma[(r-k_j+q_j)/2]} \cdot \left(\frac{R_i}{R_j}\right)^{(n-r)/2}. \tag{2.12}$$

Here, $r = m = \max\{k_j\} + 1$ will typically be chosen.

2.4 Asymptotic Methods and BIC

Laplace's asymptotic method (cf, Haughton 1988, Gelfand and Dey 1994, Kass and Raftery 1995, Dudley and Haughton 1997, and Pauler 1998) yields, as an approximation to a Bayes factor, B_{ji}, with respect to two priors $\pi_j(\boldsymbol{\theta}_j)$ and $\pi_i(\boldsymbol{\theta}_i)$,

$$B_{ji}^L = \frac{f_j(\mathbf{x}|\hat{\boldsymbol{\theta}}_j)|\hat{\mathbf{I}}_j|^{-1/2}}{f_i(\mathbf{x}|\hat{\boldsymbol{\theta}}_i)|\hat{\mathbf{I}}_I|^{-1/2}} \cdot \frac{(2\pi)^{k_j/2}\pi_j(\hat{\boldsymbol{\theta}}_j)}{(2\pi)^{k_i/2}\pi_i(\hat{\boldsymbol{\theta}}_i)}, \tag{2.13}$$

where $\hat{\mathbf{I}}_i$ and $\hat{\boldsymbol{\theta}}_i$ are the observed information matrix and m.l.e., respectively, under model M_i, and k_i is the dimension of $\boldsymbol{\theta}_i$. As the sample size goes to infinity, the first factor of B_{ji}^L typically goes to 0 or ∞, while the second factor stays bounded. The BIC criterion of Schwarz (1978) arises from choosing an appropriate constant for this second term, leading to

$$B_{ji}^S = \frac{f_j(\mathbf{x}|\hat{\boldsymbol{\theta}}_j)}{f_i(\mathbf{x}|\hat{\boldsymbol{\theta}}_i)} \cdot n^{(k_i-k_j)/2}. \tag{2.14}$$

Discussion concerning this choice can be found in Kass and Wasserman (1995) and Pauler (1998).

The BIC approximation has the advantages of simplicity and an (apparent) freedom from prior assumptions. However, it is valid only for 'nice' problems. Among the problems for which it does not apply directly are models with irregular asymptotics (see, e.g., section 4.2) and problems in which the likelihood can concentrate at the boundary of the parameter space for one of the models. Dudley and Haughton (1997) and Kass and Vaidyanathan (1992) give extensions of (2.13) to such situations, but we will not formally consider such extensions here. We note, in passing, that the approximation (A.1) in Appendix 1 is both more accurate and widely applicable than (2.13).

Illustration 1 (continued). Normal Mean

Application of (2.14) yields,

$$B_{21}^S = \frac{B_{21}^N}{\sqrt{2\pi}} = \frac{1}{\sqrt{n}}\left(1 + \frac{n\overline{x}^2}{s^2}\right)^{n/2}.$$

Illustration 2 (continued). Linear Model

The usual BIC for linear models has the simple expression

$$B_{21}^S = \left(\frac{R_1}{R_2}\right)^{n/2} \cdot n^{(k_1-k_2)/2}.$$

3 Evaluating Objective Bayesian Model Selection Methods

There are many possible methods for evaluating default Bayes factors. An obviously important criterion is studying how they work in various examples; in particular, how often they fail to apply and how often they fail to give satisfactory answers. The discussion in later sections will focus, to a large extent, on these simple criteria.

One can also study formal properties of default Bayes factors, such as consistency, compatibility with sufficiency and the likelihood principle, and various types of coherence (cf, O'Hagan 1995, 1997). These can be useful, but we feel that they are typically secondary, or can be subsumed within, discussion of the following basic principle.

Principle: *Testing and model selection methods should correspond, in some sense, to actual Bayes factors, arising from reasonable default prior distributions.*

That default Bayes factors should behave like real Bayes factors may seem almost tautological, but the principle seems to be questioned by some Bayesians. There are two major lines of argument against the principle. One is to question the value of Bayes factors as a general Bayesian measure of evidence in comparing hypotheses or models (e.g., to question whether Bayes factors have any value except as a mathematical component of posterior probabilities). For a defense of Bayes factors in this regard, see Berger (1999). The second line of argument, popularized in O'Hagan (1995), is based on a type of decomposition of the Bayes factor into components, with the argument being that one might want to 'robustify' one component (typically that arising from the prior) by borrowing information from another component (typically that arising from the likelihood). Our view of the situation is simply that only the end result, namely the overall Bayes factor, has any importance or significance, and we seek only to evaluate this end result.

One of the primary reasons that we (the authors) are Bayesians is that we believe that the best *discriminator between procedures* is study of the prior distribution giving rise to the procedures. Insights obtained from studying overall properties of procedures (e.g., consistency) are enormously crude in comparison (at least in parametric problems, where such properties follow automatically once one has established correspondence of a procedure with a real Bayesian procedure). Moreover, we believe that one of the best ways of studying any biases in a procedure is by examining the corresponding prior for biases.

A side issue, but of considerable relevance in practice, is that we know how to interpret Bayes factors. If an entirely different entity is created (i.e., a default measure that does not have an interpretation as a Bayes factor or a posterior probability), learning how it should be interpreted would be a formidable undertaking, even if it were somehow superior. We have already seen ample demonstration of this danger in our profession; for instance, regardless of the inherent value of p-values, they are a disaster in practice because the vast majority of practitioners incorrectly interpret them as posterior probabilities.

If one accepts the value (necessity?) of studying priors corresponding to a default Bayes factor, there are two hurdles in applying the above Principle. The first is that of interpreting the phrase "in some sense." Sometimes a default Bayes factor can be shown to exactly correspond to a real Bayes factor, but this can often only be done in an approximate sense. In Berger and Pericchi (1996a), we defined one useful sense of approximation, namely asymptotics: indeed, we defined an *intrinsic prior* as a prior distribution that would yield essentially the same answer as a default Bayes factor if there was a large amount of data. The reason for employing an asymptotic argument is technical: default Bayes factors often depend on "reuse" of the data, and corresponding priors will then only exist in an asymptotic sense. Appendix 1 outlines the development of asymptotic intrinsic priors.

One can, of course, question the use of an asymptotic argument here. Indeed, default Bayes factors, such as the IBF and FBF, will typically only approximate the "intrinsic prior Bayes factor" to order $O(1/\sqrt{n})$, and hence may differ substantially for smaller n. However, we have observed that any "biases" or unreasonable features of an intrinsic prior are typically also reflected in the small sample behavior of the corresponding default Bayes factor. Thus detecting unreasonable behavior of a default Bayes factor can usually be done more easily and more accurately through study of the intrinsic prior, than, say, by conducting a huge simulation study. Also, some properties, such as consistency, still follow automatically from the existence of a (proper) asymptotic intrinsic prior.

It should be noted that asymptotic determination of an intrinsic prior is related to, but distinct from, the asymptotics yielding BIC; such asymptotics at best correspond to utilization of an intrinsic prior evaluated solely at the null model (see Kass and Wasserman 1995, and Pauler 1998).

Illustration 3: As an illustration of some of the above concepts, consider the simple problem of testing a normal mean, with known variance. Suppose $X_1, ..., X_n$ are i.i.d. from the normal distribution with mean θ and variance one. It is desired to compare $M_1 : \theta = 0$ with $M_2 : \theta \neq 0$. Utilizing the standard noninformative prior, $\pi_2^N(\theta) = 1$, a minimal training sample is a single x_i, and

$$B_{21}^N = (2\pi/n)^{1/2} \exp(n\bar{x}^2/2), \quad B_{12}^N(x_i) = (2\pi)^{-1/2} \exp(-{x_i}^2/2). \tag{3.1}$$

It follows that the AIBF and MIBF are given by

$$B_{21}^{AI} = B_{21}^N \cdot (2\pi)^{-1/2} \cdot \frac{1}{n} \sum_{i=1}^{n} \exp(-\frac{1}{2}{x_i}^2), \quad B_{21}^{MI} = B_{21}^N \cdot (2\pi)^{-1/2} \exp(-\frac{1}{2}\text{Med}[{x_i}^2]). \tag{3.2}$$

The FBF, with fraction b, can also easily be shown to be

$$B_{21}^F = \sqrt{b} \cdot \exp[n(1-b)\bar{x}^2/2]. \tag{3.3}$$

The first interesting feature to note is that B_{21}^F *exactly* equals the Bayes factor arising from a $\mathcal{N}(0, n^{-1}(b^{-1}-1))$ prior, so that this prior is the exact intrinsic prior corresponding to the FBF. Furthermore, note that, as b ranges from 0 to 1, the variance of this prior ranges from ∞ to 0. Hence there is a one-to-one correspondence between the choice of b and the choice of the intrinsic prior variance. Our view is thus that, in evaluating the FBF for this situation, one need only evaluate the suitability of the prior variance corresponding to b. A choice such as $b = 1/10$ would correspond to a prior variance of $9/n$. Then, as the sample size grew, the variance would be shrinking to 0, which seems very unreasonable from a Bayesian perspective. However, a choice such as $b = 1/n$ (the choice mentioned earlier, since the minimal training sample size here is 1) would correspond to a prior variance of $(1 - n^{-1})$, which at least has a stable behavior as n grows large.

The attractive property of the FBF in this situation, of having reasonable intrinsic priors (at least for suitable choices of b), is only approximately shared by the IBFs in (3.2). Indeed, note that, as $n \to \infty$,

$$\frac{1}{n}\sum_{i=1}^{n} \exp(-\frac{1}{2}{x_i}^2) \to E_\theta[\exp(-{X_i}^2)] = 2^{-1/2}\exp(-\frac{1}{4}\theta^2), \quad \text{Med}[{x_i}^2] \to (0.46) + \theta^2,$$

the last expression only being approximate for small or moderate θ. Using (A.4), (A.5) and (A.8) in Appendix 1, it is easy to see that the intrinsic priors for the AIBF and MIBF are, respectively, the $\mathcal{N}(0,2)$ distribution and (approximately) $(1.26) \cdot \mathcal{N}(0,1)$. Thus the AIBF does correspond to use of a sensible default prior for large n; and the relative error of the approximation (at $\theta = 0$, say) is of order $(0.4)/\sqrt{n}$, indicating that the AIBF will behave like a real Bayes factor even for quite modest n. The situation for the MIBF is different, in that it fails to behave like a true Bayes factor because of the constant (1.26) multiplying the (proper) $\mathcal{N}(0,1)$ prior. We would view this as a defect of the MIBF here, amounting to a 26% "bias" in favor of M_2. Again, however, this must be kept in perspective; a 26% bias for a default Bayes factor should usually not be of much practical concern.

As a matter of definition, note that we use the term "intrinsic prior" to stand for any prior, proper or improper, to which a default Bayes factor corresponds (at least approximately). Of course, if the intrinsic prior turns out to be improper, some additional justification may be needed. Indeed, perhaps the most problematical aspect of the above Principle is interpreting the term "reasonable default prior distributions." Consider, for instance, Illustration 1 in section 2.1, namely deciding if a normal mean μ is zero or not, when the variance σ^2 is also unknown. As mentioned in section 2.1, Jeffreys (1961) devotes considerable discussion to the question of what makes a default prior "reasonable" in such a situation. His arguments involve concepts such as balance (the prior should give equal mass to the regions $\mu < 0$ and $\mu > 0$), symmetry, scaling (given σ^2, μ should have a distribution with scale σ), propriety (given σ^2, μ should have a proper distribution but the nuisance parameter, σ^2, can have an improper distribution) and appropriateness of the prior tails. While one might not agree with all of the conclusions of Jeffreys, his arguments as to what makes a default prior "reasonable" are classic.

One central issue in this regard is Difficulty 4 in section 1.5, namely the need to recognize when it is reasonable (and when it is not reasonable) to assign a 'common' parameter from different models the same prior distribution. Jeffreys proposed utilizing the same prior distribution for such a common parameter only when the parameter is orthogonal (in the sense of Fisher information) to the other parameters in the models under consideration. For a modern application of this principle, see Clyde and Parmigiani 1996.

In part because of the difficulties inherent in orthogonal paramerization, and in part because of philosophical arguments, our own preference has been to think in terms of "predictive matching" priors, rather than priors based on orthogonalization. The underlying motivation is the foundational Bayesian view that one should concentrate on predictive distributions of observables; models and priors are, at best, convenient abstractions. According to this perspective, it is a predictive distribution $m(\mathbf{y})$ that describes reality, where $\mathbf{y}$ is a variable of predictive interest. We can choose to represent $m(\mathbf{y})$ as $m(\mathbf{y}) = \int f_i(\mathbf{y}|\boldsymbol{\theta}_i)\pi_i(\boldsymbol{\theta}_i)d\boldsymbol{\theta}_i$, where f_i is a model and π_i a prior, but these are merely a convenient abstraction.

From this perspective, if one is comparing models $M_1 : f_1$ versus $M_2 : f_2$, then the priors π_1 and π_2 should be chosen so that $m_1(\mathbf{y})$ and $m_2(\mathbf{y})$ are as close as possible. Thus we think of π_1 and π_2 as being properly *calibrated* if, when filtered through the models M_1 and M_2, they yield similar predictives. This could be assessed by defining some distance measure, $d(m_1, m_2)$, and calling π_1 and π_2 calibrated if $d(m_1, m_2)$ is small.

One key issue in operationalizing this idea is that of choosing the variable $\mathbf{y}$ at which a predictive match is desired. It seems natural, in the exchangeable case, to choose $\mathbf{y}$ to be a "future" sample of data. Previously (Berger and Pericchi 1997), we suggested that this should be essentially an "imaginary minimal training sample," which would typically be the smallest set of observations for which all model parameters are identifiable, but we have lately realized that a better choice would be the imaginary minimal training sample for the *smallest* model. (This work and its implications will be reported elsewhere.)

We will utilize the notion of predictive matching in examples later in the paper. Note that these ideas are related to ideas of elicitation through predictives (cf, Kadane et.al. 1980). Also similar uses of predictive matching to define priors for model selection can be found in Suzuki (1983), Laud and Ibrahim (1995) and Ibrahim and Laud (1994).

While the above considerations are far from precise, they can frequently be implemented, at least in part. Furthermore, the idea is not to demand demonstration that a default Bayes factor has a reasonable intrinsic prior in a given situation, before it can be used (although this would certainly be reassuring); rather, the idea is to show that a default Bayes factor approach leads to reasonable intrinsic priors for a wide variety of situations, so that Bayesians will be willing to trust the methodology in new settings.

Before proceeding to the 'evaluation' examples of later sections, we look back at the illustrations from the previous section to demonstrate the use of the desiderata presented here.

Illustration 1 (continued). Normal Mean, Intrinsic Priors

Intrinsic priors for the various default approaches can be derived using the technique in Appendix 1. The needed mle's for application of that technique are $\hat{\mu} = \overline{x}$, $\hat{\sigma}_1 = (\sum_{i=1}^{n} x_i^2/n)^{1/2} = (\overline{x}^2 + s^2/n)^{1/2}$, and $\hat{\sigma}_2 = (s^2/n)^{1/2}$.

Conventional Prior Approach: Obviously the intrinsic prior would be the original conventional prior. The Zellner and Siow (1980) conventional prior is quite sensible, as has already been discussed, but the g-prior has the unpleasant feature, mentioned in section 2.1, of yielding a Bayes factor that does not converge to zero even when the evidence against M_1 is overwhelming.

Intrinsic Bayes Factor Approach: In Berger and Pericchi (1996a), it is shown that the intrinsic priors corresponding to the AIBF are given by the usual noninformative priors for the standard deviations ($\pi_i^I(\sigma_i) = 1/\sigma_i$) and

$$\pi_2^I(\mu|\sigma_2) = \frac{1 - \exp[-\mu^2/\sigma_2^2]}{2\sqrt{\pi}(\mu^2/\sigma_2)}.$$

This last conditional distribution is proper (integrating to one over μ) and, furthermore, is virtually equivalent to the Cauchy$(0, \sigma_2)$ choice of $\pi_2(\mu|\sigma_2)$ suggested by Jeffreys (1961); indeed, the two prior densities never differ by more than 15%. Note that π_2^I also obeys all of the desiderata of Jeffreys for choice of a good conventional prior for this situation. That the AIBF has a conditionally proper intrinsic prior is not specific to this example, but holds generally for nested models, as discussed in Berger and Pericchi (1996a), Dmochowski (1996), and Moreno, Bertolino and Racugno (1998a). This is a rather remarkable property of AIBFs and leads to our recommending their use in the nested situation. A final curiosity is that, even though the original noninformative prior that was used for σ_2 was $\pi(\sigma_2) = 1/\sigma_2^2$, the intrinsic prior for σ_2 is the standard reference prior $\pi(\sigma_2) = 1/\sigma_2$.

An intrinsic prior can also be found for the MIBF. It is very similar to π^I above, but differs by a moderate constant and is hence not conditionally proper. We would interpret this constant as the amount of bias inherent in use of the MIBF.

Fractional Bayes Factor Approach: Selecting $r = m = 2$ as before, it is shown in De Santis and Spezzaferri (1997) that the intrinsic priors are given by the usual noninformative priors for the standard deviations ($\pi_i^I(\sigma_i) = 1/\sigma_i$) and

$$\pi_2^{F(2/n)}(\mu|\sigma_2) = \sqrt{\pi} \cdot \text{Cauchy}(0, \sigma_2).$$

This conditional prior clearly integrates to $\sqrt{\pi}$, which we would interpret as indicating a bias of about 77% in favor of M_2. The prior is otherwise the same as that recommended by Jeffreys (1961), so that we would judge the FBF to be reasonable here, except for the bias. Note that increasing r will lead to a conditional intrinsic prior that is a Student-t density with r degrees of freedom, but will still have a biasing factor.

BIC Approach: It is easy to see that the intrinsic prior corresponding to BIC is constant in μ, which we do not feel is optimal. In Kass and Wasserman (1995), it is argued that B_{21}^S, the BIC Bayes factor, is approximately a real Bayes factor with respect to the usual noninformative prior for the variances (okay) and $\pi_2(\mu|\sigma_2) = \mathcal{N}(0, \sigma_2^2)$. For this prior, however, the Laplace approximation in (2.13) yields

$$B_{21}^L = B_{21}^S \cdot \exp[-\overline{x}^2/(2\hat{\sigma}_2^2)].$$

If M_1 is the true model and n is large, then $\overline{x}$ will be close to zero and B_{21}^L will be close to B_{21}^S. This is the basis of the Kass and Wasserman (1995) argument. Of course, ignoring $\exp[-\overline{x}^2/(2\hat{\sigma}_2^2)]$ can be bad if the sample size is not large or M_1 is not true. Indeed, we would view this quantity as the 'bias' against M_1 that is inherent in the use of BIC, and the amount of bias here could be very substantial. It is observed in Kass and Wasserman (1995) that, when n is large and the bias is extreme (because $|\overline{x}|$ is large), then this 'error' will not really matter, since B_{21}^S will overwhelm the bias term. This does require n to be large, however.

Illustration 2 (continued). Linear Models, Intrinsic Priors

Conventional Prior Approach: If only two models are under consideration in the Zellner and Siow (1980) approach, the intrinsic prior would obviously be the original conventional prior. If there are multiple models under consideration, however, intrinsic priors do not typically exist. For instance, suppose the encompassing approach (discussed in the paragraph preceding (2.1)) is utilized. The prior for the encompassing model, M_0, changes with the M_i being considered in the pairwise comparisons. This suggests that there is no single assignment of priors that would correspond to the Zellner and Siow procedure, and this can, indeed, be shown to be the case. We do not feel that this is a particularly serious flaw, however, just as we do not feel that a slight 'bias' in terms of the intrinsic prior mass is a serious flaw.

What about the conventional priors themselves? Following the arguments of Jeffreys (1961), we feel that it is very reasonable to assign 'common' parameters the usual noninformative priors and to use a proper conditional prior for the 'extra' parameters that is centered at $\mathbf{0}$ and has a Cauchy shape. Choosing the the prior scale matrix to be $\mathbf{Z}_2^t\mathbf{Z}_2/n$ is considerably less compelling, however. Indeed, in Nadal (1999), it is argued that, as the number of groups grows large in ANOVA models, this prior can become too concentrated about zero. Exploration of this important issue in other situations is needed.

For the conventional g-priors, there is no problem with multiple model coherency, as the priors are defined separately for each model. However, as indicated at the end of section 2.1, these priors have the undesirable property that the Bayes factor will not go to zero, even when the evidence against a model becomes overwhelming. These conventional priors also depend on $\mathbf{X}^t\mathbf{X}$, and can thus have the potential problem of 'over concentration about zero,' that was just mentioned.

Intrinsic Bayes Factor Approach: For the AIBF, pairwise intrinsic priors exist, and are discussed in Berger and Pericchi (1996b, 1997). We generally recommend use of the encompassing approach when dealing with linear models, so that all models are compared with the encompassing model, M_0, and Bayes factors are computed via (2.1). The main result is that the intrinsic prior for comparing M_1 (say) to M_0 is the usual reference prior for M_1, given by $\pi_1^I(\boldsymbol{\beta}_1, \sigma_1) = 1/\sigma_1$, while, under M_0, $\pi_0^I(\boldsymbol{\beta}_0, \boldsymbol{\beta}_1, \sigma_0) = \pi_0^I(\boldsymbol{\beta}_0|\boldsymbol{\beta}_1, \sigma_0)/\sigma_0$, where $\pi_0^I(\boldsymbol{\beta}_0|\boldsymbol{\beta}_1, \sigma_0)$ is a proper density when the AIBF is derived using reference priors (i.e., $q_1 = q_0 = 0$); is proper when the AIBF is derived using modified Jeffreys priors (i.e., $q_1 = 0$, $q_0 = k_0 - k_1 = p$); but is not proper when the AIBF is derived using the Jeffreys rule noninformative priors (i.e., $q_1 = k_1, q_0 = k_0$). Thus the use of Jeffreys rule priors is not recommended for deriving the AIBF.

As an example, use of the modified Jeffreys priors, in deriving the AIBF, results in the (proper) conditional intrinsic prior

$$\pi_0^I(\boldsymbol{\beta}_0|\boldsymbol{\beta}_1, \sigma_0) = \frac{\sigma_0^{-p}c^*}{L} \cdot \sum_{l=1}^{L} \frac{|\mathbf{X}_0^t(l)\mathbf{X}_0(l)|^{1/2}}{|\mathbf{X}_1^t(l)\mathbf{X}_1(l)|^{1/2}} \cdot \psi(\lambda(l), \sigma_0),$$

where $c^* = (2\pi)^{-p/2}$, $\psi(\lambda(l), \sigma_0) = 2^{-p}\exp(-\lambda(l)/2)M((p+1)/2, p+1, \lambda(l)/2)$, M is the Kummer function $M(a,b,z) = \frac{\Gamma(b)}{\Gamma(a)} \cdot \sum_{j=0}^{\infty} \frac{\Gamma(a+j)}{\Gamma(b+j)} \cdot \frac{z^j}{j!}$ and $\lambda(l) = \sigma_0^{-2}\boldsymbol{\beta}_0^t\mathbf{X}_0^t(l)(I - \mathbf{X}_1(l)[\mathbf{X}_1^t(l)\mathbf{X}_1(l)]^{-1}\mathbf{X}_1^t(l))\mathbf{X}_0(l)\boldsymbol{\beta}_0$.

This (proper) conditional intrinsic prior behaves very much like a mixture of p-variate t-densities with one degree of freedom, location $\mathbf{0}$, and scale matrices

$$\boldsymbol{\Sigma}(l) = 2\sigma_0^2[\mathbf{X}_0^t(l)(I - \mathbf{X}_1(l)[\mathbf{X}_1^t(l)\mathbf{X}_1(l)]^{-1}\mathbf{X}_1^t(l))\mathbf{X}_0(l)]^{-1}.$$

Each individual training sample captures part of the structure of the design matrix, which thus gets reflected in the corresponding scale matrix above, but the overall mixture prior does not concentrate about the nested model to nearly as great an extent as do the earlier discussed conventional priors, which effectively choose the full $\sigma_0^2(\mathbf{X}_0^t\mathbf{X}_0)^{-1}$ as the scale matrix. In this regard, the intrinsic priors seem considerably more sensible.

Fractional Bayes Factor Approach: For ease of comparison, we assume that M_i is nested in M_j, use modified Jeffreys priors to derive the FBF, and assume that $b = m/n$, where $m = k_j + 1$ is the minimal training sample. Using the expressions in De Santis and Spezzaferri (1997), one can show that the FBF has an intrinsic prior of exactly the Zellner and Siow (1980) form, except that the conditional prior of the 'extra' parameters does not integrate to one. Indeed, the biasing factor is

$$\left(\frac{m}{2}\right)^{p/2} \cdot \frac{\sqrt{\pi}}{\Gamma[(p+1)/2]},$$

which can be quite large. Thus, although the FBF has an intrinsic prior with a sensible centering and Cauchy form, it has a potentially large (constant) bias. Futhermore, the concerns expressied earlier concerning the scale matrix in the Zellner and Siow (1980) prior also clearly apply to the intrinsic prior for the FBF.

BIC Approach: The situation here is the same as in Illustration 1. Priors can clearly be found for which the BIC Bayes factor equals the actual Bayes factor when the m.l.e of the 'extra parameters' in the more complex model is at $\mathbf{0}$. But, as this m.l.e. departs from $\mathbf{0}$, the BIC Bayes factor will have an increasingly dramatic bias towards the more complex model. This bias becomes even more severe as the dimensional difference between the two models increases. See Pauler (1998) for a related discussion.

4 Study Examples for Comparison of the Objective Methodologies

Construction of counterexamples to default model selection methodologies is a time-honored tradition. These can be useful in suggesting positive alterations of the theory, and in suggesting the limits of applicability of the methodology. The situation here is complicated, however, by the fact that, almost by definition, it is possible to create counterexamples to any default Bayesian methodology (at least if one believes basic Bayesian coherency arguments). This means that the value of a counterexample is proportional to how representative it is of statistical application. Most useful are examples that are representative of large classes of statistical problems, and that indicate difficulties with the various methodologies that likely apply to the entire class of problems.

Six such examples will be studied here. In most of these examples, conventional priors do not exist, and the asymptotic approaches are of limited use; hence these examples will tend to focus on comparison of FBFs and IBFs. Furthermore, the examples mostly consider situations in which FBFs encounter certain difficulties, while IBFs do not. The intent of the examples is not, however, to suggest that the FBF approach is questionable; indeed, we are convinced that the FBF approach is very useful in default model selection. Rather, the examples are primarily studies we conducted to attempt to cause the IBF approach to "fail," in order to understand its limitations. We did not succeed in causing the IBF approach to fail, however.

4.1 Improper Likelihoods (Example 1)

There are a variety of models for which the full likelihood has infinite mass (considered as a function of $\boldsymbol{\theta}$), and these models can pose challenges in use of default Bayes factors. In a number of situations, such as exponential regression models (see Ye and Berger 1991), use of suitable noninformative priors (e.g., Jeffreys or reference priors) will correct the problem, yielding finite marginal densities (so that default Bayes factors based on utilization of noninformative priors will then exist). That Jeffreys and reference priors seem to yield finite marginal densities the great majority of the time is one of the unexplained mysteries of default Bayesian analysis, and is one of the signifiant attractions in their use.

Another class of models with improper likelihoods is considerably more problematical. These are models in which no improper prior can yield a finite marginal density. One common example is mixture models, of which the following is a simple illustration. Suppose $X_1, ..., X_n$ are i.i.d. from the mixture density

$$f(x_i|\boldsymbol{\theta}) = p \cdot (2\pi)^{-1/2} e^{-x_i{}^2/2} + (1-p) \cdot (2\pi)^{-1/2} e^{-(x_i-\mu)^2/2}, \tag{4.1}$$

where $\boldsymbol{\theta} = (p, \mu)$ is unknown. Suppose we want to compare $M_1 : \mu = 0$ versus $M_2 : \mu \neq 0$, although the following difficulty applies to virtually any Bayesian attempt to utilize this density. While p poses no problem, since either the uniform prior or Jeffreys (reference) prior for p are proper, μ poses a considerable problem. The likelihood is the product (over i from 1 to n) of the density in (4.1), and expanding the product leads to a sum of terms, the first of which is a positive constant not depending on μ. Thus using, say, the uniform prior on p, $m(\mathbf{x}) = \int f(\mathbf{x}|\mu, p)\pi(\mu)\, d\mu\, dp$ is clearly infinite, unless $\pi(\mu)$ is proper (or, at least, has finite mass). Since this is true for any sample size, it is clear that there is no minimal training sample for this problem, and hence IBFs cannot be defined. Likewise it is easy to see that FBFs do not exist.

As asymptotic methods do not operate through integration, one might hope that it would be possible to apply them to problems with improper likelihoods. While this might be so, the problem is complicated by the fact that, in situations such as this, the correct asymptotic expressions can be very difficult to compute. It would appear that the only option here is thus to use the conventional prior approach; for the mixture problem, for instance, one might choose a Cauchy(0, 1) prior for μ. Interestingly, however, one can modify the IBF approach or FBF approach to deal with mixture models. The idea in the above example would be to "pretend" that a training sample arises from the second component of the mixture in the IBF approach. It seems unnatural, however, to give all training samples equal weight; rather it is appealing to choose them in accordance with the probability that the training sample arose from the second component (the "classification probability" in mixture model language). Since these classification probabilities must be estimated, the procedure must operate iteratively. An algorithm implementing this idea in general mixture models is developed in Shui (1996), and appears to work quite successfully.

The chief limitation of this modified IBF approach is its considerable computational expense; thus, an FBF version of the analysis was also developed in Shui (1996). The key idea in this development was to notice that certain fractions of parts of the likelihood (with differing fractions for different parts of the likelihood) would have effects similar to those of the probabilistically classified training samples. The fractions still have to be determined iteratively, but a fixed point algorithm can be used to do so, rather than the stochastic iterative algorithm needed for the IBF. The FBF version is thus much more computationally efficient. Another recent modification of the IBF strategy for dealing with mixture models is the 'expected posterior prior' approach or Perez and Berger (2000); this is discussed in section 5.5.

4.2 Irregular Models (Example 2)

In a large class of statistical models, the parameter space becomes constrained by the data. Such models do not typically have regular asymptotics. One of the simplest such examples is the location exponential distribution. Thus, suppose $X_1, X_2, ..., X_n$ are i.i.d. with density

$$f(x_i \mid \theta) = \exp(-(x_i - \theta))\, 1_{(\theta,\infty)}(x_i),$$

where "1" denotes the indicator function. It is desired to compare

$$M_1 : \theta = \theta_0 \quad \text{versus} \quad M_2 : \theta > \theta_0,$$

employing the usual non-informative prior $\pi_2^N(\theta) = 1$. Computation yields

$$m_1^N(\mathbf{x}) = \exp(n\theta_0 - S) 1_{(\theta_0,\infty)}(x_{min}),$$

$$m_2^N(\mathbf{x}) = \frac{1}{n}[\exp(nx_{min} - S) - \exp(n\theta_0 - S)] \cdot 1_{(\theta_0,\infty)}(x_{min}),$$

where $S = \sum_{i=1}^n x_i$ and $x_{min} = min[x_1, ..., x_n]$. Hence,

$$B_{21}^N(\mathbf{x}) = \frac{1}{n}[\exp(n(x_{min} - \theta_0)) - 1].$$

Intrinsic Bayes Factors: Any single observation is a minimal training sample. The Arithmetic and Median IBFs are thus given by

$$B_{21}^{AI} = B_{21}^N \cdot \frac{1}{n}\sum_{i=1}^{n}[\exp(x_i - \theta_0) - 1]^{-1}, \quad B_{21}^{MI} = B_{21}^N \cdot [\exp(\text{Med}[x_i] - \theta_0) - 1]^{-1}, \quad (4.2)$$

the last following from the monotonicity of $[\exp(x - \theta_0) - 1]^{-1}$.

In Appendix 2, the intrinsic priors for the AIBF and the MIBF are computed. They are given, respectively, (on $\theta > \theta_0$) by

$$\pi_2^{AI}(\theta) = -e^{(\theta-\theta_0)}\log(1 - e^{(\theta_0-\theta)}) - 1, \quad \pi_2^{MI}(\theta) = [2e^{(\theta-\theta_0)} - 1]^{-1}. \quad (4.3)$$

While $\pi_2^{AI}(\theta)$ can be shown to integrate to 1, the mass of $\pi_2^{MI}(\theta)$ is only about 0.69. The MIBF thus appears to have a modest bias in favor of M_1. It should also be noted that the statement that $\pi_2^{AI}(\theta)$ is the intrinsic prior is something of a conjecture, due to a technical issue discussed in Appendix 2.

Fractional Bayes Factors: Application of the FBF approach to this problem, utilizing fraction b of the likelihood, results in

$$B_{21}^F = B_{21}^N \cdot bn[e^{bn(x_{min}-\theta_0)} - 1]^{-1}. \quad (4.4)$$

This is completely unsatisfactory. As one indication of this, it is easy to show that $B_{21}^F > 1$ for any $0 < b < 1$. (The proof is trivial, using the fact that $b^{-1}[\exp(bv) - 1]$ is increasing in b for any $v > 0$.) Always favoring the more complex model, no matter what the data and no matter what fraction b is used, is clearly unreasonable.

The difficulty with the FBF here is that the most important part of the likelihood is the indicator function $1_{(x_{min},\infty)}(\theta)$, giving the data-dependent region in which the parameter θ must lie. The FBF operates by attempting to use just a "fraction" of the likelihood to update the prior, but clearly a fraction of an indicator function is the indicator function itself. This difficulty with the FBF could well apply to virtually any non-regular problem in which there are data-dependent constraints on the parameter. (Note that the IBF overcomes this problem because most training samples only provide a mild constraint on the parameter space and one "averages" over these mild constraints.)

BIC and Conventional Priors: BIC is inapplicable in this problem because the asymptotics are non-regular, so that (2.13) does not apply. Note, however, that the alternative asymptotic approximation (A.1) in Appendix 1 does apply; asymptotic approximations to Bayes factors that are based on this approximation are under development. To the best of our knowledge, no conventional proper priors have been proposed for this problem.

4.3 One-Sided Testing (Example 3)

One-sided testing is, of course, a broad class of problems, especially if considered in multivariate contexts. It poses interesting issues for all default methodologies; see Berger and Mortera (1999), Lingham and Sivaganesan (1997, 1999) and Sun and Kim (1997) for general discussion of these issues. Here, we consider only one-sided testing in the scale exponential model, to illustrate some of the more basic points.

Suppose $X_1, ..., X_n$ are i.i.d. with density of the form

$$f(x_i|\theta) = \theta^{-1} e^{-x_i/\theta} \quad \text{for } x_i > 0 \text{ and } \theta > 0. \tag{4.5}$$

It is desired to test $M_1 : \theta < \theta_0$ versus $M_2 : \theta > \theta_0$.

For the IBF and FBF approaches, it is natural to utilize the standard non-informative prior $\pi_i^N(\theta) = 1/\theta$ (for both models). Computation then yields

$$B_{21}^N(\mathbf{x}) = \frac{\int_{\theta_0}^{\infty} \theta^{-(n+1)} \exp(-n\bar{x}/\theta) d\theta}{\int_0^{\theta_0} \theta^{-(n+1)} \exp(-n\bar{x}/\theta) d\theta} = ([\gamma(n, n\bar{x}/\theta_0)]^{-1} - 1)^{-1},$$

where $\gamma(\alpha, x) = (1/\Gamma(\alpha)) \int_0^x \xi^{\alpha-1} e^{-\xi} d\xi$ is the incomplete Gamma function.

IBFs and FBFs: A minimal training sample is any single observation, x_i, and

$$B_{12}^N(x_i) = (e^{x_i/\theta_0} - 1)^{-1}.$$

A basic limitation of IBFs in one-sided testing is that arithmetic IBFs typically do not have intrinsic priors (cf., Dmochowski 1995, and Moreno, Bertolino, and Racugno 1998a). The reason is that $B_{12}^N(X_i)$ typically does not have finite expectation (under one of the models), so that (A.4) in Appendix 1 may well fail. Two possible alternatives are the encompassing arithmetic IBF (and its expected version, both defined in Berger and Pericchi 1996a), and the median IBF. These are extensively studied in Berger and Mortera (1999), with the advantage seemingly belonging to the encompassing versions. While we would thus recommend these IBFs for the one-sided testing problem (in tune with our overall philosophy that different default Bayes factors can be optimal for different situations), we herein consider only the median IBF (in part, to also show its limitations).

Since $B_{12}^N(x_i)$ is clearly monotonic in x_i, the MIBF is

$$B_{21}^{MI} = B_{21}^N \cdot (\exp(\mathrm{Med}[x_i]/\theta_0) - 1)^{-1}. \quad (4.6)$$

It is easy to see that, with the choice $b = 1/n$, the FBF is given by

$$B_{21}^F = B_{21}^N \cdot (\exp(\bar{x}/\theta_0) - 1)^{-1}. \quad (4.7)$$

Proper intrinsic priors exist for the median IBF and the FBF. They are computed in Berger and Mortera (1999) and (up to an irrelevant normalizing constant) are given, respectively, by

$$\pi^{FI}(\theta) = \begin{cases} \theta^{-1}(e^{\theta/\theta_0} - 1) & \theta < \theta_0 \\ (e-1)\theta^{-1}(e^{\theta/\theta_0} - 1)^{-1} & \theta > \theta_0 \end{cases},$$
$$\pi^{MI}(\theta) = \begin{cases} \theta^{-1}(2^{\theta/\theta_0} - 1) & \theta < \theta_0 \\ \theta^{-1}(2^{\theta/\theta_0} - 1)^{-1} & \theta > \theta_0 \end{cases}. \quad (4.8)$$

This intrinsic prior for the FBF has the somewhat unappealing property of being discontinuous at θ_0, but this is unlikely to cause any serious problems in practice. A considerably more important property of intrinsic priors in one-sided testing is their degree of "balance between the hypotheses" as indicated by the prior odds ratio $Pr(M_1)/Pr(M_2)$. For the two intrinsic priors above, computations in Berger and Mortera (1999) yielded prior odds ratios of 2.67 and 1.46, respectively. That these are not equal to one indicates that both the FBF and median IBF are biased, here in favor of M_1. The 46% relative bias of the median IBF may be large enough to be of concern to some, but the situation with the FBF is quite problematical, as it effectively carries a 169% bias against M_2. One can easily see this strong bias in simple data examples also (see Berger and Mortera 1999).

Conventional Priors: In one-sided testing, it is often felt to be legitimate to perform a default Bayesian analysis directly, with standard noninformative prior distributions. Thus, in the one-sided exponential testing problem above, it would be common to use the noninformative prior $\pi^N(\theta) = 1/\theta$, and directly compute the Bayes factor of M_1 to M_2. A variety of arguments can be given which suggest that this is reasonable from a Bayesian perspective, at least as an approximation with large sample sizes. And the resulting answer coincides with the classical p-value (this is typically true for location or scale invariant models), so that it would seem that the entire profession accepts this conventional prior analysis.

Note, however, that the legitimacy of using noninformative priors directly has only been suggested in simple (invariant) situations of one-sided testing, whereas the IBF and FBF appear to be very widely applicable. Perhaps more importantly, one can question whether the conventional prior Bayesian answer is actually suitable for small or moderate sample sizes. The usual Bayesian justification for direct use of the conventional prior here is that the resulting answer (also the p-value) is the lower bound of the posterior probability of the relevant M_i over reasonable classes of prior densities (cf, Casella and Berger 1987). It is natural to ask, however, if this lower bound is really the best evidential summary to provide. In Bayesian terms, if it were the case that the posterior probability of M_1 were some number between 0.05 and 0.5, depending on assumptions, would it really be reasonable to report 0.05 as the evidential summary?

This point was dramatically illustrated in the discussion by Morris of the Casella and Berger article (Morris 1987), wherein a reasonable practical situation was considered and it was demonstrated that the lower bound is an unreasonable measure of the evidence against M_1. From a Bayesian perspective, Morris's argument was essentially that typical prior beliefs will concentrate closer to the dividing line between the hypotheses (θ_0 in the one-sided testing problem mentioned above), and that using a prior distribution which is extremely diffuse is thus unreasonable, at least for small or moderate sample sizes.

Interestingly, the IBFs and FBF do produce answers which are not as extreme as the standard (classical or Bayesian) answers, often being 2 to 10 times larger than the p-values against a hypothesis (equivalently, the conventional posterior probability of the hypothesis) when the sample size is small; see Berger and Mortera (1999).

4.4 Increasing Multiplicity of Parameters (Example 4)

In numerous models in use today, the number of parameters is increasing with the amount of data. The original example of this is the Neyman-Scott problem, the testing version of which we will consider here. Suppose we observe

$$X_{ij} \sim \mathcal{N}(\mu_i,\ \sigma^2), \quad i = 1, \ldots, n;\ j = 1, 2. \tag{4.9}$$

We are interested in comparing the two models $M_1 : \sigma^2 = \sigma_0^2$ versus $M_2 : \sigma^2 \neq \sigma_0^2$. Defining $\bar{x}_i = (x_{i1} + x_{i2})/2$, $\bar{\mathbf{x}} = (\bar{x}_1, \ldots, \bar{x}_n)$, $S^2 = \sum_{i=1}^n (x_{i1} - x_{i2})^2$, and $\boldsymbol{\mu} = (\mu_1, \ldots, \mu_n)$, the likelihood function (under M_2) can be written

$$L(\boldsymbol{\mu}, \sigma) \propto \sigma^{-2n} \cdot \exp\left[-\frac{1}{\sigma^2}(|\bar{\mathbf{x}} - \boldsymbol{\mu}|^2 + \frac{S^2}{4})\right]. \tag{4.10}$$

The feature of this problem that is most relevant to the following analysis is that the information available concerning each μ_i is limited (only two observations for each), while the information concerning σ^2 is increasing with n. This is indicated by the Fisher information matrix: it is diagonal, with diagonal elements $(2/\sigma^2, \ldots, 2/\sigma^2, 2n/\sigma^2)$, the last entry corresponding to the information for σ^2. This great difference in information can pose difficulties for default Bayes factors. Furthermore, techniques such as BIC, which are based on a fixed number of parameters and asymptotics in the sample size, cannot be used here.

The reference prior for the problem is

$$\pi_1^N(\boldsymbol{\mu}) = 1, \quad \text{and} \quad \pi_2^N(\boldsymbol{\mu}, \sigma) = 1/\sigma.$$

Note that the noninformative Jeffreys rule prior for M_2, namely $\pi_2^N(\boldsymbol{\mu}, \sigma) = 1/\sigma^n$, typically gives very bad performance for any inferences involving the Neyman-Scott problem. For the reference prior, computation yields

$$B_{21}^N = \frac{1}{2}\Gamma(\frac{n}{2})\frac{\sigma_0^n \exp[S^2/(4\sigma_0^2)]}{(S^2/4)^{n/2}}. \tag{4.11}$$

Intrinsic Bayes Factors: It is easy to see that a minimal training sample consists of both observations corresponding to one of the μ_i, and one observation corresponding to each of the others. (Using all the observations corresponding to one of the μ_i for the training sample should be cause for worry, but the IBFs will be seen to be fine.) Computation for a training sample shows that $B_{12}^N(\mathbf{x}(l))$ depends only on the two observations corresponding to the same μ_i, and it follows easily that the AIBF is given by

$$B_{21}^{AI} = B_{21}^N \cdot \frac{1}{n}\sum_{i=1}^{n} \frac{|x_{i1} - x_{i2}|}{\sqrt{\pi}\sigma_0} \exp[-\frac{(x_{i1} - x_{i2})^2}{4\sigma_0{}^2}]. \tag{4.12}$$

In Appendix 3, it is shown that the intrinsic prior corresponding to the AIBF is

$$\pi_1^{AI}(\boldsymbol{\mu}) = 1, \text{ and } \pi_2^{AI}(\boldsymbol{\mu}, \sigma) = \frac{2}{\pi\sigma_0(1 + \sigma^2/\sigma_0{}^2)}. \tag{4.13}$$

Thus (for M_2) given σ, $\boldsymbol{\mu}$ has the usual constant noninformative prior, while, marginally, σ has the half-Cauchy distribution, with median equal to σ_0. This density for σ is not only proper, but it has the appealing property of being "balanced," in the sense of giving equal weight to $[\sigma < \sigma_0]$ and $[\sigma > \sigma_0]$. (Since π_2^{AI} is improper, it is not technically correct to talk about conditional and marginal densities; but such statements can be justified here in various limiting senses. Also, because of the "impropriety in the intrinsic prior for $\boldsymbol{\mu}$," properties such as consistency do not automatically hold, but consistency, at least, is easy to prove directly.)

Fractional Bayes Factors: For $b \leq 1/2$, the fractional Bayes factor does not exist. For $b > 1/2$, an easy computation yields

$$B_{21}^{F,b} = \frac{b^{n(b-1/2)}\Gamma(n/2)}{\Gamma(n(b-1/2))} \cdot [\frac{S^2}{4\sigma_0{}^2}]^{n(b-1)} \cdot \exp[\frac{(1-b)S^2}{4\sigma_0{}^2}]. \tag{4.14}$$

This is not a satisfactory default Bayes factor. Indeed it need not even be consistent, as shown by the following lemma (whose proof is given in Appendix 3).

Lemma 1 *Consider any monotonic sequence of fractions* $(b_1, b_2, \ldots)$. *Then* B_{21}^{F,b_n} *is not necessarily consistent as* $n \to \infty$; *for instance, if* $\sigma^2 = 2{\sigma_0}^2$, *then* $\lim_{n\to\infty} B_{21}^{F,b_n} \le 1$, *even though* M_2 *is true. Furthermore, if* $n(1-b_n) \to \infty$, *then* $B_{21}^{F,b_n} \to 0$, *so that the FBF will eventually be certain to select the wrong model.*

The failure of the FBF in this example is due to the fact that (in intuitive terms) it was forced to take the same fraction, b, of the likelihood for each μ_i and the likelihood for σ, even though the amount of information in the likelihood for each is vastly different. (An earlier example of this phenomenon, based on comparison of two exponential means with vastly different sample sizes, was given in Iwaki 1997.) Note that the IBF was fine in this regard, effectively taking just one observation for training purposes for each of the μ_i and σ. The obvious suggestion is thus to attempt to write the likelihood function in terms of likelihoods for μ and for σ, and then take different "fractions" of each. The natural way to write the likelihood function in this regard is

$$L(\mu,\sigma) \propto [\frac{1}{\sigma^n} e^{-|\bar{\mathbf{x}}-\mu|^2/\sigma^2}] \cdot [\frac{1}{\sigma^n} e^{-S^2/(4\sigma^2)}]. \tag{4.15}$$

Taking fraction b_1 of the first component of the likelihood, and fraction b_2 of the second, leads to a "training likelihood"

$$L^{b_1,b_2}(\mu,\sigma) \propto [\frac{1}{\sigma^n} e^{-|\bar{\mathbf{x}}-\mu|^2/\sigma^2}]^{b_1} \cdot [\frac{1}{\sigma^n} e^{-S^2/(4\sigma^2)}]^{b_2}. \tag{4.16}$$

Inserting this into (2.8), in place of L^b. defines a more general class of fractional Bayes factors.

As an example, consider the choices $b_1 = (n-1)/n$ and $b_2 = 2/n$ (roughly reflecting the fact that there is n times as much information in the likelihood about σ^2 as about each μ_i). Computation yields that the resulting FBF is

$$B_{21}^{F,b_1,b_2} = B_{21}^N \cdot [\frac{2S^2}{n\sigma_0\pi}]^{1/2} e^{-S^2/(2n\sigma_0^2)}. \tag{4.17}$$

This is quite sensible; indeed, it is shown in Appendix 3 that there is a corresponding intrinsic prior, given by

$$\pi_1^{FI}(\mu) = 1, \text{ and } \pi_2^{FI}(\mu,\sigma) = \frac{2}{\sqrt{\pi}\sigma_0} \cdot e^{-\sigma^2/{\sigma_0}^2}, \tag{4.18}$$

which is constant in μ and (under M_2) gives σ the half-Normal distribution, with scale $\sigma_0/\sqrt{2}$. This density for σ is proper, but is not quite as "balanced" as was π_2^{AI}, in that the median of the prior is roughly $\sigma_0/2$, not σ_0. Nevertheless, we would view this as a very satisfactory intrinsic prior, and hence an appealing default Bayes factor.

While this generalized "fractionalization" appears to solve the problem here, note that the solution comes with the price of introducing additional complexity into the

FBF approach. Indeed, it is not at all apparent, in general, how one should go about choosing the factorization of the likelihood (as in (4.15)), or which fractions should be used for each component. For interesting ideas in this direction, see De Santis and Spezzaferri (1998b).

4.5 Group Invariant Models (Example 5)

A very "nice" class of model comparisons is that which consists of comparisons between models having a given invariance structure. For general discussion of this class of comparisons, see Berger, Pericchi, and Varshavsky (1998); here we content ourselves with a simple example, which is also the final example in Bertolino and Racugno (1997) (also discussed from a different perspective in Berger and Pericchi 1997). Of interest is comparison between three separate scale models: thus assume $X_1, ..., X_n$ are i.i.d., with $(X_i - \mu_0)/\sigma$ (μ_0 known) having either a standard

$$M_1 : \textit{Normal}, \quad M_2 : \textit{Laplace} \text{ (double exponential)}, \quad \text{or } M_3 : \textit{Cauchy}$$

density. For the three models, Bertolino and Racugno (1997) consider noninformative priors of the form $\pi_j(\sigma) = \sigma^{-\alpha_j}$, with various choices of α_1, α_2, and α_3.

We should begin by noting that, for comparing models with the same group invariance structure (the multiplicative group for the above scale invariant models), there is a very strong reason to use the reference prior (given by $\alpha_j = 1$ here). This reason can be stated in two ways. First, the reference prior is typically exactly predictively matched for imaginary minimal training samples (in the sense discussed in section 3) for *all* models with the same group structure. This is a powerful inherent justification for its use, suggesting that one can directly use the B^N_{ji} in such a situation. The second way in which this can be viewed is to note that, when one has exact predictive matching for all minimal training samples, then all versions of IBFs reduce to simply B^N_{ji}. To see this explicitly in the scale parameter case, note that the marginal density of a minimal training sample (a single x_i here) is

$$m^N_j(x_i) = \frac{1}{2|x_i - \mu_0|}, \; j = 1, 2, 3.$$

Since these are equal for all three models, the "correction terms" in IBFs are all equal to one, and the IBFs reduce to B^N_{ji}. (See Berger, Pericchi and Varshavsky 1998, Berger and Pericchi 1997, and Sansó, Pericchi and Moreno 1996, for general discussion of this phenomenon. Note, in particular, that the predictive matching actually occurs for the *right invariant Haar measure* of the group action on the parameter space; this virtually always equals the so-called one-at-a-time reference prior, however.)

Table 1: *MIBF and FBF for Separate Scale Models, with reference and other priors.*

α_1	α_2	α_3	B_{21}^M	B_{21}^F	B_{31}^M	B_{31}^F	B_{32}^M	B_{32}^F
1	1	1	2.64	1.86	1.88	0.90	0.71	0.48
1	1.5	1.5	2.38	2.06	1.52	0.46	0.64	0.22

For priors other than the reference prior, IBFs will not simplify and so, to further explore the example of Bertolino and Racugno (1997), we will have to utilize an IBF. The three models here are separate models of equal complexity, and so it is not clear which to place in the numerator of the AIBF. Indeed, Bertolino and Racugno (1997) show that this ambiguity can lead to problems; they chose different values of the hyperparameters to show that the AIBF changes with the priors and that, for certain values of the hyperparameters, the AIBF can be incoherent in the sense that, simultaneously, $B_{12}^{AI} > 1$, and $B_{21}^{AI} > 1$. This issue is explored in Berger and Pericchi (1997), with the resulting recommendation that the MIBF be employed for separate models. (Note that the AIBF was, from the beginning, only recommended for nested models.) Hence we consider the MIBF in the following.

The FBF does not simplify in this situation when reference priors are used; comparison of invariant models is thus a situation where the IBF is actually simpler than the FBF. Furthermore, the FBF utilizes a fraction of the likelihood for updating the prior even though the predictive matching argument suggests that this is unnecessary. In computing the FBF below we used the fraction $b = 1/n$, since the minimal training sample is of size one.

In spite of our strong preference for the reference prior in this problem, we put the MIBF and the FBF to the test, not only with the reference prior ($\alpha_1 = \alpha_2 = \alpha_3 = 1$), but also with the prior specified by ($\alpha_1 = 1,\ \alpha_2 = \alpha_3 = 1.5$). (The behavior we observed for the other priors considered in Bertolino and Racugno 1997, was not substantially different, so we confined attention to the above two priors.) In Table 1, we present the Bayes factors arising from the data considered in Bertolino and Racugno (1997), namely $\{x_i - \mu_0\} = \{-1, -0.4, -0.2, 0.001, 0.01, 0.1, 0.3, 1\}$. Notice that the MIBF is quite robust to the choice of noninformative prior. The FBF seems somewhat more sensitive to the prior but, of more concern, gives answers which are considerably different than those provided by the IBF with the reference prior (which we are arguing is the correct default answer for this problem). Of course, this is but one example and one data set, but it does reinforce the earlier message that IBFs are most well suited to group invariant situations.

Finally, observe that we essentially argued above that there is a highly acceptable *conventional prior* for model comparisons among models with the same group invariance structure, namely the reference prior (or, more precisely, the right invariant Haar measure). We have not studied the accuracy of BIC in this situation, but no obvious difficulties in its application are apparent, as long as the models are regular. (BIC would not apply, for instance, if one of the models were the scale uniform model.)

4.6 When Neither Model is True (Example 6)

In practice, it is probably rare for *any* of the entertained models to be true. (See Key, Pericchi and Smith 1999, for general discussion.) How this affects the various default Bayes factors is only in the preliminary stages of investigation, but the issue is important enough to deserve mention.

Note, first of all, the interesting result in Smith (1995), that the *geometric* intrinsic Bayes factor (which replaces the arithmetic average in the AIBF with a geometric average), is an estimated version of the optimal model selector under a prequential scenario in which neither of the models being considered is the true model. The geometric IBF (GIBF) surfaces again in the following example, arising from O'Hagan (1997).

Consider the scenario of the Illustration 3 in section 3, where $X_1, \ldots, X_n$ are i.i.d. from the normal distribution with mean θ and variance one, and it is desired to compare $M_1 : \theta = 0$ with $M_2 : \theta \neq 0$. Recall that the FBF (for $b = 1/n$) is given by

$$B_{21}^F = \frac{1}{\sqrt{n}} \exp\left(\frac{n-1}{2} \bar{x}^2\right).$$

Simple computation shows that the GIBF is given by

$$B_{21}^{GI} = \frac{1}{\sqrt{n}} \exp\left(\frac{n-1}{2}[\bar{x}^2 - \frac{S^2}{n}]\right),$$

where $S^2 = \sum_{i=1}^n (x_i - \bar{x})^2/(n-1)$. Concerning S^2, O'Hagan (1997) says: "The appearance of the sample variance in this formula represents the effect of the variability in the partial Bayes factor. The more the variability in the data, the more the IBF favours the simpler model, which is not intuitively reasonable behaviour."

O'Hagan's conclusion is far from clear. Indeed, if S^2 is large, most statisticians would begin to doubt the model assumption that $\sigma^2 = 1$. To explore the situation, let us assume that the *true* model for the data is $M_3 : \mathcal{N}(0, \sigma^2)$, with $\sigma^2 > 1$. Note that M_1 is clearly closer to M_3 than is M_2, so that we should still prefer M_1 to M_2. (We are not viewing this problem from the viewpoint of diagnostics or model elaboration; the scenario is still that we must choose between M_1 and M_2 using a default Bayes factor, and all we are trying to ascertain is which default Bayes factors will do a good job of choosing the model closest to M_3.)

It is useful to define $Z = \sqrt{n}\bar{x}/\sigma$, which clearly has a $\mathcal{N}(0,1)$ distribution under M_3. In terms of Z, we can write the FBF and the GIBF as

$$B_{21}^F = \frac{1}{\sqrt{n}} \exp\left(\frac{(n-1)}{2n} Z^2\sigma^2\right), \quad B_{21}^{GI} = \frac{1}{\sqrt{n}} \exp\left(\frac{(n-1)}{2n}[Z^2\sigma^2 - S^2]\right).$$

If σ^2 is large enough, B_{21}^F will clearly typically favor the "worst" model M_2. For instance, even in the moderate case that $\sigma = 2$ and $Z = 1$, it is easy to check that B_{21}^F will exceed 1 (thus favoring M_2) unless the sample size is greater than 50. In contrast, the GIBF appears to be trying to compensate for M_3 by subtracting S^2 from the exponent. (Note that $E[Z^2\sigma^2 - S^2] = 0$.) Even for very large σ^2, the GIBF will typically favor M_1.

A different view of the situation can be obtained by studying the intrinsic priors that result from this "wrong models" scenario. Since the FBF was earlier seen to be the exact Bayes factor for a $\mathcal{N}(0,\ 1 - n^{-1})$ prior, we have no work to do there. Currently, we have only preliminary results on intrinsic priors for IBFs under a "wrong models" scenario but, for the AIBF in this situation, it is easy to show that its intrinsic prior is $\mathcal{N}(0,\ \sigma^2 + 1)$ (assuming M_3 is true). Thus the AIBF is "obtaining the scale for the prior" from the true model, not the assumed (incorrect) models.

The phenomenon here also relates to the criticism of IBFs that they typically do not depend solely on the sufficient statistics for the models, in that the training samples are not typically sufficient statistics. We have previously answered this criticism with the responses "it usually makes little difference in practice;" and "if desired, the matter is often easy to resolve, by either employing the *expected* IBF (which only depends on the sufficient statistics), or simply by taking an appropriate expectation of any IBF, conditional on the sufficient statistics (since, by definition, such an expectation does not depend on the unknown parameters)." We have always been reluctant to apply either of these "corrections," however, precisely because it seemed quite possible that IBFs were somehow employing information about the true model that was contained in the training samples. Note that any default Bayes factors which depend only on sufficient statistics (e.g., the FBF, BIC, and even conventional prior Bayes factors) cannot adapt to the true model in a "wrong models" scenario. Obviously much more investigation needs to be done to ascertain the success with which IBFs adapt to the true model, and it would be foolish to suggest that IBFs can adapt to all possible model deviations, but the indications are tantalizing.

5 Summary Comparisons

We do not repeat all of the comparisons made in section 4, but it is useful to summarize certain basic points. Also, we include some comparative comments obtained from other studies or examples.

5.1 Clarity of Definition

Comparison of the approaches is made more difficult by the fact that the approaches are, in various senses, not uniquely defined. Furthermore, as we are searching for an "automatic" procedure, we deem it to be negative if a default methodology requires considerable user input.

Conventional Prior Approach: The conventional prior approach suffers from the obvious problem of having a large possible multiplicity of conventional priors for a given situation. Part of the problem here is the brutal honesty of the approach: the conventional prior is clearly visible for all to see and criticize. In contrast, while BIC, IBFs, and FBFs may (at best) correspond to Bayes factors with intrinsic priors, these priors are not immediately visible. While statisticians would generally prefer the honesty of the conventional prior approach, many users of statistics prefer to "sweep such issues under the rug" (to borrow from I. J. Good), so that the conventional prior approach may well suffer in practical usage. (Let us be clear here: we are not happy with this state of affairs, but would much prefer to see BIC, IBFs, or FBFs used in practice by someone who, for this reason alone, would reject the conventional prior approach and use, say, a p-value.)

Intrinsic Bayes Factors: For the intrinsic Bayes factor, indeterminacy arises partly because of the existence of a variety of different IBFs. Our view of the IBF approach is that it is a *strategy* for approaching model selection problems, and that different IBFs will be "optimal" in different situations. (It is arguably naive to think that any single default Bayes factor will be optimal for all situations.) Thus, in Berger and Pericchi (1996a), we recommended that one should use the "expected IBF" if the sample size is small; the "arithmetic IBF" for two nested models; and the "encompassing IBF" for multiple linear models. We gave no clear recommendation for non-nested models. Recently (Berger and Pericchi 1997b), we added the "median IBF" to our recommended list, in part to fill the gap for non-nested models; in part to serve as an alternative to the encompassing approach for multiple models; and possibly to serve as a "single default tool" for users who do not wish to vary the tool with the problem. (While not necessarily always optimal from our perspective, the median IBF is almost always reasonable - though see Ghosh and Samanta 2001.)

As pointed out in Berger and Pericchi (1996a) (and emphasized in O'Hagan 1997), IBFs also have the indeterminacy of being dependent on the way in which the data is presented. For instance, in O'Hagan (1997) it is observed that whether one presents the data as $(x_1, x_2, \ldots, x_n)$ or as $(x_1 - \bar{x}, \ldots, x_n - \bar{x}, \bar{x})$ can have a pronounced effect on IBFs if one were to blindly apply the formulae. Of course, one can always reconstruct the original data from $(x_1 - \bar{x}, \ldots, x_n - \bar{x}, \bar{x})$ but this requires some additional knowledge on the part of the user about IBFs. An obvious suggestion is to recommend that IBFs be applied to raw, not processed, data (although if the data could be processed to make them more nearly exchangeable, this would probably be advantageous; see section 4.5 in Iwaki 1997, for an artificial, but interesting, example of how non-exchangeable data can affect IBFs). Sometimes, however, the notion of 'raw' data is not even well-defined. If the data consists of a realization of a stochastic process, for instance, there is no natural notion of a single 'raw' piece of data. For suggestions concerning the application of IBFs in such situations, see Sivaganesan and Lingham (1998, 1999). Even more problematical is the situation in which only sufficient statistics are available. Then the only IBFs available are the *expected* IBF and direct use of the intrinsic prior, since these are not based on resampling the actual data. Note that FBFs are essentially immune to these difficulties.

Fractional Bayes Factor: For the fractional Bayes factor, the main lack of definition is in the choice of the fraction b. Indeed, without a "recipe" for choosing b, one might question whether the FBF approach is actually a "default" Bayes factor. In Illustration 3 in section 3, for instance, we saw that choice of b is equivalent to choice of the (normal) prior variance; thus, suggesting that the choice of b be left to the investigator is equivalent to suggesting that the investigator subjectively choose the prior variance, i.e., adopt the subjective Bayesian approach. In practice, the most commonly used choice of b is m/n, where m is the minimal training sample size (when it exists).

We also saw that the standard FBF approach was unable to deal with various important classes of problems, such as those discussed in sections 4.1 and 4.4, unless it was modified to allow for using differing "fractions" of parts of the likelihood. If one embarks upon this generalization of the FBF approach (as we have found it essential to do), then the definitional issue appears to become even more involved; we have found, however, that choosing the (multiple) fractions by following IBF insight still succeeds.

BIC: Indeterminacy in asymptotic approximations such as BIC arises from several sources. First, one can utilize different asymptotic approximations. For instance, expression (A.1) in Appendix 1 typically gives a much more accurate approximation to a Bayes factor than does (2.13), and can be used as the basis for asymptotic approximations. (This is being explored elsewhere.) Even with the standard Laplace approximation, there are a variety of possible choices of the constant which replaces $\pi_j(\hat{\boldsymbol{\theta}}_j)$ and $\pi_i(\hat{\boldsymbol{\theta}}_i)$ in (2.13). And, even with a supposedly well-specified method such as BIC, which depends only on the likelihood, the dimensions of the parameters, and the sample size, there is considerable uncertainty as to how to define the sample size once one departs from i.i.d. situations. Pauler (1998) has begun the enterprise of determining "effective sample size" to deal with more complex situations. Finally, if the model sizes grow with the sample size, BIC can be extremely inadequate; see Berger, Ghosh and Mukhopadhyay (1999).

5.2 Computational Simplicity

The computational edge is typically with BIC, as it requires only standard m.l.e. computations in its calculation. (Note that computation of Bayes factors is not a domain in which MCMC algorithms have an edge over m.l.e. computations.)

In general, use of conventional priors and FBFs is roughly equal in computational complexity although, for many "standard" problems, FBFs are available in closed form while conventional prior Bayes factors may require numerical integration. Normal linear models is one such scenario, where FBFs (and even IBFs) are available in closed form, but the standard conventional priors are Cauchy or t-priors, which require numerical integration.

IBFs are typically the most difficult default Bayes factors to compute (comparison of invariant models being an exception). Also, with large sample sizes, computation of IBFs is only possible with use of suitable schemes for sampling from the training samples, since the number of training samples might be enormous. These issues are discussed in Berger and Pericchi (1996a) and Varshavksy (1995). An interesting result from the latter work, based on the theory of U-statistics, is that utilizing only a small multiple of n training samples, where n is the overall sample size, is essentially as accurate as utilizing all training samples. Of course, even computing n of the $B_{12}^N(\mathbf{x}(l))$ might appear to be computationally imposing, but note that these are very frequently available in closed form, even when B_{21}^N is not. Finally, it should be mentioned that Expected IBFs and even direct use of the intrinsic prior may be computationally more attractive in certain scenarios since they involve no training sample computations (cf, Berger and Pericchi 1996a).

5.3 Domain of Applicability

Conventional priors have been developed primarily on a case-by-case basis, starting with the situations considered in Jeffreys (1961). There are, however, two general classes of problems for which satisfactory conventional priors can be said to exist. The first is the comparison of models having the same invariance structure (as in section 4.5 above). Then, using the notion of predictive matching, Berger, Pericchi, and Varshavsky (1998) argue that model selection can be done using the right invariant Haar measure corresponding to the group action for the models. Typically, this prior is simply the reference prior. Note that these priors are usually improper, but use of such is eminently defensible in this situation.

The second class of problems in which general conventional priors can be said to exist is nested models. The idea here is simply to use the intrinsic prior corresponding to the arithmetic IBF (as given, say, by (A.4) and (A.8) in Appendix 1). Our experience is that this intrinsic prior is an excellent conventional prior, typically being predictively matched for nuisance parameters and proper for the parameter under test. Of course, one can question the utility of computing and using the intrinsic prior, compared with simply using the arithmetic IBF to compute the Bayes factor directly; but, for small sample sizes, or for those who feel more comfortable using a conventional prior, this approach to construction of conventional priors deserves very serious consideration. Note that this is the first general approach to the construction of conventional priors in nested models.

BIC is actually quite limited in applicability, requiring larger sample sizes, models with regular asymptotics, models in which the likelihood does not concentrate on a boundary (as in one-sided testing), and determination of "effective sample size" in non-i.i.d. situations. Again, however, there is ongoing work attempting to resolve some of these limitations.

FBFs are impressively general in applicability, especially if the modifications suggested in sections 4.1 and 4.4 are considered. Irregular models (section 4.2) and one-sided testing (section 4.3) seem to cause problems for which there is no clear remedy. Also, there are domains, such as comparison of invariant models, where FBFs seem to introduce an unneeded (and somewhat detrimental) complication.

IBFs have the widest range of applicability, modulo possible computational limitations. The main non-computational limitation of the more common IBFs is very small sample sizes; and "small" must be interpreted sensibly, remembering that the IBF is roughly "getting the prior from the data." Thus, if one were considering a parameterized outlier model, with a data set having only one outlier, it would not be reasonable to use an IBF based on training samples, since there is not enough data to effectively obtain the prior distribution for the parameters of the outlier distribution. (This was illustrated in Example 3 of O'Hagan, 1997.) For some "small sample" situations of this type, IBFs seem to work fine (e.g., the Neyman-Scott problem in section 4.4), but caution should be the rule. (See Beattie, Fong, and Lin, 2001, for another example.) For extremely small sample sizes, expected IBFs or direct use of the intrinsic prior are probably at least as sensible as anything else of a default nature, but it can be argued that no default Bayes factors are really likely to be sensible for extremely small sample sizes. (There will typically be extreme sensitivity to the conventional prior chosen, to the choice of b in the FBF, etc.).

The various types of IBFs each have their own range of applicability, with the AIBF possibly being the most limited, and the MIBF being the most general. The AIBF is essentially limited to nested models (wherein it seems to perform the best). We have not found any serious limitations to applicability of the MIBF, except for situations, such as that in section 4.1, where no sample will yield a finite marginal if improper priors are used. (Recall, however, that, even with mixture models, it was possible to alter the IBF strategy to yield successful default Bayes factors.)

5.4 Correspondence with Reasonable Intrinsic Priors

Conventional Prior Approach: It may seem somewhat odd to ask whether the conventional prior approach yields Bayes factors which correspond to reasonable intrinsic priors; indeed, what we are simply concerned with here are the implications of "reasonable." Some of the difficulties inherent in the approach were discussed in sections 2 and 3, and numerous - perhaps most - of the constructions of conventional priors in the literature have violated one or more of the basic considerations mentioned therein. This should serve as a warning that the conventional prior approach must be viewed with the same scrutiny as other methods. Again, it is instructive to read Jeffreys (1961), and see the care with which he constructed conventional priors.

BIC Approach: The most basic fact in consideration of asymptotic approaches, such as BIC, is that they cannot correspond to an actual Bayesian analysis, since they replace the $\pi_j(\hat{\boldsymbol{\theta}}_j)$ and $\pi_i(\hat{\boldsymbol{\theta}}_i)$ in (2.13) with a constant, independent of the data. An interesting argument can be given in defense of BIC, however, in nested model scenarios. Indeed, Kass and Wasserman (1995) and Pauler (1998) argue that BIC then does correspond to an actual Bayes factor under the nested model; the "constant" used in the approximation can essentially be chosen to be the constant that would arise from a default conventional prior (which they call the "unit information prior") in the Laplace approximation under the nested model. The argument proceeds by observing that, under the more complex model, the Bayes factor (of the complex to the nested model) will typically go to infinity at an exponential rate, and hence that the inaccuracy then induced by replacing $\pi_j(\hat{\boldsymbol{\theta}}_j)$ and $\pi_i(\hat{\boldsymbol{\theta}}_i)$ with the "wrong" constant is of limited practical concern.

Our view of this argument is positive, in that we feel it does cleverly justify use of BIC (or other such approximations), if a quick approximation is needed. One should not read too much into the argument, however. With moderate amounts of data, it is very frequently the case that the Bayes factor is not conclusively in favor of either hypothesis and, when this is the case, one cannot trust the approximation; virtually by definition, one is then unsure as to which model is correct and, if the complex model is correct, the approximation can be bad. (And this is even ignoring the question of the accuracy of the Laplace approximation for smaller data sets.)

Another set of issues surrounds the choice of the "unit information prior," which is the intrinsic prior behind the BIC justification. Our original plan to also discuss this set of issues has run aground due to the length of the chapter, and so we will have to delay discussion to another forum.

FBF Approach: FBFs do seem to typically have reasonable intrinsic priors, as long as b is chosen to roughly reflect the same fraction of the likelihood as is a minimal training sample of the data. Problems in which there is concern include irregular models (section 4.2), where FBFs can fail to behave at all like Bayes factors; one-sided testing (section 4.3), where FBFs often correspond to intrinsic priors which are quite biased in favor of one of the hypotheses; and problems of increasing parameter multiplicity or highly varying parameter information content (section 4.4), where only FBFs which are modified to allow for multiple fractionation may be sensible. In spite of these serious omissions, we are generally quite positive about the true Bayesian nature of FBFs, and do not hesitate to use them in straightforward situations (or "tune" them with IBF reasoning in more complicated situations).

IBF Approach: From the beginning, our interest in IBFs has been motivated by the amazing ability they seem to possess to correspond to Bayes factors with reasonable intrinsic priors in even highly challenging situations, such as those in sections 4.2, 4.3, and 4.4. And in "nice" situations, such as that of section 4.5 where the natural default priors are eminently sensible for direct use, IBFs do the "right thing" and leave the default priors alone. Time and again we have also observed strong indications that intrinsic priors from IBFs have the key property of predictive matching (discussed in section 3), but we have not been able to formulate a general result in this direction. (Part of the difficulty is the technical complexity of having to deal with improper intrinsic priors; see Moreno, Bertolino, and Racugno 1998a, for possible tools to use in this regard.)

The situation with IBFs is certainly not perfect in regards to intrinsic priors, however. Only for comparing two nested models, and using the AIBF, are reasonable intrinsic priors almost guaranteed to exist. And the AIBF typically does not have intrinsic priors outside of nested models. The other IBFs (such as the median IBF) typically yield reasonable, but not ideal, intrinsic priors. Thus, in section 4.2, the intrinsic prior for the MIBF only had mass 0.69 while, in section 4.3, it had a modestly unbalanced prior odds ratio; such problems indicate that the MIBF is typically modestly "biased" in favor of one the models.

5.5 Comparisons with Other Recent Approaches

Expected posterior prior approach: This is a recent highly promising approach, based on use of 'imaginary training samples,' developed in Pérez (1998) and Pérez and Berger (2000) (and, independently in a special case, in Schluter, Deely and Nicholson 1999). This approach utilizes imaginary training samples to directly develop default conventional priors for use in model comparison. Letting $\mathbf{x}^*$ denote the imaginary training sample (usually taken to be a minimal training sample) and starting with noninformative priors $\pi_i^N(\boldsymbol{\theta}_i)$ for M_i as usual, one first defines the posterior distributions, given $\mathbf{x}^*$,

$$\pi_i^*(\boldsymbol{\theta}_i|\,\mathbf{x}^*) = \frac{f_i(\mathbf{x}^*|\,\boldsymbol{\theta}_i)\pi_i^N(\boldsymbol{\theta}_i)}{\int f_i(\mathbf{x}^*|\,\boldsymbol{\theta}_i)\pi_i^N(\boldsymbol{\theta}_i)d\boldsymbol{\theta}_i}. \tag{5.1}$$

Since $\mathbf{x}^*$ is not actually observed, these posteriors are not available for use in model selection. However, we can let $m^*(\mathbf{x}^*)$ be a suitable predictive measure for the (imaginary) data $\mathbf{x}^*$, and define priors for model comparison by

$$\pi_i^*(\boldsymbol{\theta}_i) = \int \pi_i^N(\boldsymbol{\theta}_i|\,\mathbf{x}^*)m^*(\mathbf{x}^*)d\mathbf{x}^*. \tag{5.2}$$

These are called *expected posterior priors* (or *EP priors* for short) because this last integral can be viewed as the expectation (with respect to m^*) of the posteriors in (5.1).

EP priors can successfully deal with the four difficulties discussed in section 1.5. Of particular note is that, because they can be viewed as an expectation of training sample posteriors, it is usually possible to utilize MCMC computational schemes to compute Bayes factors. Also noteworthy is that, since the $\pi_i^*(\boldsymbol{\theta}_i)$ are actual conventional priors, they inherit the attractive properties of that approach.

The key issue in utilization of EP priors is that of appropriate choice of $m^*(\mathbf{x}^*)$. Two choices that are natural are (i) the empirical distribution of the actual data (so that the EP prior approach can then be viewed as a resampling approach); and (ii) the (usually improper) predictive arising from an *encompassed* model (i.e., a model that is nested within all others under consideration) and under the usual nonformative prior. Of considerable interest is that this latter approach results in the $\pi_i^*(\boldsymbol{\theta}_i)$ being the intrinsic priors corresponding to the AIBF for nested models. Thus the EP prior approach can be viewed as an outgrowth of the IBF approach (and, indeed, arose in that way). Alternatively, it can be viewed as a method of actually implementing use of IBF intrinsic priors.

Posterior expected marginal likelihood approach: This is an approach recently developed in Iwaki (1997), which also contains extensive discussion and comparison with other methods. This approach is very similar to the EP prior approach but, instead of using $m^*(\mathbf{x}^*)$ in (5.2), it uses the predictive distribution of $\mathbf{x}^*$, given the actual data $\mathbf{x}$. This approach thus also generates priors, but they are data-dependent. For instance, in comparison of nested models, these priors seem to (roughly) be centered at the m.l.e. under the more complex model, rather than centered at the nested model. We view this particular type of data-dependency in the prior as inducing a bias in favor of the more complex model, and hence prefer the EP prior approach. (Note, however, that the empirical version of the EP prior approach can be similarly criticized.) The degree of this bias, however, is modest compared to most other methodologies, and so we view the approach of Iwaki (1997) as promising and deserving serious study.

Robust Bayesian analysis approach: Another approach that can be used as a default model selection method is robust Bayesian analysis (see Berger 1994, and Ríos Insua and Ruggeri, 2000, for reviews and references). In particular, one can often find upper and/or lower bounds on Bayes factors over wide classes of prior distributions. These bounds can be the basis of a default analysis. The three limitations of the approach are: (i) often only an upper or a lower bound, but not both, are available; (ii) computations can be formidable; and (iii) conclusions often cannot be reached solely from a bound on the Bayes factor. In spite of these limitations, it should be recognized that, when the approach can be applied, one obtains an answer that is considerably more compelling than answers from other default methodologies. Note, also, that there have been several interesting recent developments utilizing robust Bayesian methodology in concert with either the fractional or intrinsic Bayes factor approaches. References include De Santis and Spezzaferri (1996, 1999), Moreno (1997), Moreno, Bertolino, and Racugno (1998b), Sansó, Pericchi, and Moreno (1996).

6 Recommendations

We end with a (provisional) answer to the question that motivated this paper: When should each of the default Bayes factor methodologies be used, and when should they not? In one sense, this question cannot be answered without asking another: Who is to be the user of the methodology, and in what fashion will it be used? BIC, the FBF (with, say, $b = m/n$ and constant priors) and the MIBF (with constant priors) could each be used as a single fairly general tool by unsophisticated users. At the other extreme, highly sophisticated users can view all default model selection methods as simply a collection of available tools, with specific strengths and weaknesses, that can be modified or adapted to deal with highly complex situations. A middle ground, which we find it useful to imagine, is that of a good Bayesian statistical computer package, in which a set of rules could be encoded to employ the default model selection strategy most suited for the situation being analyzed. In this spirit, we conclude with a few such comments about each of the four approaches we have considered.

The conventional prior approach should be used if there is a reasonable (and well-studied) conventional prior available for the given situation, and if operating in an environment where use of a specific conventional prior is sociologically acceptable. The approach is particularly valuable if the sample size is small (in which case all of the other approaches become suspect), or in problems (such as mixture models) where marginal densities with respect to noninformative priors are not finite.

The asymptotic approaches (such as BIC) can be used with confidence in situations with large sample sizes (relative to the number of parameters). For moderate sample sizes, there is a justification for their use when calculational necessity precludes utilization of any of the other methodologies; if one of the other approaches can be implemented, however, that would typically be preferred. Note, also, that the standard BIC has considerable limitations in its domain of applicability.

The intrinsic Bayes factor approach is the most generally applicable approach, if it can be implemented computationally. IBFs based on training samples can be used with considerable confidence (unless the sample size is very small) in (i) nested model comparisons (AIBF or encompassing IBF preferred); (ii) comparison of non-nested models of roughly equal size (MIBF preferred); and (iii) situations where reference priors are available for computing the IBFs. If the models are of highly varying dimension and reference priors are not available, we know of nothing better to use than the MIBF, but we do not have enough experience to be overly comfortable. When the sample size is very small, we recommend use of the expected IBF (or direct use of the intrinsic prior as a conventional prior). One should also be aware that, viewed as a *strategy*, IBFs can often be modified to handle difficult situations, as in the situation with mixture models discussed in section 4.1.

The recommended domain of the fractional Bayes factor approach is a large region somewhere between the recommended domains of the IBF and BIC. The FBF is typically easier to compute than the IBF, but it is considerably more difficult to compute than BIC. On the other hand, the range of applicability of the FBF is considerably greater than that of BIC, but is more limited than that of IBFs. Note that, if one generalizes the FBF to allow use of varying fractions in different parts of the likelihood, with the fractions chosen to mimic the effect of minimal training samples on IBFs (or, alternatively, chosen to reflect the varying amount of information about the parameters in the likelihood), the generality of the approach can be considerably increased. FBFs are typically available even for very small sample sizes, but their utility in such situations is tempered by the fact that the answer will typically then be highly sensitive to the choice of the fraction b, and no reasonable automatic choices for this fraction seem possible in such situations.

With judicious application of the above four methodologies, we are, for the first time, at the point where we can tackle most of the model selection and hypothesis testing problems with which we are presented.

Appendix 1: Intrinsic Prior Equations

The formal definition of an intrinsic prior, given in Berger and Pericchi (1996a), was based on an asymptotic analysis, utilizing the following approximation to a Bayes factor associated with priors π_j and π_i:

$$B_{ji} = B_{ji}^N \cdot \frac{\pi_j(\hat{\boldsymbol{\theta}}_j)\pi_i^N(\hat{\boldsymbol{\theta}}_i)}{\pi_j^N(\hat{\boldsymbol{\theta}}_j)\pi_i(\hat{\boldsymbol{\theta}}_i)}(1+o(1)); \tag{A.1}$$

here $\hat{\boldsymbol{\theta}}_i$ and $\hat{\boldsymbol{\theta}}_j$ are the m.l.e.s under M_i and M_j. (The approximation in (A.1) holds in considerably greater generality than does the Schwarz approximation in (2.13).)

Most default Bayes factors (certainly FBFs and IBFs) can be written in the form

$$B_{ji} = B_{ji}^N \cdot \tilde{B}_{ij}, \tag{A.2}$$

where $\tilde{B}_{ij}$ is often called the *correction factor.* To define intrinsic priors, equate (A.1) with (A.2), yielding

$$\frac{\pi_j(\hat{\boldsymbol{\theta}}_j)\pi_i^N(\hat{\boldsymbol{\theta}}_i)}{\pi_j^N(\hat{\boldsymbol{\theta}}_j)\pi_i(\hat{\boldsymbol{\theta}}_i)}(1+o(1)) = \tilde{B}_{ij}. \tag{A.3}$$

We next need to make some assumptions about the limiting behavior of the quantities in (A.3). The following are typically satisfied, and will be assumed to hold as the sample size grows to infinity:

(i) Under M_j, $\hat{\boldsymbol{\theta}}_j \to \boldsymbol{\theta}_j$, $\hat{\boldsymbol{\theta}}_i \to \psi_i(\boldsymbol{\theta}_j)$, and $\tilde{B}_{ij} \to B_j^*(\boldsymbol{\theta}_j)$.

(ii) Under M_i, $\hat{\boldsymbol{\theta}}_i \to \boldsymbol{\theta}_i$, $\hat{\boldsymbol{\theta}}_j \to \psi_j(\boldsymbol{\theta}_i)$, and $\tilde{B}_{ij} \to B_i^*(\boldsymbol{\theta}_i)$.

When dealing with the AIBF, it will typically be the case that, for $k = i$ or $k = j$,

$$B_k^*(\boldsymbol{\theta}_k) = \lim_{L\to\infty} E_{\boldsymbol{\theta}_k}^{M_k}\left[\frac{1}{L}\sum_{l=1}^{L} B_{ij}^N(l)\right]; \tag{A.4}$$

if the $\mathbf{X}(l)$ are exchangeable, then the limits and averages over L can be removed. For the MIBF, it will simply be the case that, for $k = i$ or $k = j$,

$$B_k^*(\boldsymbol{\theta}_k) = \lim_{L\to\infty} \text{Med}[B_{ij}^N(l)]. \tag{A.5}$$

For the FBF, De Santis and Spezzaferri (1997) show, for the i.i.d. situation and with $b = m/n$ (where m is fixed) that, for $k = i$ or $k = j$,

$$\begin{aligned} B_k^*(\tilde{\boldsymbol{\theta}}_k) &= \lim_{n\to\infty} \frac{\int L^b(\boldsymbol{\theta}_i)\pi_i^N(\boldsymbol{\theta}_i)d\boldsymbol{\theta}_i}{\int L^b(\boldsymbol{\theta}_j)\pi_j^N(\boldsymbol{\theta}_j)d\boldsymbol{\theta}_j} \\ &= \frac{\int \exp\left(m\, E_{\tilde{\boldsymbol{\theta}}_k}^{M_k}[\log(f_i(X_l\,|\,\boldsymbol{\theta}_i))]\right)\pi_i^N(\boldsymbol{\theta}_i)d\boldsymbol{\theta}_i}{\int \exp\left(m\, E_{\tilde{\boldsymbol{\theta}}_k}^{M_k}[\log(f_j(X_l\,|\,\boldsymbol{\theta}_j))]\right)\pi_j^N(\boldsymbol{\theta}_j)d\boldsymbol{\theta}_j}. \end{aligned} \tag{A.6}$$

(We have written $\tilde{\boldsymbol{\theta}}_k$ instead of $\boldsymbol{\theta}_k$ above to distinguish this variable from the dummy variables of integration; also, X_l stands for a single observation.)

Passing to the limit in (A.3), first under M_j and then under M_i, results in the following two equations which define the *intrinsic prior* (π_j^I, π_i^I):

$$\begin{aligned} \frac{\pi_j^I(\boldsymbol{\theta}_j)\pi_i^N(\psi_i(\boldsymbol{\theta}_j))}{\pi_j^N(\boldsymbol{\theta}_j)\pi_i^I(\psi_i(\boldsymbol{\theta}_j))} &= B_j^*(\boldsymbol{\theta}_j), \\ \frac{\pi_j^I(\psi_j(\boldsymbol{\theta}_i))\pi_i^N(\boldsymbol{\theta}_i)}{\pi_j^N(\psi_j(\boldsymbol{\theta}_i))\pi_i^I(\boldsymbol{\theta}_i)} &= B_i^*(\boldsymbol{\theta}_i). \end{aligned} \tag{A.7}$$

The motivation, again, is that priors which satisfy (A.7) would yield answers which are asymptotically equivalent to use of the given default Bayes factors. We note that solutions are not necessarily unique, do not necessarily exist, and are not necessarily proper (cf, Dmochowski 1996, and Moreno, Bertolino, and Racugno 1998a).

In the nested model scenario (M_i nested in M_j), and under mild assumptions, solutions to (A.7) are trivially given by

$$\pi_i^I(\boldsymbol{\theta}_i) = \pi_i^N(\boldsymbol{\theta}_i), \quad \pi_j^I(\boldsymbol{\theta}_j) = \pi_j^N(\boldsymbol{\theta}_j)B_j^*(\boldsymbol{\theta}_j). \tag{A.8}$$

Typically there are also many other solutions, but the solutions in (A.8) are the simplest. See Dmochowski (1996) and Moreno, Bertolino, and Racugno, (1998a), for interesting characterizations of intrinsic priors in nested problems.

Appendix 2: Technical details from Section 4.2

We seek to establish (4.3). Even though the situation is highly nonregular, (A.1) can still be shown to apply, providing $\pi_2^N(\theta)$ and $\pi_2(\theta)$ are continuous and bounded at θ_0.

Furthermore, this can be viewed as a nested problem, so that the intrinsic priors are given by (A.8) after determining the limits in (A.4) and (A.5).

For the MIBF, note that $\text{Med}[X_i] \to \theta_0 + \log 2$. The intrinsic prior in (4.3) follows immediately. For the AIBF, computation yields

$$E_\theta^{M_2}[(e^{(X_i-\theta_0)} - 1)^{-1}] = -e^{(\theta-\theta_0)} \log(1 - e^{(\theta_0-\theta)}) - 1. \qquad (A.9)$$

This would indeed seem to establish the validity of (4.3). There is, however, an interesting technical difficulty, namely that (A.4) does not hold under M_1 (the expectation being infinite). While this might seem to doom the analysis, the intrinsic prior can be seen to be unbounded at θ_0, behaving as $-(\log(\theta - \theta_0) + 1)$. This means that (A.1) does not directly apply either, but it seems quite possible that generalized versions of (A.1), (A.3), and (A.4) would hold, since all quantities seem to be going to infinity at the same rate at θ_0. Unfortunately, we have not yet completed this generalization. The issue is of more than technical interest in that proper behavior of the AIBF, in such a difficult situation, can yield considerably increased confidence in the procedure.

Appendix 3: Technical details from Section 4.4

Intrinsic Priors: The intrinsic prior equations in Appendix 1 do not apply directly to this situation, because their asymptotic motivation is clearly inapplicable to the μ_i. Note, however, that the priors in (4.13) and (4.18) are both constant in the μ_i. Hence we can directly integrate the likelihood in (4.10) over the μ_i, before applying the results in Appendix 1. The resulting marginal likelihood of σ is

$$L_n(\sigma) \propto \sigma^{-n} e^{(-S^2/(4\sigma^2))}. \qquad (A.10)$$

Note also that S^2 is a sum of the i.i.d. $S_i = (X_{i1} - X_{i2})^2$, having density

$$f(s_i \mid \sigma) \propto \sigma^{-1} e^{(-s_i/(4\sigma^2))},$$

and that

$$B_{21}^N = \frac{\int L_n(\sigma) \cdot \sigma^{-1}\, d\sigma}{L_n(\sigma_0)}.$$

Hence the results in Appendix 1 apply directly to this "marginal" problem.

Since the problem is comparison of nested models, (A.8) applies. For the AIBF, (A.4) yields

$$B_2^*(\sigma) = E_\sigma^{M_2}\left[\frac{|X_{i1} - X_{i2}|}{\sqrt{\pi}\sigma_0} \exp\left(-\frac{(X_{i1} - X_{i2})^2}{4{\sigma_0}^2}\right)\right] = \frac{2\sigma}{\pi\sigma_0}(1 + \frac{\sigma^2}{{\sigma_0}^2})^{-1}.$$

Inserting this into (A.8) yields (4.13). For the FBF in (4.17), one can (in the marginal problem) simply observe that $S^2/(2n) \to \sigma^2$. Hence (A.8) directly yields (4.18) as the intrinsic prior.

Proof of Lemma 1: Note first that $S^2/(2\sigma^2)$ has a chi-squared distribution with n degrees of freedom, from which it follows that, when $\sigma^2 = 2\sigma_0^2$ and $n \to \infty$,

$$\frac{S^2}{4n\sigma_0^2} = 1 + Z(\frac{2}{n})^{1/2} + \gamma_n, \tag{A.11}$$

where Z is standard normal and $\gamma_n = O(1/n)$. Note next that

$$\begin{aligned} \log(B_{21}^{F,b_n}) &= n(1-b_n)\cdot\frac{S^2}{4n\sigma_0^2} + n(b_n-1)[\log(\frac{S^2}{4n\sigma_0^2}) - \log n] \\ &\quad + n(b_n - \frac{1}{2})\log b_n + \log\Gamma(\frac{n}{2}) - \log\Gamma(n(b_n - \frac{1}{2})). \end{aligned} \tag{A.12}$$

Case 1. Assume that $n(b_n - 1/2) \to \infty$, and that b_n is bounded away from 1.

Using (A.11) in (A.12), expanding $\log(1 + Z\sqrt{2/n} + \gamma_n)$, and using Stirling's approximation with the Γ functions, results in the expression

$$\log(B_{21}^{F,b_n}) = (1-b_n)Z^2 + \frac{1}{2}\log(2b_n - 1) + n[(b_n - \frac{1}{2})\log(b_n/(b_n - \frac{1}{2})) - \frac{1}{2}\log 2] + O(\frac{1}{\sqrt{n}}). \tag{A.13}$$

It can be shown that $(b - 0.5)\log(b/(b - 0.5))$ is an increasing function of b on $(0.5, 1)$, so that the term in square brackets in (A.13) is negative (since b_n is bounded away from 1). It is immediate that the FBF is then tending to 0.

Case 2. Assume that $n(b_n - 1/2) \to \infty$ and that $b_n \to 1$.

Expanding the log terms in (A.13) and taking limits yields

$$\lim_{n\to\infty} \log(B_{21}^{F,b_n}) = \lim_{n\to\infty} n(1 - b_n)(\frac{1}{2} - \log 2).$$

Since $(0.5 - \log 2)$ is negative, the conclusion that the FBF is ≤ 1 in the limit is immediate. Also, if $n(1 - b_n) \to \infty$, the conclusion that the FBF $\to 0$ follows.

Case 3. Assume that $\rho_n = n(b_n - 1/2) < K < \infty$.

Inserting ρ_n into (A.12) and performing a similar expansion to that leading to (A.13), results in the expression

$$\log(B_{21}^{F,b_n}) = \frac{1}{2}Z^2 + \rho_n(\log\frac{n}{2} - 1) - \frac{\log n}{2} - \frac{(n-1)}{2}\log 2 + \frac{\log(2\pi)}{2} - \log\Gamma(\rho_n) + O(\frac{1}{\sqrt{n}}).$$

The only positive unbounded term, $\rho_n \log n$, is clearly dominated by $-(n/2)\log 2$, so that the expression goes to $-\infty$, establishing that the FBF goes to 0. The case where

ρ_n is not bounded above (and does not go to ∞ in limit) is handled by a more tedious version of Cases 1 and 3. We omit the details.

REFERENCES

Barbieri, M. and Berger, J. (2001). Optimal predictive variable selection. ISDS Discussion Paper, Duke Univ.

Berger, J. (1994). An overview of robust Bayesian analysis (with Discussion). *Test* **3**, 5-124.

Berger, J. (1999). Bayes Factors. In the *Encyclopedia of Statistical Sciences*, Update Volume 3 (S. Kotz, et al., eds.) 20-29, Wiley, New York.

Berger, J. and Bernardo, J. M. (1992). On the development of the reference prior method. In *Bayesian Statistics 4* (J. M. Bernardo, J. O. Berger, A. P. Dawid and A. F. M. Smith, eds.) 35-60, Oxford Univ. Press.

Berger, J., Boukai, B. and Wang, Y. (1997). Unified frequentist and Bayesian testing of a precise hypothesis (with discussion). *Statist. Sci.* **12**, 133-160.

Berger, J., Boukai, B. and Wang, Y. (1999). Simultaneous Bayesian - frequentist sequential testing of nested hypotheses. *Biometrika* **86**, 79-92.

Berger, J., Brown, L. and Wolpert, R. (1994). A unified conditional frequentist and Bayesian test for fixed and sequential hypothesis testing. *Ann. Statist.* **22**, 1787-1807.

Berger, J. and Delampady, M. (1987). Testing precise hypotheses. *Statist. Sci.* **3**, 317-352.

Berger, J., Ghosh, J.K. and Mukhopadhyay, N. (1999). Approximations and consistency of Bayes factors as model dimension grows. Technical Report, Dept. Stat., Purdue Univ.

Berger, J. and Mortera, J. (1999). Default Bayes factors for one-sided hypothesis testing. *J. Amer. Statist. Asso.* **94**, 542-554.

Berger, J. and Pericchi, L. (1996a). The intrinsic Bayes factor for model selection and prediction. *J. Amer. Statist. Asso.* **91**, 109–122.

Berger, J. and Pericchi, L. (1996b). The intrinsic Bayes factor for linear models. In *Bayesian Statistics 5* (J. M. Bernardo, J. O. Berger, A. P. Dawid and A. F. M. Smith, eds.) 23-42, Oxford Univ. Press.

Berger, J. and Pericchi, L. (1997). On the justification of default and intrinsic Bayes factors. In *Modeling and Prediction* (J. C. Lee et al., eds.) 276–293, Springer-Verlag, New York.

Berger, J. and Pericchi, L. (1998). Accurate and stable Bayesian model selection: the median intrinsic Bayes factor. *Sankhyā Ser. B* **60**, 1-18.

Berger, J., Pericchi, L., and Varshavsky, J. (1998). Bayes factors and marginal distributions in invariant situations. *Sankhyā Ser. A* **60**, 307-321.

Berger, J. and Sellke, T. (1987). Testing a point null hypothesis: the irreconcilability of P-values and evidence. *J. Amer. Statist. Asso.* **82**, 112-122.

Berk, R. (1966). Limiting behavior of posterior distributions when the model is incorrect. *Ann. Math. Statist.*, **37**, 51-58.

Bernardo, J. M. (1979). Reference posterior distributions for Bayesian inference. *J. Roy. Statist. Soc. Ser. B* **41**, 113-147.

Beattie, S., Fong, D., and Lin, D. (2001). A two-stage Bayesian model selection strategy for supersaturated designs. To appear in *Technometrics.*

Bertolino, F. and Racugno, W. (1997). Is the intrinsic Bayes factor intrinsic? *Metron,* **LIV**, 5-15.

Carlin, B.P. and Chib, S. (1995). Bayesian model choice via Markov Chain Monte Carlo. *J. Roy. Statist. Soc. Ser. B* **57**, 473-484.

Casella, G., and Berger, R. (1987). Reconciling Bayesian and frequentist evidence in the one-sided testing problem (with discussion). *J. Amer. Statist. Asso.* **82**, 106-111.

Chib, S. and Jeliazkov, I. (2001). Marginal likelihood from the Metropolis-Hastings output. *J. Amer. Statist. Asso.* **91**, 270-281.

Clyde, M. (1999). Bayesian model averaging and model search strategies. In *Bayesian Statistics 6*, (J.M. Bernardo, A.P. Dawid, J.O. Berger, and A.F.M. Smith, eds.) 157-185, Oxford Univ. Press.

Clyde, M., DeSimone, H. and Parmigiani, G. (1996). Prediction via orthogonalized model mixing. *J. Amer. Statist. Asso.* **91**, 1197-1208.

Clyde, M. and George, E.I. (2000). Flexible empirical Bayes estimation for wavelets. *J. Roy. Statist. Soc. Ser. B* **62**, 681-698.

Cox, D. R. (1961). Tests of separate families of hypotheses. In *Proc. Fourth Berkeley Symp. Math. Statist. Probab.* **1** 105-123, Univ. of California Press, Berkeley.

Dass, S. and Berger, J. (1998). Unified Bayesian and conditional frequentist testing of composite hypotheses. ISDS Discussion Paper 98-43, Duke Univ.

Dellaportas, P., Forster, J. and Ntzoufras, I. (2001). On Bayesian model and variable selection using MCMC. To appear in *Statist. Comput.*.

De Santis, F., and Spezzaferri, F. (1996). Comparing hierarchical models using Bayes factor and fractional Bayes factors: a robust analysis. In *Bayesian Robustness* (J. Berger, et al., eds.) **29** IMS Lecture Notes, Hayward, California.

De Santis, F., and Spezzaferri, F. (1997). Alternative Bayes factors for model selection. *Canad. J. Statist.* **25**, 503-515.

De Santis, F., and Spezzaferri, F. (1998a). Consistent fractional Bayes factor for linear models. *Pubblicazioni Scientifiche del Dipartimento di Statistica, Probab. e Stat. Appl.* **19.**, Univ. di Roma, "La Sapienza", Serie a,

De Santis, F., and Spezzaferri, F. (1998b). Bayes factors and hierarchical models. *J. Statist. Plann. Inference* **74**, 323-342.

De Santis, F., and Spezzaferri, F. (1999). Methods for robust and default Bayesian model selection: the fractional Bayes factor approach. *Internat. Statist. Rev.* **67**, 1-20.

de Vos, A. F. (1993). A fair comparison between regression models of different dimension. Technical Report, The Free University, Amsterdam.

Dmochowski, J. (1996). Intrinsic priors via Kullback-Leibler Geometry. In *Bayesian Statistics 5* (J. M. Bernardo, J. O. Berger, A. P. Dawid and A. F. M. Smith, eds.) 543-549, Oxford Univ. Press.

Draper, D. (1995). Assessment and propagation of model uncertainty. *Jour. of the Roy. Statist. Soc. Ser. B* **57**, 45-98.

Dudley, R. and Haughton, D. (1997). Information criteria for Multiple Data Sets and Restricted Parameters. *Statist. Sinica* **7**, 265-284.

Dupuis, J.A. and Robert, C.P. (1998). Bayesian variable selection in qualitative models by Kullback-Leibler projections. In *Proceedings of the Workshop on Model Selection* (W. Racugno, ed.) 275-305, CNR, Collana Atti di Congressi, Pitagora, Editrice, Bologna.

Edwards, W., Lindman, H, and Savage, L.J. (1963). Bayesian statistical inference for psychological research. *Psycological Review* **70**, 193-242.

Findley, D.F. (1991). Counterexamples to Parsimony and BIC. *Ann. Inst. Statist. Math.* **43**, 505-514

Gelfand, A. E. and Dey, D. K. (1994). Bayesian model choice: asymptotics and exact calculations. *J. Roy. Statist. Soc. Ser. B* **56**, 501-514.

Gelfand, A. E., Dey, D. K. and Chang, H. (1992). Model determination using predictive distributions with implementations via sampling-based methods. In *Bayesian Statistics 4* (J. M. Bernardo, J. O. Berger, A. P. Dawid and A. F. M. Smith, eds.) 147-167, Oxford Univ. Press.

Gelfand, A. E. and Ghosh, S. K. (1998). Model choice: a minimum posterior predictive loss approach. *Biometrika* **85**, 1-11.

George, E. I. and Foster, D. P. (2000). Empirical Bayes variable selection. *Biometrika* **87**, 731-747.

George, E. I. and McCulloch, R. E. (1993). Variable selection via Gibbs sampling. *J. Amer. Statist. Asso.* **88**, 881-889.

Ghosh, J.K. and Samanta, T. (2001). Nonsubjective Bayesian testing - an overview. To appear in *J. Statist. Plann. Inference.*

Godsill, S. J. (2001). On the relationship between Markov Chain Monte Carlo methods for model uncertainty. *J. Comput. Graph. Statist.* **10**, 230-248.

Goutis, C. and Robert, C.P. (1998). Model choice in generalized linear models: a Bayesian approach via Kullback-Leibler projections. *Biometrika* **85**, 29-37.

Green, P. (1995). Reversible jump Markov Chain Monte Carlo computation and Bayesian model determination. *Biometrika* **82**, 711-732.

Han, C. and Carlin, B. (2001). Markov Chain Monte Carlo methods for computing Bayes factors: a comparative review. *J. Amer. Statist. Asso.* **96**, 1122-1132.

Haughton, D. (1988). On the choice of a model to fit data from an exponential family. *Ann. Statist.* **16**, 342-355.

Ibrahim, J. and Laud, P. (1994). A predictive approach to the analysis of designed experiments. *J. Amer. Statist. Asso.* **89**, 309-319.

Iwaki, K. (1997). Posterior expected marginal likelihood for testing hypotheses. *J. Economics, Asia Univ.* **21**, 105-134.

Jefferys, W. and Berger, J.O. (1992). Ockham's razor and Bayesian analysis. *American Scientist* **80**, 64-72.

Jeffreys, H. (1961). *Theory of Probability*, Oxford Univ. Press.

Kadane, J.B., Dickey, J., Winkler, R., Smith, W., and Peters, S. (1980). Interactive elicitation of opinion for a normal linear model. *J. Amer. Statist. Asso.* **75**, 845-854.

Kass, R. E. and Raftery, A. (1995). Bayes factors. *J. Amer. Statist. Asso.* **90**, 773-795.

Kass, R. E. and Vaidyanathan, S. K. (1992). Approximate Bayes factors and orthogonal parameters with application to testing equality of two binomial proportions. *J. Roy. Statist. Soc. Ser. B* **54**, 129-144.

Kass, R. E. and Wasserman, L. (1995). A reference Bayesian test for nested hypotheses and its relationship to the Schwarz criterion. *J. Amer. Statist. Asso.* **90**, 928-934.

Key, J. T., Pericchi, L. R., and Smith, A. F. M. (1999). Bayesian model choice: what and why? In *Bayesian Statistics 6* (J.M. Bernardo, A.P. Dawid, J.O. Berger and A.F.M. Smith, eds.) 343-370, Oxford Univ. Press.

Kim, S. and Sun, D. (2000). Intrinsic priors for model selection using an encompassing model. *Life Time Data Analysis* **6**, 251-269.

Laud, P.W. and Ibrahim, J. (1995). Predictive model selection. *J. Roy. Statist. Soc. B* **57**, 247-262.

Lavine, M. and Schervish, M. J. (1999). Bayes factors: what they are and what they are not. *Amer. Statist.* **53**, 119-122.

Lingham, R. and Sivaganesan, S. (1997). Testing hypotheses about the power law process under failure truncation using intrinsic Bayes factors. *Ann. Inst. Statist. Math.* **49**, 693-710.

Lingham, R. and Sivaganesan, S. (1999). Intrinsic Bayes factor approach to a test for the power law process. *J. Statist. Plann. Inference* **77**, 195-220.

Moreno, E. (1997). Bayes factors for intrinsic and fractional priors in nested models: Bayesian robustness. Technical Report, Univ. of Granada.

Moreno, E., Bertolino, F., and Racugno, W. (1998a). An intrinsic limiting procedure for model selection and hypothesis testing. *J. Amer. Statist. Asso.* **93**, 1451-1460.

Moreno, E., Bertolino, F., and Racugno, W. (1998b). Model selection and hypothesis testing. Technical Report, Univ. of Granada.

Moreno, E., Bertolino, F., and Racugno, W. (1999). Default Bayesian analysis of the Behrens-Fisher problem. *J. Statist. Plann. Inference* **81**, 323-333.

Morris, C. (1987). Comment to "Testing a point null Hypothesis: the irreconcilability of p-values and evidence." *J. Amer. Statist. Asso.* **82**, 112-139.

Nadal, N. (1999). El Análisis de Varianza basado en los Factores de Bayes Intrínsecos. Ph.D dissertation, Universidad Simón Bolívar, Venezuela.

O'Hagan, A. (1995). Fractional Bayes factors for model comparisons. *J. Roy. Statist. Soc. Ser. B* **57**, 99-138.

O'Hagan, A. (1997). Properties of intrinsic and fractional Bayes factors. *Test* **6**, 101-118.

Pauler, D. (1998). The Schwarz criterion and related methods for normal linear models. *Biometrika* **85**, 13-27.

Pérez, J. M. (1998). Development of conventional prior distributions for model comparisons. Ph.D. dissertation, Purdue Univ.

Pérez, J. M. and Berger, J. (2000). Expected posterior prior distributions for model selection. ISDS Discussion Paper 00-08, Duke Univ.

Pericchi, L. R. and Pérez, M. E. (1994). Posterior robustness with more than one sampling model. *J. Statist. Plann. Inference* **40**, 279-294.

Raftery, A., Madigan, D. and Hoeting, J. (1997). Bayesian model averaging for linear regression models. *J. Amer. Statist. Assoc.* **92**, 179-191.

Ríos Insua, D. and Ruggeri, F. (2000). *Robust Bayesian Analysis.* Springer-Verlag, New York.

Sansó, B., Pericchi, L. R., and Moreno, E. (1996). On the robustness of the intrinsic Bayes factor for nested models. In *Bayesian Robustness* (J. O. Berger, et al., eds.) **29**, 157-176, IMS Lecture Notes, Hayward, California.

Schluter, P.J., Deely, J.J. and Nicholson, A.J. (1999). The averaged Bayes factor: A new method for selecting between competing models. Technical Report, Univ. of Canterbury.

Schwarz, G. (1978). Estimating the dimension of a model. *Ann. Statist.* **6**, 461-464.

Sellke, T., Bayarri, M.J. and Berger, J. (2001). Calibration of P-values for testing precise null hypotheses. *Amer. Statist.* **55**, 62-71.

Shao, J. (1997). An asymptotic theory for linear model selection. *Statist. Sinica* **7**, 221-264.

Shibata, R. (1981). An optimal selection of regression variables. *Biometrika*, **68**, 45-54.

Shively, T. S., Kohn, R., and Wood, S. (1999). Variable selection and function estimation in additive nonparametric regression using a data-based prior (with discussion). *J. Amer. Statist. Asso.* **94**, 777-806.

Shui, C. (1996). Default Bayesian analysis of mixture models. Ph.D. dissertation, Dep. of Statistics, Purdue Univ.

Sivaganesan, S. and Lingham, R. (1998). Bayes factors for a test about the drift of Brownian motion under noninformative priors. Technical Report, Div. of Statistics, Northern Illinois Univ.

Sivaganesan, S. and Lingham, R. (1998). Bayes factors for model selection for some diffusion processes under improper priors. Technical Report, Div. of Statistics, Northern Illinois Univ.

Smith, A. F. M. (1995). Discussion of O'Hagan, A. *J. Roy. Statist. Soc. Ser. B* **57**, 99-138.

Smith, A. F. M. and Spiegelhalter, D. J. (1980). Bayes factors and choice criteria for linear models. *J. Roy. Statist. Soc. Ser. B* **42**, 213-220.

Spiegelhalter, D. J. and Smith, A. F. M. (1982). Bayes factors for linear and log-Linear models with vague prior information. *J. Roy. Statist. Soc. Ser. B* **44**, 377-387.

Stone, M. (1979). Comments on "Model Selection Criteria of Akaike and Schwarz." *J. Roy. Statist. Soc. Ser. B* **41**, 276-278.

Sun, D. and Kim, S. (1997). Intrinsic priors for testing ordered exponential means. Technical Report, Univ. of Missouri.

Suzuki, Y. (1983). On Bayesian approach to model selection. In *Proc. Internat. Statist. Inst.* 288-291, Voorburg, ISI Publications.

Varshavsky, J. (1995). On the development of intrinsic Bayes factors. Ph.D. dissertation, Purdue Univ.

Verdinelli, I. and Wasserman, L. (1995). Computing Bayes factors using a generalization of the Savage-Dickey density ratio. *J. Amer. Statist. Asso.* **90**, 614-618.

Ye, K. and Berger, J. (1991). Noninformative priors for inferences in exponential regression models. *Biometrika* **78**, 645-656.

Young, K. and Amiss, J. (1995). Comparisons of Bayes factors with non-informative priors. Technical Report, Univ. of Surrey.

Zellner, A. and Siow (1980). Posterior odds for selected regression hypotheses. In *Bayesian Statistics* (J. M. Bernardo, M. H. DeGroot, D. V. Lindley and A. F. M. Smith, eds.) 585-603, Valencia Univ. Press, Valencia.

Zellner, A. (1984). Posterior odds ratios for regression hypothesis: general considerations and some specific results. In *Basic Issues in Econometrics* 275-305, Univ. of Chicago Press.

Zellner, A. (1986). On assessing prior distributions and Bayesian regression analysis with g-prior distributions. In *Bayesian Inference and Decision Techniques: Essays in Honor of Bruno de Finetti* (P.K. Goel and A. Zellner, eds.) 233-243, North-Holland, Amsterdam.

DISCUSSION

J. K. Ghosh and Tapas Samanta

Indian Statistical Institute

A number of objective Bayesian methods of model selection and testing a sharp null hypothesis have been proposed and developed to deal with the difficulties with improper noninformative priors. Professors Berger and Pericchi have been the leading contributors to this area. Their present work is an illuminating overview of the subject as it stands now. As part of the overview they take up the important problem of evaluating these methods and give some final recommendations. We generally agree with them but feel many points need further study.

1. *Intrinsic Priors.* Berger and Pericchi are guided by the principle that a good method should produce a Bayes factor that is equal up to $o_p(1)$ to a Bayes factor with "reasonable default prior". The default Bayes factors like the intrinsic Bayes factor (IBF) and the fractional Bayes factor (FBF) seem to have this correspondence (at least asymptotically) as illustrated with a number of examples.

A potential problem is that it may not be easy to agree to a "reasonable default prior" in examples which have not been enriched by contextual discussions like that of Jeffreys. In such cases we would suggest that one can go a step further and argue as follows. If there is a prior which gives rise to a Bayes factor that is well approximated by an appealing data analytic procedure, then each of the two – the prior and the method – lends support to the other. We feel the above default Bayes factors are naturally developed, intuitively very appealing and may be considered to be good default Bayes factors in their own merits. The intrinsic priors may therefore be considered to be natural default priors as they correspond to naturally developed good default Bayes factors. In turn this argument strengthens the default Bayes factors as argued by Berger and Pericchi.

Thus these automatic methods represent one way of generating (conventional) default priors for hypothesis testing and model selection problems at least in the cases where the model dimension is small compared with sample size.

It is interesting to observe that the Cauchy prior recommended by Jeffreys for the problem considered in Illustration 1 can be obtained in this way, vide Ghosh and Samanta (2001, Section 2.5).

'J. K. Ghosh is Jawharlal Nehru Professor and Tapas Samanta is Professor, Indian Statistical Institute, 203 B.T. Road, Calcutta 700 037, India; email: tapas@isical.ac.in. J. K. Ghosh is also Professor, Department of Statistics, Purdue University, U.S.A; email: ghosh@stat.purdue.edu

The case of large model dimension may need further study, since the approximation by BIC breaks down, vide Berger et al. (1999).

A related technical question is whether an intrinsic prior is a probability measure. The authors have an interesting earlier result on this (Berger and Pericchi 1996a, Theorem 1) which is, however, based on the assumption that $\pi_1^N(\boldsymbol{\theta}_1)$ is proper. If one proceeds in the same way for two parameters or parameter vectors (ξ, η) where η is the parameter of interest, the conditional distribution $\pi_2^I(\eta|\xi)$ will be proper if

$$\int \pi_2^I(\xi, \eta) d\eta < \infty.$$

Using the argument in the proof of Theorem 1 of Berger and Pericchi (1996a), this condition is equivalent to the integrability of

$$\frac{m_1^N(x(l))}{m_2^N(x(l))} m_2^N(x(l), \xi)$$

with respect to $dx(l)$, where

$$m_2^N(x(l), \xi) = \int \pi_2^N(\xi, \eta) f_2(x(l)|\xi, \eta) d\eta.$$

This condition cannot be simplified any further. So a more general result than Theorem 1 of Berger and Pericchi (1996a) would not be easy to obtain. Hence it would be good to have a catalogue of examples where the intrinsic prior is a probability measure or close to being one in some sense. This has been verified by the authors and their students for a number of examples including normal linear models and some exponential models but other examples are worth studying.

2. *Effect of Using Minimal Training Samples.* IBF uses minimal sample for training, leaving as much of the data as possible for model comparison. Use of larger training samples is expected to correspond to Bayes factors with more peaked (intrinsic) priors. This can be easily seen in simple examples like those in Illustrations 1 and 3. In Illustration 3, for example, the intrinsic prior based on minimal training samples (of size 1) is a $N(0, 2)$ prior, whereas training samples of size 2 will lead to a $N(0, 1)$ prior. An argument as to why it is expected to be so in the general nested case is given in Ghosh and Samanta (2001, Section 3).

3. *Scale of Priors.* To fix ideas consider only a one-dimensional parameter θ. Most Bayesian tests and model selection procedures use a prior under the more complex model which has a scale comparable with the scale of variation of a single random variable X_i, namely, σ. This idea goes back to Jeffreys and is usually right. But one can think of special situations where other scales may be more appropriate. For example, if the sample size is chosen carefully at the planning stage, the alternatives like $\theta = \delta/\sqrt{n}$ may

be of special importance. A prior that does not recognize this will put too small a weight on such θ's. This point has been argued in more detail in Ghosh and Samanta (2001, Section 5.2) where it has also been suggested that the apparently dramatic difference between common classical and common Bayesian tests is really a matter of judgement on the scale in a particular problem and that there is no single scale that is right in all problems.

For simplicity, let θ be a location parameter and suppose there is some reason to believe that the prior under M_2 should have a scale comparable with $\sigma/\sqrt{c}$. In this case one may first find an intrinsic prior π with training sample size equal to one and then choose π_c as the distribution of $\theta = \theta'/\sqrt{c}$ where $\theta' \sim \pi$. The prior π_c will be the appropriate prior to use for θ.

If the intrinsic prior π is not easy to find, how can one directly modify the algorithm of IBF's to utilize the information that $\sigma/\sqrt{c}$ is the right scale for θ under M_2? Changing the parameter but keeping the training samples independent of c doesn't seem to help. At least for c growing at a sufficiently slow rate with the sample size n the following recipe should work. The parameter isn't changed but the training sample is of size $[c]$, the integral part of c. For $N(\mu, 1)$ it can be verified this works for fixed c as well as for $c = o(n)$. If $c = O(n)$, one can easily see that this method does not work.

Another different way of handling this problem is to start with a data dependent prior (Ghosh and Samanta 2001, Section 2.5) and make a suitable change there. We have not explored the consequences of doing this.

4. *Examples.* Since this is an exciting area of new research, there is a need to try out these tools in many examples and examine carefully the inference resulting from these tools. There are several examples of this sort in the paper. They are all quite interesting. However, the conclusions regarding the IBF's and FBF that are drawn from them need further study and verification. Justifying these new Bayes factors is difficult enough. To compare them or choose one from among them is even more difficult. It is true that the IBF's are closer cousins of the non-controversial conditional BF's than the FBF and so seem easier to interpret and accept. However, some of the examples provide at best ambiguous evidence about which of these BF's one should use. Some of our reservations are discussed below.

To start with, it seems unclear whether one should compare a single criterion like the FBF with all the IBF's together.

Secondly, in some examples there are aspects that are missed. For example, the fact that the GIBF is smaller than the FBF, vide Section 4.6, doesn't seem to be a strong argument against the FBF. There are several ways of seeing this. In Ghosh and Samanta (2001) we derive the FBF as a GIBF adjusted for the fact that the GIBF uses a data dependent prior that integrates to less than one. Also, if one criterion is always smaller than the other, then it is clear that none of them can be better than the other in all circumstances. It is easy to provide numerical evidence of this sort in Example 6.

Finally, in Example 6, the advantage that the AIBF has over the FBF is bought at a price. For, being exactly equal to a BF with respect to a prior may be preferable to being so only asymptotically.

We too have our own preferences, namely, the AIBF, a trimmed AIBF and the median IBF but believe it's too early to come to any definite conclusion.

Example 1 (Section 4.1) illustrates the difficulties with the class of models with "improper likelihoods" such as the mixture models for which the IBF and FBF cannot be directly employed. It refers to an unpublished work of Shui (1996) that considers modifications of the IBF and the FBF approaches to deal with the mixture models. Hopefully, the interesting work of Shui would be published soon.

5. *Teaching Nonsubjective Bayes Testing.* How should one teach nonsubjective Bayes testing in an undergraduate course? What would be the best way to communicate these ideas to the students who cannot be expected to understand all the subtleties of an IBF? At least in the classical examples ($N(\mu, 1)$ or $N(\mu, \sigma^2)$) it may be easier to motivate and use a BF based on a default or intrinsic prior but one would still have to motivate the prior. Do Berger and Pericchi have any suggestions?

Fulvio De Santis
Università di Roma, "La Sapienza"

1. Introduction.

Model selection and hypothesis testing are difficult topics. In these problems, as we depart from the usual assumption of explicitly stated underlying model, fundamental statistical principles (e.g., likelihood, sufficiency, etc.) begin to fade and we are left with no clear direction. Substantial debate over the appropriate model selection and hypothesis testing problems has taken place inside and outside the Bayesian community. Within the Bayesian approach, controversies arise on how model selection should be performed, even in the *ideal* situation where prior information is available. For example, the recent renewed interest in the development of *default* model selection methods has

[0]Fulvio De Santis is Assistant Professor, Dipartimento di Statistica, Probabilità e Statistiche Applicate , Università di Roma, "La Sapienza", P.le A. Moro, 5, 00185 - Roma, Italy; email: fulvio.desantis@uniroma1.it

witnessed many conflicts: not only is the statistical problem hard, but some statisticians make it even harder by the controversial use of improper priors! The objective Bayesian approach is of great theoretical and applied importance, not only for its connections to non-Bayesian analysis, but because it has been able to produce original, useful and sensible statistical tools. Furthermore, until a few years ago, little had been done for default model selection, compared to standard estimation. Nevertheless, in the last few years, important advances have been made towards developing useful strategies for objective model selection, as the present paper by Berger and Pericchi (B&P, from now on) demonstrates. This paper is an important piece of work for at least three reasons. First, it summarises the extensive experience and important contributions of the two authors and their collaborators. Furthermore, the paper critiques most of the methods currently available for default model selection: formal properties, behaviour in important classes of problems, compatibility with subjective Bayesian methods, difficulties in implementation and computation. Third, and most importantly, the authors provide a fair comparison of different model selection methods. This allows them to recognize limits of the IBF approach and to acknowledge merits of competitive methods.

The goal of this discussion is to outline possible strategies for selecting among default Bayes factors (DBFs) or for comparing such methods. Next sections propose a possible complement to the authors' presentation, not an alternative.

This note consists of two sections. Section 2 is devoted to two possible ways for comparing DBFs. Namely, the discussion focuses on finite-sample properties of DBFs (Section 2.1) and on the use of a frequentist pre-experimental analysis to perform a *neutral* comparison between competing methods (Section 2.2). Section 3 considers the more specific issue of the selection of the fraction(s) in the FBF approach.

2. Strategies for comparing and selecting default Bayes factors.

The rich literature on DBFs has given a lot of attention to comparative issues, and the paper of B&P reviews the main approaches to this goal. Comparative analyses of DBFs have focused, in general, on two aspects: *a*) coherence of the methods, mainly thought of as the ability of DBFs to satisfy some typical properties of *true* BFs; and *b*) asymptotic correspondence to real BFs (intrinsic prior theory). The next two subsections propose to widen the comparison of DBFs from two different points of view.

2.1. Ordinary and default Bayes factors: a finite-sample analysis of compatibility:

According to B&P (see Section 3), DBFs must be judged in the light of their correspondence to actual Bayes factors. Since, in most of the cases, correspondence cannot be established for finite sample sizes, the authors argue that such a correspondence might be established asymptotically. This consideration motivates the intrinsic prior methodology. However, in a finite sample setting, it can be of interest to evaluate how "far" a DBF is from a real BF. For a given set of data, we can evaluate the compatibility of a DBF with an ordinary BF computed with a proper prior. This is of particular interest in the presence of weak prior information. Specifically, suppose that we want to compare a fully specified model $f_1(\cdot)$ with a second model $f_2(\cdot|\theta_2)$, with unknown θ_2. Let us assume that, in the presence of partial prior information, we succeed in eliciting a class Γ of priors for θ_2, but we are not able to select any specific prior in this set. (Note that, if prior information is totally lacking, Γ is the class of all the distributions for θ_2). In this context the most natural approach is to look at the range of the standard BF over Γ but, as it is often the case, such a range might be so large not to lead to decisive evidence in favour of either one of the two models. This is a context in which DBF methodology can be of some help, even in the presence of partial prior information.

A first way to resort to a DBF, B^D_{21}, is to look at its range over Γ, rather than considering the range of the standard BF. Note that we are suggesting to compute B^D_{21}, originally proposed to be used with improper priors, using the proper priors in Γ. Often the range of B^D_{21} is more informative than the range of B_{21}, as it would be the case in the example under consideration, if Γ contained flat priors for an unbounded parameter space for θ_2. This fact was extensively discussed in De Santis and Spezzaferri (1997), among others. See Liseo (2000) for a recent discussion on robustness isssues in Bayesian model selection.

A second approach is the following: Given the class Γ of priors for θ_2, we can decide to use a DBF with a non-informative prior, π^N. One may study the compatibility of the DBF with the class Γ in order to determine if the method is *sensible.* Given a set of data $\mathbf{x}_n$ of size n, we say that B^D_{21} is Γ−compatible if there is at least one prior π^* in Γ such that B^D_{21} equals the BF computed with π^*. Hence, using either the standard BF with π^* or the DBF with π^N, we would anyhow be using a true Bayesian method. Of course, in most of the cases such a prior π^* does not exist. It is, however, interesting to establish how far we are from the class Γ, when we use B^D_{21}. As noted above, the difficulty in determining a π^* in Γ for finite n is the motivation for looking at intrinsic priors. In this case Γ is the class of all the distributions and the compatibility between B^D_{21} and Γ is established only approximatively.

To illustrate how the approach outlined above can be used for comparing alternative DBFs, let us get back to our finite sample set-up and let $d[B^D_{21}(\pi^N), B_{21}(\pi)]$ be some sort of *distance* between the DBF and the true BF for $\pi \in \Gamma$ (here we stress the dependence of BF and DBF on the respective priors). Also, let $\bar{B}^D_{21}$ be an alternative DBFs for the same problem. Then, for a given data set $\mathbf{x}_n$, we prefer B_{21}^D to $\bar{B}_{21}^D$ if

$$\inf_{\pi\in\Gamma} d[B_{21}^D(\pi^N), B_{21}(\pi)] < \inf_{\pi\in\Gamma} d[\bar{B}_{21}^D(\pi^N), B_{21}(\pi)].$$

If $\inf_{\pi\in\Gamma} d[B_{21}^D(\pi^N), B_{21}(\pi)] = 0$, B_{21}^D is Γ-compatible; otherwise it is to be preferred to the alternative DBF since it is closer to the true BF.

For example, consider the simple testing problem of Illustration 3 in the paper. In this problem, the point null hypothesis, $\theta = 0$, for a normal mean with variance equal to one (model M_1) is tested against a two-sided alternative (model M_2). Consider for the prior under the alternative the standard class of conjugate priors, Γ_{Con}, with mean zero and variance $\tau^2 \in I\!R^+$. Suppose we are interested in comparing 4 typical DBFs: fractional BF (with $b = 1/n$), expected arithmetic IBF, BIC and posterior BF (POBF, Aitkin, 1991). For example, suppose that $n = 10$ and $\sqrt{n}\bar{x} = 1.96$ (corresponding to the classical .05 p-value). It is easy to check that in this case the values of the 4 DBFs are 1.78, 1.53, 2.15 and 4.83 respectively, but the range of the standard BF in the conjugate class is $(0, 2.11)$ Therefore, with respect to the observed data, FBF and *expected* IBF are both $\Gamma_{Con}-$compatible, while BIC and POBF are not. However, being BIC "closer" to the possible values of BF in Γ_{Con}, it is preferable to POBF.

Of course, compatibility of a DBF with a class of priors is not necessarily guaranteed over the sample space and for any given sample size. In the example under consideration, it is easy to check that FBF is uniformly $\Gamma_{Con}-$compatible, regardless of n, but the remaining DBFs (*expected* IBF, BIC and POBF) are $\Gamma-$compatible only if the sample mean is less than $\sqrt{2}e^{-1/2}$, $e^{-1/2}$ and $\sqrt{2/n}e^{-1/2}$, respectively.

Let us extend this idea. In the pre-experimental set-up, it might be of some interest to look at the probability of obtaining a DBF that is $\Gamma-$compatible. In the previous simple example, let us focus on *expected* IBF and BIC. It is easy to check that, under the null hypothesis, the probabilities that such DBFs are $\Gamma_{Con}-$ compatible are $\Phi(\sqrt{2n}e^{-1/2})$ and $\Phi(\sqrt{n}e^{-1/2})$, respectively, where $\Phi(\cdot)$ is the c.d.f. of a standard normal. Therefore, in this case, *expected* IBF is uniformly more likely to be $\Gamma_{Con}-$compatible than BIC, under the null. This fact implies that, had we to choose the sample size in order to be guaranteed to have a $\Gamma_{Con}-$compatible DBF at a given probability level, less data are needed for *expected* IBF than for BIC.

This analysis is admittedly crude since the frequentist pre-experimental behaviour of DBFs should also be studied under the alternative. However, it might give an idea of how to bridge the asymptotic analysis of compatibility between DBFs and real BFs, represented by the intrinsic prior theory, to the finite-sample necessity of evaluating DBFs, in the presence of partial prior information

2.2. A neutral comparison of default Bayes factors in the presence of proper priors: a frequentist analysis:

As mentioned by B&P (Section 5.5) and also discussed in the previous section, DBFs can be used with proper priors as robust methods. It is intuitively easy to understand that gain in robustness of the prior results in loss of discriminatory power (d.p., in the following) of the model choice criteria. Of course, between two fairly robust methods, the one whose loss in d.p. is less should be preferred. Loss in d.p. of a BF can be quantified by extending ideas given in Verdinelli (1996) and Royall (1997). Following Verdinelli (1996), we say that a BF (ordinary or default) is *decisive* if it provides clear evidence in favor of M_1 or M_2. Given some data and chosen a suitable threshold $k > 1$, a BF is decisive if it is greater than k or smaller than $1/k$. However, discriminating ability of BFs must be established before the experiment is performed. Hence, the idea is to evaluate the frequentist probabilities that such criteria are decisive. Two alternative DBFs can then be compared as follows: if, in order to achieve a certain probability of being decisive, a DBF requires less data than another, the former has a greater d.p. than the latter. In the standard test of a normal mean, already considered in the previous section, assuming equal probabilities for the two hypotheses and a prior variance $\tau^2 = 1.5$, the mimimal sample sizes required to have decisive ordinary, fractional and *expected* intrinsic BFs are 11, 27 and 18 respectively when $k = 3$. Therefore, even though both FBF and *expected* IBF's d.p. are less than ordinary BF's d.p. as expected, FBF seems to be more conservative than *expected* IBF as a choice criterion. Of course, that the choice of k is crucial and calibration of thresholds for DBFs deverves investigation.

In principle, the above analysis might be extended to the comparison of DBFs defined with improper priors. However, problems arise in the computation of the frequentist probabilities since these require the use of the marginals of the data that are not defined when improper priors are used. This problem is considered in De Santis (2000).

3. On the choice of the fraction(s) for FBFs.

B&P clearly point out that in the FBF approach the choice of the fraction b is crucial. They show that in the Neyman and Scott testing problem, illustrated in Section 4.4, the basic definition of FBF must be extended in order to achieve consistency. A similar problem has been pointed out by Iwaki (1997). In the specific context of linear models, even though the same can be proved in more general set-ups, whenever the data are not exchangeable, De Santis and Spezzaferri (1999) show how the use of a unique fraction for the likelihoods in the FBF's correction factor might lead to inconsistencies. They also propose a constructive method to derive a *multiple-fractions*, consistent FBF. In my opinion, FBF as well as IBFs can simply be seen as the result of a suitable combination of partial Bayes factors: in the FBF, a geometric mean of the likelihoods for all the different training samples is performed, while, in the IBF approach, suitable averages of the entire correction terms are computed. In this way, at least in relevant problems, the selection of b, or of multiple fractions b_i's, is automatically made once partial BFs are defined. In fact, the choice of the fraction(s) is the automatic result of the likelihood-combination process. From this perspective, the computation fo the fraction(s) is, at most, as hard as it is the determination of the terms to average in the IBF approach.

A further approach is to simply regard b (let us consider now, for simplicity, the simple case of a unique fraction) as a constant in $(0,1)$, not necessarily related to the size of the training sample. It has been often noted that the choice of b has an effect on both the sensitivity to the priors and on the d.p. of the criterion (Gilks 1995, O'Hagan 1995). Conigliani and O'Hagan (2000) study the effect of the choice of b on the sensitivity of the standard FBF to both proper and improper priors. They conclude that, on the grounds of sensitivity to the prior, the choice $b = \ell_0/n$, where ℓ_0 is the minimal training sample size, is often appropriate, but not necessarily the unique. The authors point out correctly that, in addition to the effect on the sensitivity to the priors, the effect on the d.p. must be taken into account in the choice of b. This last aspect is however hard to quantify. A possible, natural way to pursue this is again based on a pre-experimental analysis of FBF. We can look at the choice of b as a design problem, and dicriminatory power of the FBF can then be quantified by the pre-experimental probability of observing decisive evidence in favour of M_1 or M_2. The idea is to determine the probability of having decisive evidence (i.e. strong discriminatory power) as a function of b to be used in order to select, before performing the experiment, the optimal fraction. As a simple example, suppose again that, under M_1, $X \sim N(0,1)$, and, under M_2, $X \sim N(\theta, 1)$. In this case, if $\pi_2^N(\theta) \propto 1$, the resulting FBF corresponds to a Bayes factor obtained using, as a prior under M_2, a $N(0,(b-1)/bn)$ density, with $b \in (0,1)$. Therefore, noting that, marginally, $\bar{x} \sim N(0,1/n)$ under M_1 while, under M_2, $\bar{x} \sim N(0,(1-b)/bn)$, computation of the probability of observing a decisive FBF, as a function of b, is straightforward. Table 1 shows the probabilities of obtaining decisive evidence with FBF, when $k = 3$ and equal prior probabilities for the two hypotheses are assumed, for some values of n and b.

Table 1. *Probabilities of Decisive FBF*

n	10	20	30	50	100	200	500
$b = 1/n$	0.429	0.712	0.787	0.849	0.904	0.937	0.964
$b = 2/n$	0.179	0.429	0.634	0.757	0.849	0.904	0.945

It is clear that reduction in discriminatory power of FBF depends strongly on the sample size: it is substantial for small sample sizes ($n = 10, 20$) but less and less influent as the sample size increases.

Beyond the above stardard oversimplistic example, such an analysis might be the starting point to develop an objective quantitative measure of discriminatory power of the FBF, as a function of b. This measure could be combined with measures of sensitivity of the FBF to the prior, such the ones proposed in Conigliani and O'Hagan (2000), in a unifying tool to be used to choose b.

Two final comments are in order. First, in principle the above analysis can be also performed in the presence of *multiple fractions* FBF. Secondly, and more importantly, as noted above computation of the probabilities to be used to set the fraction(s) requires the knowledge of the marginal distributions of the data under the two models, and this is, in general, much more complicated than it is in this problem. The use of fractional priors might be, at least in some cases, of help (De Santis, 2000).

ADDITIONAL REFERENCES

Conigliani, C. and O'Hagan, A. (2000). Sensitivity measures of the fractional Bayes factor to prior distributions. *Canad. J. Statist.* **28**.

De Santis, F. (2000). Statistical evidence and sample size for robust and default Bayes testing. Technical Report, Univ. of Rome, "La Sapienza".

Gilks, W.R. (1995). Discussion of O'Hagan. *J. Roy. Statist. Soc. Ser. B* 57 118-120.

Liseo, B. (2000). Robustness issues in Bayesian model selection. In: *Robust Bayesian Analysis.* Lectures Notes in Statistics, 152, 197-222. Springer-Verlag.

Royall, M.R. (1997). *Statistical Evidence: A Likelihood Paradigm.* Chapman and Hall, London.

Verdinelli, I. (1996). Bayesian designs of experiments for the linear model. PhD dissertation, Dept. of Statistics, Carnegie Mellon Univ.

REJOINDER

J. O. Berger and L. R. Pericchi

We thank the discussants for their very interesting comments and viewpoints. We respond to each in turn, using the numbering scheme of the discussants. If we do not mention a section of a discussion, it is because we appreciate and agree with the points mentioned therein.

Reply to Professor De Santis:

2.1. De Santis rightly observes that it would be nice to have formal ways of evaluating objective Bayes factors in small sample settings. To date, our efforts in this direction have been limited to determining the intrinsic prior (based on asymptotics) and then investigating, for small samples, the extent to which the objective Bayes factor is close to the Bayes factor from the intrinsic prior. If the two Bayes factors are close, one can rest easy. But if they differ, one is not sure what to conclude.

De Santis investigates this issue within the context of partial prior information, as specified by a class Γ of prior distributions. He considers several possible ways of measuring compatibility of objective Bayes factors with the information in Γ.

A variant of this idea that is in tune with our strategy for development of objective Bayes factors is to directly utilize the partial prior information in the construction of the Bayes factor. One natural approach is to first calculate the reference prior, subject to the restriction of being in the class Γ (cf. Sun and Berger 1998). If the reference prior is proper, it can be immediately used to calculate the BF. Otherwise, one could use the constrained reference prior to compute a default Bayes factor (e.g., an IBF). Of course, if the prior is unconstrained, this reduces to the ordinary definition of an objective Bayes factor.

2.2. As argued in the chapter, our recommendation for evaluating an objective Bayes factor method is simply: discover which prior is effectively being used when applying the method, and informally judge whether or not this prior is reasonable. Our experience is that, by looking at this intrinsic prior, one can gain a great deal of insight into possible biases or inadequacies of the corresponding objective Bayes factor method.

In contrast, comparison of operating characteristics of objective Bayes factors rarely seems to yield clear insights. The problem is that the results are, of necessity, highly dependent on the particular operating characteristic that one considers. And even then, one objective Bayes factor will rarely uniformly dominate another. After all, virtually any Bayesian procedure in testing is formally admissible from a frequentist perspective, meaning that uniform domination cannot be attained.

Consider, for instance, the use of 'discriminatory power' as discussed by De Santis. For the particular situation he considered, the IBF happened to have higher discriminatory power than the FBF. But both the IBF and the FBF in this example have proper intrinsic priors, so that presumably a different choice of the 'design' prior for θ (the prior under which discriminatory power is computed) or a different choice of k (or allowing different k for choice of different models) could easily reverse this finding. Another issue here is that we feel the 'design' prior must be fixed in carrying out pre-experimental comparisons and, in particular, should not equal the intrinsic prior for a procedure (as is apparently done in section 3 of the discussion). In other words, one must fix the pre-experimental 'truth' and then judge various procedures against this truth, rather than allowing the truth to shift with the procedure.

Of course, part of the message of De Santis is that one should formally consider experimental design, with the goal of ensuring that the discriminatory power of the procedure to be used is adequate, and we completely agree. It is just that we do not feel that generic comparison of objective Bayes factors can easily be carried out in this way.

3. De Santis reaffirms the need for modifying the FBF to allow for multiple fractions. Indeed, De Santis and Spezzaferri (1999) propose a quite compelling method for determining the multiple fractions. We have two comments. First, no method can overcome the difficulties of FBFs in irregular models, such as our Example 2. Second, when observations can be dependent, it is something of a misnomer to call the method an FBF method, since it produces a prior that cannot then be written in terms of fractions of multiplicative parts of the full likelihood. Indeed, their method is more closely related to what is known as use of the *empirical expected posterior prior* (use of (5.2) with m^* chosen to be the empirical distribution of minimal training samples); (5.2) can then be viewed as the arithmetic average of minimal training sample posteriors, while the approach of De Santis and Spezzaferri leads to a geometric average of these training sample posteriors. These connections are all very interesting and affirm the basic point made by De Santis that all these objective Bayes factors are based on much the same principle.

Reply to Professors Ghosh and Samanta:

1. We agree with Ghosh and Samanta that the direct intuitive appeal of certain of the objective Bayes factors can actually lend support to use of the corresponding intrinsic prior. It is indeed useful to think of the justification as a two-way street. We also agree that, in situations in which the number of parameters is allowed to grow with the sample size, the existing theory of Intrinsic Priors need not apply (although it can sometimes be directly modified in an appropriate fashion, as was done in our Example 4).

Ghosh and Samanta raise the interesting issue of propriety of the conditional Intrinsic Distribution $\pi_2^I(\eta|\psi)$. We actually seek much more than just propriety of this distribution; we also want the (typically improper) marginal intrinsic priors for the nuisance parameters under the two models to be properly 'calibrated.' Dass (2000) shows that this can be done in problems with a suitable group structure, as long as the initial noninformative priors that are used to derive the IBF are the right Haar priors.

2. We agree with the observation that increasing the size of the training sample will imply more peaked Intrinsic Priors. This is of particular relevance because of the next comments of the discussants.

3. The discussion of the 'Scale of the Priors' is fascinating. It may, indeed, frequently be the case that, in well-designed experiments, the pre-experimentally chosen sample size, n_0, is such that 'local alternatives' like $\theta = \delta/\sqrt{n_0}$ are those that are apriori viewed to be likely, and objective Bayes factors would need to adjust to this scale. The various technical mechanisms discussed by Ghosh and Samanta for achieving this adjustment (such as increasing the training sample size in IBFs) are quite clever.

There remains, however, the outstanding practical issue of determining when a scale adjustment is necessary. One possibility - seemingly that envisaged by Ghosh and Samanta - is to subjectively elicit the appropriate scale, and then embed this scale in a suitable default procedure. This is entirely reasonable, but does require some subjective thinking.

One might, of course, begin by computing the answers arising from both a 'local alternative' scale and the usual scale; it is only if these yield contrasting conclusions that one would need to make a subjective decision as to which scale is most appropriate. By the 'local alternative' scale here we effectively mean that which would arise as a lower bound from a robust Bayesian analysis with respect to a reasonable class of priors. Unfortunately, it is not clear how one can automatically find this scale through use of modifications of IBF type procedures.

This discussion is also related to the idea of 'local' vs 'global' alternatives in Smith and Spiegelhalter (1980). Also of interest, from that paper, is the observation that use of local alternatives can lead to AIC type approximations to Bayes factors; this is related to the observation of Ghosh and Samanta that local scales can bring Bayesian and frequentist answers closer together. While this is true asymptotically, it should be pointed out that, for moderate sample sizes, a significant discrepancy remains between, say, p-values and Bayes factors (at any scale).

4. *Examples:* As pointed out by the discussants, one always has to consider comparison-by-example with caution. The main point of the examples in the chapter was to indicate the types of things that could go wrong with default procedures (so that one could be properly cautious in their use) rather than to try to 'prove' which procedures were better. Our surprise that the IBF seemed to automatically overcome all obstacles probably led us to emphasize the comparison aspect a bit too much.

Concerning the FBF, we should mention that we have always thought of the FBF as a multitude of procedures, especially when multiple fractions are allowed. Thus we introduced multiple factions in examples (such as Example 4) where it seemed to be necessary.

Our purpose in Example 6 *When Neither Model is True* was apparently not stated very clearly. We certainly did not mean to suggest that the GIBF is better than the FBF because it is smaller. In the example we were, instead, reacting to the comment in O'Hagan (1997) to the effect that the appearance of the sample variance in the GIBF leads to *"...not intuitively reasonable behaviour"*. We were simply suggesting that the situation is far from clear, and that the appearance of the sample variance in the GIBF can be motivated through consideration of robustness to the assumption $\sigma^2 = 1$. The effect, on the FBF, of violation of this assumption is quite serious, while the GIBF seems to compensate rather well to its violation. We also wanted to present the example to point out that IBFs may well have advantages over, say, the use of the corresponding intrinsic prior, when it is suspected that none of the models under consideration may be true.

5. *Teaching Non-Subjective Bayes Testing.* We have certainly been thinking about possible ways to teach this material. Part of the problem is that one might choose to emphasize quite different methods for different audiences. In a Bayesian course emphasizing MCMC, it would be natural to focus on objective testing and model comparison based on use of expected posterior priors (often equivalent to intrinsic priors from AIBFs), since they can usually be directly incorporated into MCMC schemes. For a low level undergraduate service course, one might settle for simply teaching students to calibrate p-values via the $BF = -ep\log(p)$ formula of Sellke, Bayarri and Berger (2001). At a higher level undergraduate course, one might emphasize the idea of training samples and present the median IBF as a general purpose testing and model selection tool. There is surely also a role for approximations, such as BIC and its possible generalizations or reformulations.

ADDITIONAL REFERENCES

Dass, S.C. (2000). Propriety of intrinsic priors in invariant testing situations. Technical Report, Michigan State Univ.

Sun, D. and Berger, J. (1998). Reference priors with partial information. *Biometrika* **85**, 55-71.

Model Selection
IMS Lecture Notes - Monograph Series (2001) Volume 38

Scales of Evidence for Model Selection: Fisher versus Jeffreys

Bradley Efron and Alan Gous

Stanford University and Cariden Technologies, Inc.

Abstract

Model selection refers to a data-based choice among competing statistical models, for example choosing between a linear or a quadratic regression function. The most popular model selection techniques are based on interpretations of p-values, using a scale originally suggested by Fisher: .05 is moderate evidence against the smaller model, .01 is strong evidence, etc. Recent Bayesian literature, going back to work by Jeffreys, suggests a quite different answer to the model selection problem. Jeffreys provided an interpretive scale for Bayes factors, a scale which can be implemented in practice by use of the BIC (Bayesian Information Criterion.) The Jeffreys scale often produces much more conservative results, especially in large samples, so for instance a .01 p-value may correspond to barely any evidence at all against the smaller model. This paper tries to reconcile the two theories by giving an interpretation of Fisher's scale in terms of Bayes factors. A general interpretation is given which works fine when checked for the one-dimensional Gaussian problem, where standard hypothesis testing is seen to coincide with a Bayesian analysis that assumes stronger (more informative) priors than those used by the BIC. This argument fails in higher dimensions, where Fisher's scale must be made more conservative in order to get a proper Bayes justification.

Bradley Efron is Max H. Stein Professor of Statistics, Department of Statistics, Sequia Hall, Stanford University, Paolo Alto, CA 94305-4065, U.S.A; email: brad@stat.Stanford.EDU. Alan Gous is Director of Research & Development, Cariden Technologies, Inc., 1733 Woodside Road, Suite 310, Redwood City, CA 94061, U.S.A.; email: alan@cariden.com.

Contents

1 Introduction

In model selection problems the statistician must use the data to choose between discrete alternative models, for instance which explanatory variables to include in a regression analysis. A particularly simple example involves a single Gaussian observation

$$x \sim N(\theta, 1). \tag{1.1}$$

and the choice between two models, a smaller one inside a bigger one,

$$\mathcal{M}_0 : \theta = 0 \quad \text{versus} \quad \mathcal{M} : \theta \neq 0. \tag{1.2}$$

This example shows up frequently in the model selection literature, and we will use it here for comparison of different methods.

By far the most widely-used model selection methods are based on hypothesis tests. A test statistic $S(\mathbf{x})$ depending on the observed data $\mathbf{x}$ is evaluated, $S(\mathbf{x}) = s$, and the *critical level* α is calculated,

$$\alpha = \text{prob}_{\mathcal{M}_0}\{S(\mathbf{x}) < s\} \tag{1.3}$$

(so $1 - \alpha$ equals the p-value or significance level). Here we suppose that larger values of S indicate stronger evidence against the smaller hypothesis $\mathcal{M}_0$, as with the optimal test statistic $S(\mathbf{x}) = |x|$ for situation (1.1)-(1.2). This leaves us with the problem of evaluating the strength of evidence for $\mathcal{M}$ and against $\mathcal{M}_0$. Frequentists use a scale of evidence set down by Fisher in the 1920's. Table 1 gives Fisher's scale as it is commonly interpreted: $\alpha = .99$ is strong evidence in favor of $\mathcal{M}$ versus $\mathcal{M}_0$, .95 is moderate evidence, etc. The borderline of neutral evidence is somewhere around $\alpha = .90$. Fisher, discussing chi-square tests, states it this way, in terms of $P = 1 - \alpha$: "If P is between .1 and .9 there is certainly no reason to suspect the hypothesis tested. If it is below .02 it

critical level α:	**.90**	**.95**	**.975**	**.99**	**.995**	**.999**
strength of evidence for $\mathcal{M}$:	borderline	moderate	substantial	strong	very strong	overwhelming

Table 1. Fisher's scale of evidence against hypothesis $\mathcal{M}_0$. The critical level (one minus the p-value) is the $\mathcal{M}_0$ probability of the test statistic being smaller than the value actually observed.

is strongly indicated that the hypothesis fails to account for the whole of the facts. We shall not often be astray if we draw a conventional line at .05, and consider that higher values of χ^2 indicate a real discrepancy." Section 20, Fisher (1954).

Recent Bayesian literature provides a different answer to the model selection problem. Attention is focussed on the *Bayes Factor* $B(\mathbf{x})$,

$$\frac{\text{prob}\{\mathcal{M}|\mathbf{x}\}}{\text{prob}\{\mathcal{M}_0|\mathbf{x}\}} = B(\mathbf{x})\frac{\text{prob}\{\mathcal{M}\}}{\text{prob}\{\mathcal{M}_0\}}, \tag{1.4}$$

the ratio between the aposteriori and apriori odds in favor of the bigger model $\mathcal{M}$, defined more carefully in Section 2. Jeffreys (1961) suggested a scale of evidence for interpreting Bayes factors, presented in slightly amended form in Table 2. For example $B(\mathbf{x}) = 10$ indicates positive but not strong evidence in favor of $\mathcal{M}$.

The calculation of $B(\mathbf{x})$ requires a full Bayesian prescription of the prior distribution on $\mathcal{M}$ and $\mathcal{M}_0$, as described in Section 2.1. This is not available in most situations, but Jeffreys suggested rules for objectively selecting priors in the absence of apriori knowledge, see Kass and Wasserman (1996). These rules have led to convenient data-based estimates of the Bayes factor $B(\mathbf{x})$, the most popular of which is the BIC (Bayesian Information Criterion), Schwarz (1978). Let $\widehat{B}(\mathbf{x})$ be Wilks' maximum likelihood ratio statistic, the ratio of the maximized likelihoods for $\mathcal{M}$ compared to $\mathcal{M}_0$. In a repeated sampling situation, where the data $\mathbf{x}$ comprise a random sample of size n, the BIC approximates the Bayes factor by $B_{\text{BIC}}(\mathbf{x})$,

$$\log\, B_{\text{BIC}}(\mathbf{x}) = \log\, \widehat{B}(\mathbf{x}) - \frac{d}{2}\, \log(n), \tag{1.5}$$

where d is the difference in the number of free parameters between $\mathcal{M}$ and $\mathcal{M}_0$. The rationale for this formula is discussed in Section 2.

As an example, suppose $\mathbf{x}$ is an i.i.d. (independent and identically distributed) Gaussian sample of size n,

$$x_1, x_2, \ldots x_n \overset{iid}{\sim} N(\theta_1, 1). \tag{1.6}$$

Bayes Factor $B(\mathbf{x})$:	$< \mathbf{1}$	**1 — 3**	**3 — 20**	**20 – 150**	$> \mathbf{150}$
Evidence for $\mathcal{M}$:	negative	not worth more than a bare mention	positive	strong	very strong

Table 2. Jeffreys' scale of evidence for the interpretation of Bayes factors, as amended by Kass and Raftery (1995).

and we wish to select between

$$\mathcal{M}_0 : \theta_1 = 0 \quad \text{versus} \quad \mathcal{M} : \theta_1 \neq 0. \tag{1.7}$$

This problem reduces to (1.1), (1.2) by defining

$$x = \sqrt{n}\,\bar{x} = \sum_1^n x_i/\sqrt{n} \quad \text{and} \quad \theta = \sqrt{n}\,\theta_1, \tag{1.8}$$

since x is a sufficient statistic. In this case $d = 1$, $\widehat{B}(\mathbf{x}) = x^2/2$, and

$$\log B_{\text{BIC}}(\mathbf{x}) = \frac{x^2}{2} - \frac{\log(n)}{2}. \tag{1.9}$$

We can now make a disturbing comparison, of a type first emphasized in Lindley's 1957 paper: Jeffreys' scale of evidence, as implemented by the BIC, often leads to much more conservative decisions than Fisher's scale. A data set with p-value .01 (critical level .99) may be "barely worth mentioning" on Jeffreys' scale, a shocking assertion to the medical scientist for whom a .01 significance level settles the issue. Figure 1 illustrates the comparison for situation (1.6)-(1.7).

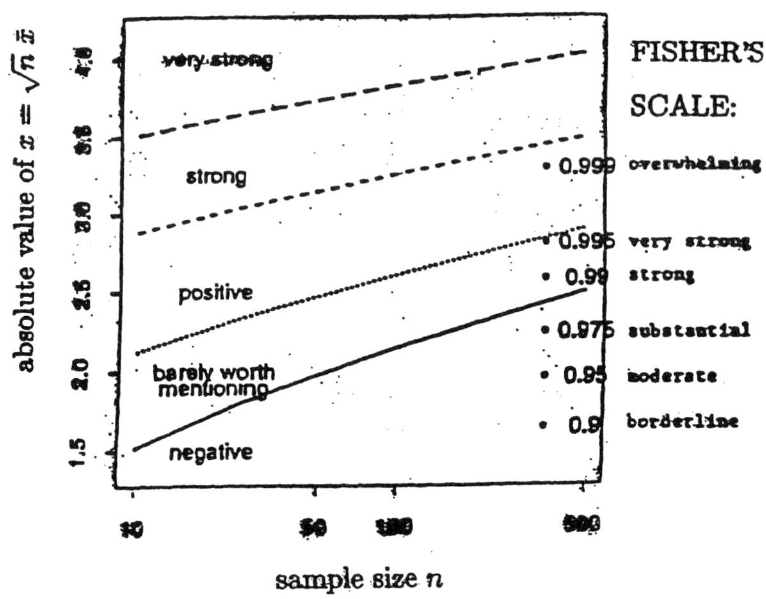

Figure 1 Jeffreys' scale of evidence for $\mathcal{M}$ as implemented by BIC; for Gaussian samples as in (1.6)-(1.9); Fisher's scale, shown at right, does not depend on n. Jeffreys' scale is much more conservative for large sample sizes. For example, $x = 2.58$ with $n = 100$ is strong Fisherian evidence, but barely worth mentioning for Jeffreys.

How can the two scales of evidence give such different results? We will try to answer the question by showing that frequentist model selection can be understood from a Bayesian point of view, but one where the prior distribution favors the bigger model more strongly than with BIC or other objectivist Bayes techniques. The reasons for this are partly historical, perhaps relating to Fisher's and Jeffreys' different scientific environments and different attitudes concerning the "null hypothesis" $\mathcal{M}_0$. More of the difference though arises from the Bayesian notion of sample size coherency, discussed in Section 4, which implies that evidence against the smaller model must be judged more cautiously in larger samples, as seen in Figure 1.

Section 2 reviews Bayes factors and develops a useful approximation formula for their comparison,

$$B(\mathbf{x}) \doteq \widehat{B}(\mathbf{x})/\widehat{B}(\mathbf{y}), \tag{1.10}$$

where $\mathbf{y}$ is a "breakeven point", that is a data set for which the Bayes factor $B(\mathbf{y})$ equals 1. Fisher's scale is interpreted as putting the breakeven point at the 90th percentile of the test statistic, e.g., at $y = 1.645$ in situation (1.1), (1.2), while the BIC puts it at $\sqrt{\log(n)}$, (1.9).

There is a healthy literature on the comparison of Bayesian and frequentist model selection. Some of this concerns the interpretation of a p-value as $\text{prob}\{\mathcal{M}_0|\mathbf{x}\}$. Berger and Selke (1987) show that this interpretation must be wrong for two-sided testing situations like (1.1)-(1.2), with the p-value underestimating $\text{prob}\{\mathcal{M}_0|\mathbf{x}\}$ as in Lindley's paradox, while Casella and Berger (1987) show that it is reasonable for one-sided problems. Berger, Boukai, and Wang (1997) reconcile p-values with Bayesian posterior probabilities using a conditional inference approach.

Andrews' (1994) results are closer to the point of view taken here. He shows that in a certain asymptotic sense there is a monotone relationship between p-values and Bayes factors. Section 3 gives some comparisons with Andrews' results. Section 5 concerns a small but illustrative example of model selection, combining the Bayesian and frequentist points of view. The paper ends with a series of remarks in section 6, and a brief summary.

Our main goal here is to reconcile Fisherian hypothesis testing with Jeffreys' Bayes-factor theory, at least as far as reconciliation is possible, and then to pinpoint the nature of the remaining differences. Technical issues will be kept to a minimum, with simple examples used to make many of the main points.

2 Approximations for Bayes Factors

The BIC formula (1.5) provides an objective approximation to the Bayes factor $B(\mathbf{x})$, objective in the sense that it is entirely data-based and does not involve subjective appraisals of prior probabilities. This section discusses a class of such approximation formulas based on the convenient representation (1.10). Our comparison of Fisher's and Jeffreys' scales of evidence will take place within this class.

2.1 Bayes Factors and the BIC

Suppose that we observe data $\mathbf{x}$ distributed according to a parametric family of densities $f_\theta(\mathbf{x})$,

$$\mathbf{x} \sim f_\theta(\mathbf{x}), \tag{2.1}$$

θ being an unknown parameter vector. We wish to choose between a smaller model and a bigger model for θ,

$$\theta \in \mathcal{M}_o \quad \text{versus} \quad \theta \in \mathcal{M}, \quad \text{with} \quad \mathcal{M}_o \subset \mathcal{M}. \tag{2.2}$$

A complete Bayesian analysis begins with prior probabilities for $\mathcal{M}_o$ and $\mathcal{M}$,

$$\pi_o = \text{prob}\{\theta \in \mathcal{M}_o\} \quad \text{and} \quad \pi = \text{prob}\{\theta \in \mathcal{M}\}, \tag{2.3}$$

$\pi_o + \pi = 1$, and also with prior conditional densities for θ given each model,

$$g_o(\theta) \quad \text{and} \quad g(\theta) \tag{2.4}$$

for the densities of θ given $\mathcal{M}_o$ and $\mathcal{M}$ respectively. We assume that $g(\theta)$ puts probability zero on $\mathcal{M}_o$ so that it is not necessary to subtract $\mathcal{M}_o$ from $\mathcal{M}$ as was done in (1.2). Sometimes it is slightly more convenient to take $\mathcal{M}_o \subset \mathcal{M}$ as in (2.2).

Letting $\pi(\mathbf{x})$ and $\pi_o(\mathbf{x})$ be the aposteriori probabilities for $\mathcal{M}$ and $\mathcal{M}_o$ having observed $\mathbf{x}$, Bayes' rule gives

$$\frac{\pi(\mathbf{x})}{\pi_o(\mathbf{x})} = \frac{\pi}{\pi_0} \cdot \frac{f(\mathbf{x})}{f_0(\mathbf{x})}, \tag{2.5}$$

where $f(\mathbf{x})$ and $f_o(\mathbf{x})$ are the two marginal densities

$$f(\mathbf{x}) = \int_{\mathcal{M}} f_\theta(\mathbf{x}) g(\theta) d\theta \quad \text{and} \quad f_o(\mathbf{x}) = \int_{\mathcal{M}_o} f_\theta(\mathbf{x}) g_o(\theta) d\theta. \tag{2.6}$$

The ratio

$$B(\mathbf{x}) = \frac{f(\mathbf{x})}{f_o(\mathbf{x})} \tag{2.7}$$

is called the *Bayes factor* in favor of $\mathcal{M}$ compared to $\mathcal{M}_o$. Kass and Raftery (1995) provide a nice overview of Bayes factors, including their origins in the work of Jeffreys, Good, and others.

Bayes' rule (2.5) relates the posterior and prior odds ratio through the Bayes factor,

$$\frac{\pi(\mathbf{x})}{\pi_o(\mathbf{x})} = \frac{\pi}{\pi_o} \cdot B(\mathbf{x}). \tag{2.8}$$

The Bayesian model selection literature tends to focus on the case $\pi/\pi_o = 1$ of equal prior odds, with the presumption that π/π_o will be appropriately adjusted in specific applications. This paper proceeds in the same spirit, except for some of the specific frequentist/Bayesian comparisons of Section 5.

In most situations there will be no obvious choice for the prior densities $g_o(\theta)$ and $g(\theta)$ in (2.4). Jeffreys (1935, 1961) suggested objective rules for choosing priors in model selection problems. This has led to a substantial current literature, Kass and Raftery's (1995) bibliography listing more than 150 papers. The BIC approximation (1.5) can be thought of as a Bayes factor based on Jeffreys' objective priors, but we will see that other approximations, closer in effect to Fisher's scale of evidence can be described in the same way.

A clear derivation of the BIC appears in Section 2 of Kass and Wasserman (1995). They follow Smith and Spiegelhalter's (1980) approach in which Bayesian objectivity is interpreted to mean a prior distribution having the same amount of information for θ as one observation out of a random sample of size n. As an example, consider the repeated-sampling Gaussian problem (1.6)-(1.7), with $x_1, x_2, \dots x_n \overset{\text{i.i.d.}}{\sim} N(\theta_1, 1)$. If we use a Gaussian prior $g(\theta_1) = N(0, A)$ then Smith and Spiegelhalter's approach suggests taking $A = 1$. The exact Bayes factor is easily calculated,

$$\log B(\mathbf{x}) = \frac{n}{n+1} \frac{x^2}{2} - \frac{\log(n+1)}{2}, \quad [x = \sqrt{n}\,\bar{x}] \tag{2.9}$$

which approaches $\log B_{\text{BIC}}(\mathbf{x}) = [x^2 - \log(n)]/2$, (1.9), as n grows large.

There seems to be an obvious objective Bayesian prior density for this situation: $g(\theta_1) = c$, a constant. This gives a Bayes factor $\int cf_{\theta_1}(\mathbf{x})d\theta_1/f_o(\mathbf{x})$ of the form

$$\log B(\mathbf{x}) = \frac{x^2}{2} - \frac{1}{2}\log(\frac{n}{2\pi c^2}). \tag{2.10}$$

However because we are dealing with an improper density function we still have to select a value for the constant c. The BIC chooses $c = 1/\sqrt{2\pi}$. The breakeven point for (2.10), where $B(\mathbf{x}) = 1$, occurs at $x = [\log(n/2\pi c^2)]^{1/2}$, so the BIC choice amounts to setting the breakeven point at $[\log(n)]^{1/2}$.

Suppose we believe entirely in model $\mathcal{M}$ in (1.6)-(1.7), and wish only to estimate θ. Then the constant prior density $g(\theta) = c$ gives answers agreeing with the usual frequentist confidence intervals, and the Bayesian objectivist does not have to worry about the choice of c. Model selection is inherently more awkward and difficult than estimation, part of the trouble coming from the different dimensionalities of $\mathcal{M}$ and $\mathcal{M}_o$. A prior distribution for situation (2.2) is inherently bumpy around the smaller model $\mathcal{M}_o$, rather than smooth as in estimation problems, making the Bayesian analysis more delicate. See Remark D in section 6.

2.2 A Useful Approximation Lemma

Our comparison of frequentist and Bayesian model selection relies on formula (1.10), $B(\mathbf{x}) \doteq \widehat{B}(\mathbf{x})/\widehat{B}(\mathbf{y})$, $\mathbf{y}$ a breakeven point, which will now be derived in somewhat more general form. We will work in an exponential family setting where $\mathcal{M}$ is a multiparameter exponential family of densities $f_\theta(\mathbf{x})$, and $\mathcal{M}_o \subset \mathcal{M}$ is also an exponential family. For example, $\mathcal{M}$ might be a logistic regression model with m predictors while $\mathcal{M}_0$, is a submodel in which $m_o < m$ of the predictors appear.

Besides the observed data point $\mathbf{x}$ consider a second point $\mathbf{x}^\dagger$, which for now can be any point in the domain of the exponential family. Definition (2.7) gives

$$B(\mathbf{x}) = \frac{B(\mathbf{x})}{B(\mathbf{x}^\dagger)}B(\mathbf{x}^\dagger) = \frac{R}{R_o}B(\mathbf{x}^\dagger),$$

where R and R_o are ratios of marginal densities (2.6),

$$R = \frac{f(\mathbf{x})}{f(\mathbf{x}^\dagger)} \quad \text{and} \quad R_o = \frac{f_o(\mathbf{x})}{f_o(\mathbf{x}^\dagger)}. \tag{2.11}$$

Section (1.2) of O'Hagen (1995) reviews some of the history of equation (2.10).

In what follows we will show that $R/R_o = B(\mathbf{x})/B(\mathbf{x}^\dagger)$ is well-approximated by the ratio of likelihood ratio statistics,

$$B(\mathbf{x})/B(\mathbf{x}^\dagger) \doteq \widehat{B}(\mathbf{x})/\widehat{B}(\mathbf{x}^\dagger), \tag{2.12}$$

where

$$\widehat{B}(\mathbf{x}) = \frac{f_{\widehat{\theta}}(\mathbf{x})}{f_{\widehat{\theta}_o}(\mathbf{x})} \quad \text{and} \quad \widehat{B}(\mathbf{x}^\dagger) = \frac{f_{\widehat{\theta}^\dagger}(\mathbf{x}^\dagger)}{f_{\widehat{\theta}_o^\dagger}(\mathbf{x}^\dagger)}. \tag{2.13}$$

Here $(\widehat{\theta}, \widehat{\theta}_o)$ are the MLEs under $\mathcal{M}$ and $\mathcal{M}_o$ for data point $\mathbf{x}$, and similarly $(\widehat{\theta}^\dagger, \widehat{\theta}_o^\dagger)$ for $\mathbf{x}^\dagger$. Notice that R involves only densities from $\mathcal{M}$, and R_o involves only densities from $\mathcal{M}_o$, which makes the approximation theory easy: we need never compare densities from spaces of different dimensions, which is particularly helpful in working with improper priors such as $g(\theta) = \text{constant}$. This tactic is an example of Good's (1947, 1951) "device of imaginary results" used, differently, by Smith and Spiegelhalter (1980, 1982), Kass and Wasserman (1995) and Pettit (1992).

Information matrices play a central role in the derivation of (2.12). The observed and expected Fisher information matrices for θ are defined in terms of the log likelihood second derivative matrix $\ddot{\ell}_{\mathbf{x}}(\theta) = \frac{\partial^2}{\partial\theta^2} \log f_\theta(\mathbf{x})$,

$$i_{\mathbf{x}}(\theta) = -\ddot{\ell}_{\mathbf{x}}(\theta) \quad \text{and} \quad I(\theta) = E_\theta\{i_{\mathbf{x}}(\theta)\}. \tag{2.14}$$

In exponential families $i_{\mathbf{x}}(\theta) = I(\theta)$, so $i_{\mathbf{x}}(\theta)$ does not depend upon $\mathbf{x}$. In repeated sampling situations such as (1.6), $I(\theta)$ is proportional to the sample size n. The order of magnitude error rates that follow refer to n, but they are valid beyond the repeated sampling framework. In a logistic regression situation, for instance, we could take n to be trace $\{I(\theta)\}$ and the bounds would remain valid under mild regularity conditions on the choice of the covariate vectors. In practice "sample size" is a difficult concept to define and causes trouble in applications of the BIC, see Section 5 here and Section 5 of Kass and Wasserman (1995).

For estimation problems (but not for model selection) Jeffreys suggested using the invariant prior density, usually improper,

$$g(\theta) = c \cdot |I(\theta)|^{\frac{1}{2}}, \tag{2.15}$$

c being an arbitrary positive constant, Kass and Wasserman (1996). This is particularly convenient for deriving (1.10) as the following lemma shows.

Lemma Suppose $\mathcal{M}_o \subset \mathcal{M}$ are exponential families and that the prior densities (2.4) are

$$g_o(\theta) = c_o \cdot |I_o(\theta_o)|^{\frac{1}{2}} \quad \text{and} \quad g(\theta) = c \cdot |I(\theta)|^{\frac{1}{2}}, \tag{2.16}$$

$I_o(\theta_o)$ being the $\mathcal{M}_o$ Fisher information matrix at point θ_o in $\mathcal{M}_o$. Then the ratio $R/R_o = B(\mathbf{x})/B(\mathbf{x}^\dagger)$ (2.11) is approximated to second order by $\widehat{B}(\mathbf{x})/\widehat{B}(\mathbf{x}^\dagger)$, (2.13),

$$\frac{B(\mathbf{x})}{B(\mathbf{x}^\dagger)} = \frac{\widehat{B}(\mathbf{x})}{\widehat{B}(\mathbf{x}^\dagger)}[1 + O(n^{-1})]. \tag{2.17}$$

Proof For any smooth prior density $g(\theta)$, Laplace's method, Tierney, Kass and Kadane (1989), gives

$$f(\mathbf{x}) = (2\pi)^{m/2} f_{\widehat{\theta}}(\mathbf{x}) g(\widehat{\theta})\, |I(\widehat{\theta})|^{-\frac{1}{2}}(1 + O(n^{-1})), \tag{2.18}$$

$m = \dim(\mathcal{M})$, and similarly for the other combinations of parameter values and data points appearing in (2.13). Therefore

$$R = \frac{f_{\widehat{\theta}}(\mathbf{x}) g(\widehat{\theta})\, |I(\widehat{\theta})|^{-\frac{1}{2}}}{f_{\widehat{\theta}^\dagger}(\mathbf{x}^\dagger) g(\widehat{\theta}^\dagger)\, |I(\widehat{\theta}^\dagger)|^{\frac{1}{2}}}\,[1+O(n^{-1})] \quad \text{and} \quad R_o = \frac{f_{\widehat{\theta}_o}(\mathbf{x}) g_o(\widehat{\theta}_o)\, |I_o(\widehat{\theta}_o)|^{-\frac{1}{2}}}{f_{\widehat{\theta}_o^\dagger}(\mathbf{x}^\dagger)\, g_o(\widehat{\theta}_o^\dagger)\, |I_o(\widehat{\theta}_o)\,|^{-\frac{1}{2}}}\,[1+O(n^{-1})]. \tag{2.19}$$

Jeffreys' priors (2.16) result in $R \doteq f_{\widehat{\theta}}(\mathbf{x})/f_{\widehat{\theta}^\dagger}(\mathbf{x}^\dagger)$ and $R_o \doteq f_{\widehat{\theta}_o}(\mathbf{x})/f_{\widehat{\theta}_o^\dagger}(\mathbf{x}^\dagger)$, regardless of the two constants c_o and c in (2.16), verifying (2.17).

Combining (2.10) with the lemma shows that

$$B(\mathbf{x}) \doteq \frac{\widehat{B}(\mathbf{x})}{\widehat{B}(\mathbf{x}^\dagger)} B(\mathbf{x}^\dagger), \tag{2.20}$$

"$\doteq$" indicating second-order accuracy as in (2.17). Various choices of $\mathbf{x}^\dagger$ are made in Section 2.3, leading to convenient approximations for $B(\mathbf{x})$. For all such choices we will insist that $\mathbf{x}^\dagger$ have the same $\mathcal{M}_o$ MLE as $\mathbf{x}$,

$$\widehat{\theta}_o^\dagger = \widehat{\theta}_o. \tag{2.21}$$

This has the following benefit: *if $\mathcal{M}_o$ is an exponential family and $\widehat{\theta}_o^\dagger = \widehat{\theta}_o$, then $R_o = f_o(\mathbf{x})/f_o(\mathbf{x}^\dagger)$ exactly equals $f_{\widehat{\theta}_o}(\mathbf{x})/f_{\widehat{\theta}_o}(\mathbf{x}^\dagger)$, no matter what the prior $g_o(\theta_o)$ may* *be*, reducing assumptions (2.16) to $g(\theta) = c\ |I(\theta)|^{\frac{1}{2}}$, and removing part of the error in approximation (2.17). This follows easily from the exponential family fact that (2.21) makes

$$\frac{f_{\theta_o}(\mathbf{x})}{f_{\theta_o}(\mathbf{x}^{\dagger})} = \frac{f_{\widehat{\theta}_o}(\mathbf{x})}{f_{\widehat{\theta}_o}(\mathbf{x}^{\dagger})} \qquad \text{for all } \ \theta_o \in \mathcal{M}_o. \tag{2.22}$$

Approximation (2.12), $B(\mathbf{x})/B(\mathbf{x}^{\dagger}) \doteq \widehat{B}(\mathbf{x})/\widehat{B}(\mathbf{x}^{\dagger})$, tends to be highly accurate even in small samples. Table 3 shows part of a regression example with $0-1$ response data taken from Finney (1947), and used by Kass and Wasserman (1995). A total of $n = 39$ (predictor, response) pairs (v_i, x_i) are available, and we consider testing model $\mathcal{M}$, that the probability p_i of $x_i = 1$ follows the linear logistic form

$$\text{logit}(p_i) = \theta_o + \theta_1 v_i, \tag{2.23}$$

versus $\mathcal{M}_o:\ \theta_1 = 0$. The MLE of $\mathbf{p} = (p_1, p_2, \ldots, p_{39})$ under $\mathcal{M}_o$ is

$$\widehat{\mathbf{p}}_o = (.513, .513, \ldots, .513), \tag{2.24}$$

and we will take

$$\mathbf{x}^{\dagger} = .365\ \widehat{\mathbf{p}}_o + .635\ \mathbf{x}. \tag{2.25}$$

This is a breakeven point for testing $\theta_1 = 0$, as defined in Section 2.3.

v:	157	154	110	88	90	85	78	104	95	95	90	74	78
x:	1	1	1	1	1	1	0	0	0	0	0	0	0
v:	115	88	136	151	93	123	126	60	98	113	118	120	78
x:	1	1	1	1	1	0	1	0	0	0	0	1	0
v:	126	98	128	120	143	137	104	104	108	90	98	88	111
x:	1	0	1	0	1	0	0	1	1	1	0	0	1

Table 3 Logistic regression example; $n = 39$ cases of predictor v and dichotomous response x. From Finney (1947) in an experiment concerning vasoconstriction; v is x_1 from his table 1.

We can evaluate $B(\mathbf{x})$ and $B(\mathbf{x}^\dagger)$ directly by numerical integration of (2.6) over the Jeffreys' prior (2.15), obtaining

$$\frac{B(\mathbf{x})}{B(\mathbf{x}^\dagger)} = 8.888 \quad \text{compared to} \quad \frac{\widehat{B}(\mathbf{x})}{\widehat{B}(\mathbf{x}^\dagger)} = 8.841, \tag{2.26}$$

an error of only half a percent. This impressive accuracy reflects the fact that in a certain practical sense discussed in Remark I of Section 6, approximation (2.12) is third-order accurate.

The combination of Jeffreys' prior densities with exponential families makes the approximation $B(\mathbf{x})/B(\mathbf{x}^\dagger) \doteq \widehat{B}(\mathbf{x})/\widehat{B}(\mathbf{x}^\dagger)$ at least second-order accurate. Less restrictive assumptions lead to less accurate versions of (2.17) and (2.20). If we allow $\mathcal{M}_o$ to be a *curved* exponential subfamily of $\mathcal{M}$ then (2.17) may only be first-order accurate, erring by factor $1 + O(n^{-1/2})$, and similarly if Jeffreys' prior is replaced by some other slowly varying function $g(\theta)$. All of this is of more theoretical than practical importance. Much bigger differences between the frequentist and Bayesian methods arise from their different choices of the breakeven point $\mathbf{y}$ in (1.10).

2.3 The Breakeven Point

Definition A *breakeven point* $\mathbf{y}$ is a data set having the same $\mathcal{M}_o$ MLE as the observed data set $\mathbf{x}$, (2.21), and satisfying $B(\mathbf{y}) = 1$.

It follows from (2.20) that

$$B(\mathbf{x}) \doteq \widehat{B}(\mathbf{x})/\widehat{B}(\mathbf{y}) \tag{2.27}$$

so if we can find $\mathbf{y}$ we can compute a good approximation to the Bayes factor $B(\mathbf{x})$. This is especially convenient in Fisher's framework. A critical level of $\alpha_o = .90$ corresponds to the Fisherian breakeven point, so we can take $\mathbf{y}$ to be a point such that the test statistic $S(\mathbf{y})$ equals its $\mathcal{M}_o$ 90th percentile,

$$S(\mathbf{y}) = S^{(.90)}. \tag{2.28}$$

Things are particularly simple if the test statistic $S(\mathbf{x})$ is the likelihood ratio statistic $\widehat{B}(\mathbf{x})$ itself. Wilks' theorem says that $2\log(S^{.90})$ is approximated by the 90th percentile of a χ^2_d random variable, $d = \dim(\mathcal{M}) - \dim(\mathcal{M}_o)$,

$$S^{(.90)} \doteq \exp\{\chi_d^{2(.90)}/2\} \tag{2.29}$$

Then (2.27) provides an estimate of "$B_{\text{freq}}(\mathbf{x})$", the effective frequentist Bayes factor,

$$\log\ B_{\text{freq}}(\mathbf{x}) \doteq \log\ \widehat{B}(\mathbf{x}) - \chi_d^{2(.90)}/2. \tag{2.30}$$

Notice that formula (2.30) does not require explicit calculation of the breakeven point $\mathbf{y}$. It is less accurate than (2.27), see Remark E, but the difference is small in practice.

For the one-dimensional Gaussian situation (1.6)-(1.7), relation (2.27) says that an objective Bayes factor in favor of $\mathcal{M}$ should be of the form

$$\log\ B(\mathbf{x}) \doteq (x^2 - y^2)/2 \qquad (x = \sqrt{n}\ \bar{x}) \tag{2.31}$$

for some choice of the breakeven value y. The frequentist choice is $y_{\text{freq}} = 1.645$, the 90th percentile of $|x|$, while the BIC formula (1.9) uses $y_{\text{BIC}} = \sqrt{\log\ (n)}$. Table 4 shows that y_{BIC} crosses the frequentist value at sample size $n = 15$, growing larger at a rate that causes the dramatic differences seen on the right side of Figure 1. Section 4 discusses the Bayesian rationale for increasing the breakeven point as n gets bigger.

n:	10	15	100	1000	10000
$\sqrt{\log\ (n)}$:	1.52	1.645	2.15	2.63	3.04

Table 4 Breakeven point for BIC in Gaussian problem (1.6)-(1.7), as a function of sample size n. It equals frequentist value 1.645 at $n = 15$.

For another comparison consider selecting between $\mathcal{M}_o$ and $\mathcal{M}$ in the logistic regression situation of Table 3 , (2.23), which has $\widehat{B}(\mathbf{x}) = 34.28$. Applying definitions (2.30) and (1.5) gives

$$B_{\text{freq}}(\mathbf{x}) = 8.84 \quad \text{and} \quad B_{\text{BIC}}(\mathbf{x}) = 5.49, \tag{2.32}$$

showing substantial disagreement even at sample size $n = 39$.

There is another interesting choice for $\mathbf{x}^\dagger$ in the approximation formula $B(\mathbf{x}) \doteq [\widehat{B}(\mathbf{x})/\widehat{B}(\mathbf{x}^\dagger)]\ B(\mathbf{x}^\dagger)$:

Definition: A *least favorable point* $\mathbf{z}$ is a data set having its MLE under both $\mathcal{M}_o$ and $\mathcal{M}$ equal $\widehat{\theta}_o$, the $\mathcal{M}_o$ MLE for the actual data set $\mathbf{x}$. The point $\mathbf{z}$ is least favorable to the bigger hypothesis in the sense that the availability of $\mathcal{M}$ does not change the MLE from its $\mathcal{M}_o$ value. In examples (1.1), (1.2), and (1.6), (1.7), $z = 0$.

We have $\widehat{B}(\mathbf{z}) = f_{\widehat{\theta}_o}(\mathbf{z})/f_{\widehat{\theta}_o}(\mathbf{z}) = 1$, so (2.20) gives $B(\mathbf{x}) \doteq \widehat{B}(\mathbf{x}) \cdot B(\mathbf{z})$, or equivalently

$$\log\ B(\mathbf{x}) \doteq \log\ \widehat{B}(\mathbf{x}) - \log\ B^{-1}(\mathbf{z}), \tag{2.33}$$

where $B^{-1}(\mathbf{z}) = 1/B(\mathbf{z})$ is the Bayes factor for $\mathcal{M}_o$ compared to $\mathcal{M}$. This has the following interpretation: to obtain the log Bayes factor in favor of $\mathcal{M}$, *penalize* the corresponding log likelihood ratio by the log of the Bayes factor in favor of $\mathcal{M}_o$ at the least favorable point. Comparing (2.33) with (2.27) shows that

$$B^{-1}(\mathbf{z}) \doteq \widehat{B}(\mathbf{y}), \tag{2.34}$$

so the penalty against log $\widehat{B}(\mathbf{x})$ also equals the log of the likelihood ratio statistic at the breakeven point.

From (1.5) and (2.30) we see that the BIC and frequentist penalties are

$$\frac{d}{2}\log\ (n) \quad \text{and} \quad \log\ S^{(.90)} \doteq \chi_d^{2(.90)}/2 \tag{2.35}$$

respectively. The BIC penalty amounts to taking $B^{-1}(\mathbf{z}) = n^{d/2}$ in (2.33), so that at the least favorable point for $\mathcal{M}$ there can be a large Bayes factor in favor of $\mathcal{M}_o$. By contrast, in the frequentist framework their can never be a large Bayes factor for $\mathcal{M}_o$, the maximum possible factor being $\exp\{\chi_d^{2(.90)}/2\}$, equaling 3.87 for $d = 1$.

3 Frequentists As Bayesians

Section 2 argues that objective Bayes factors should be of the form $B(\mathbf{x}) \doteq \widehat{B}(\mathbf{x})/\widehat{B}(\mathbf{y})$, and that Fisher's scale of evidence amounts to putting the breakeven point $\mathbf{y}$ at a value $\mathbf{y}^{(.90)}$ such that $S(\mathbf{y}^{(.90)}) = S^{(.90)}$, the 90th percentile of the test statistic S. Taken together these arguments suggest defining the frequentist Bayes factor to be $B_{\text{freq}}(\mathbf{x}) \doteq \widehat{B}(\mathbf{x})/\widehat{B}(\mathbf{y}^{(.90)})$, or for convenience the cruder approximation (2.30).

For the one-dimensional Gaussian situation (1.6)-(1.7) with $x = \sqrt{n}\ \bar{x}$, we have

$$B_{\text{freq}}(\mathbf{x}) = \exp((x^2 - y^2)/2), \quad y = 1.645. \tag{3.1}$$

Table 5 shows $B_{\text{freq}}(\mathbf{x})$ for values of x corresponding to Fisher's scale in Table 1. For example, $|x| = 2.58$, which is strong evidence against $\mathcal{M}_o$ on Fisher's scale, corresponds to Bayes factor 7.13, giving aposteriori probability .88 for $\mathcal{M}$ assuming equal prior probabilities on $\mathcal{M}$ and $\mathcal{M}_o$. Expressing the frequentist results in terms of $B_{\text{freq}}(\mathbf{x})$, instead of comparing than with $B_{\text{BIC}}(\mathbf{x})$ as in Figure 1, reduces the discrepancy between Fisher's

Fisher:	borderline	moderate	substantial	strong	very strong	overwhelming
α:	**.90**	**.95**	**.975**	**.99**	**.995**	**.999**
$\lvert x \rvert$:	1.645	1.96	2.24	2.58	2.81	3.29
$B_{\text{freq}}(\mathbf{x})$:	1	1.76	3.19	7.13	13.29	58.03
$\pi(\mathbf{x})$:	.50	.64	.76	.88	.93	.98

Table 5. Frequentist Bayes factors corresponding to critical levels on Fisher's scale; one-dimensional Gaussian case, (3.1); $\pi(\mathbf{x})$ is prob$\{\mathcal{M}|\mathbf{x}\}$ assuming equal prior probabilities for $\mathcal{M}$ and $\mathcal{M}_o$.

and Jeffreys' scales of evidence, though Jeffreys' scale remains somewhat more favorable to $\mathcal{M}_o$.

$B_{\text{freq}}(\mathbf{x})$ is an answer to the question "what kind of Bayesians are frequentist model selectors?" We can fortify our belief in this answer by finding a genuine prior density $g(\theta)$ on $\mathcal{M}$, as opposed to Jeffreys' improper prior, that gives Bayes factors close to $B_{\text{freq}}(\mathbf{x})$. This section shows that such priors exist in the one-dimensional Gaussian situation (1.6)-(1.7), and in its one-sided version, but not in higher dimensional Gaussian problems. These results have a close connection to the work of Andrews (1994), discussed below.

3.1 The One-Dimensional Gaussian Case

We consider the simple problem (1.1), (1.2), which is equivalent to the repeated sampling version (1.6), (1.7). Figure 2 compares $B_{\text{freq}}(\mathbf{x})$, (3.1), with $B_{4.85}(\mathbf{x})$, the actual Bayes factor(2.7) when $g(\theta) = \mathcal{U}[\pm\, 4.85]$, the uniform density for θ on $[-4.85, 4.85]$. (Since $\mathcal{M}_o$ consists of the single point $\theta = 0$, the other conditional density $g_o(\theta_o)$ in (2.4) must put all of its probability on zero). We see that the match is excellent. The average absolute error over the range of Fisher's scale,

$$Q = \int_{-3.29}^{3.29} |B_{4.85}(\mathbf{x})/B_{\text{freq}}(\mathbf{x}) - 1| \; dx/(2 \cdot 3.29) \tag{3.2}$$

is only 0.011. Section 3.3 motivates the choice 4.85 and shows that still better matching priors are possible. The optimum choice among symmetric priors $g(\theta)$ is supported on six points,

$$\begin{array}{cccc} \theta: & \pm.71 & \pm2.18 & \pm3.86 \\ g: & .147 & .159 & .194 \end{array} \tag{3.3}$$

and gives $Q = .00061$, the minimum possible Q value according to the linear programming theory of Section 3.3.

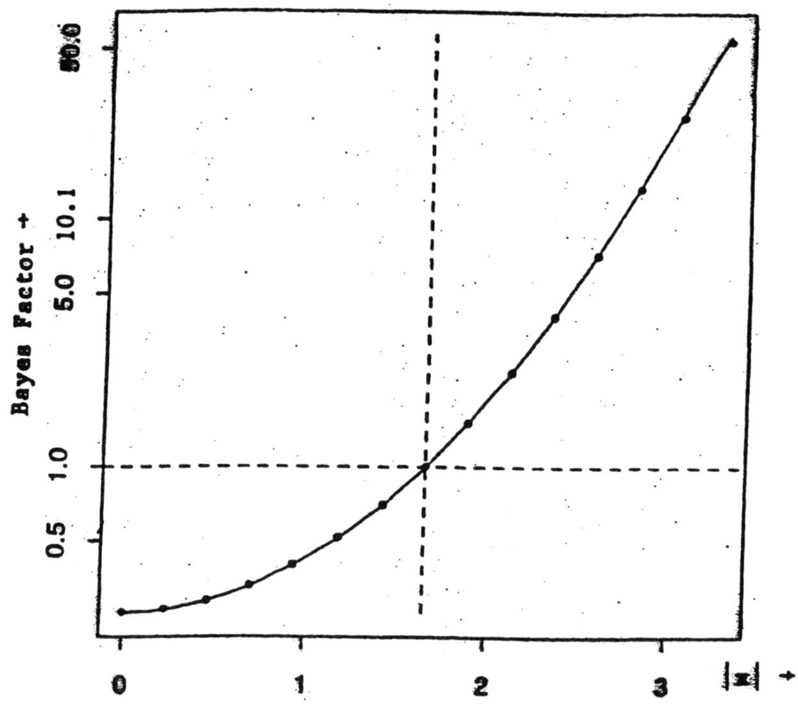

Figure 2 Comparison of log $B_{\text{freq}}(\mathbf{x})$, (solid curve, from (3.1)), with log $B_{4.85}(\mathbf{x})$ (dots) the genuine Bayes factor for the one-dimensional Gaussian situation if $g(\theta)$ is uniform on $[-4.85, 4.85]$.

All of this says that the frequentist is behaving like a somewhat unobjective Bayesian: the prior distribution $g(\theta)$ on the alternative hypothesis $\mathcal{M}$ is entirely supported within a few standard errors of the null hypothesis $\mathcal{M}_o$. By contrast, the BIC criterion is nearly equivalent to using a prior density $g(\theta)$ uniformly distributed over $\pm\ (\pi \cdot n/2)^{1/2}$, see formula (3.17) of Section (3.3),

$$\begin{array}{rccccc} n: & 10 & 15 & 100 & 1000 & 10000 \\ (\pi \cdot n/2)^{1/2}: & 3.96 & 4.85 & 12.53 & 39.63 & 125.33 \end{array} \tag{3.4}$$

The $n^{1/2}$ growth in the range of support for $g(\theta)$ is rooted in notions of Bayesian coherency as discussed in Section 4.

As in Table 4, the BIC and frequentist results agree at $n = 15$. To put it another way, the frequentist is effectively using a prior density $g(\theta)$ having about 1/15th of the data's information for estimating θ within model $\mathcal{M}$, while the BIC prior has only 1/nth of the information. See Remark H of Section 6.

The results in this section are closely related to those of Andrews (1994). Andrews considers the asymptotic analysis of model-selection problem (2.1), (2.2) when the prior density $g(\theta)$ in (2.8) is shrinking toward $\mathcal{M}_o$ at rate $1/\sqrt{n}$. For a class of elliptically symmetric $g(\theta)$ densities chosen to match the hypothesis-testing situation, he shows (in our notation) that $B(\mathbf{x})$ is asymptotically a function of $\widehat{B}(\mathbf{x})$. This amounts to defining a frequentist Bayes factor, like our $B_{\text{freq}}(\mathbf{x})$ though the specific form is a little different.

In situation (1.6)-(1.8) Andrews' theory would suggest

$$\log\ B_{\text{freq}}(\mathbf{x}) = \frac{A}{A+1}\,\frac{x^2}{2} - \frac{1}{2}\ \log\ (A+1), \tag{3.5}$$

as in his equation (2.6), this being the actual Bayes factor $B(\mathbf{x})$ starting from the Gaussian prior $g(\theta) \sim N(0, A)$. He recommends choosing the prior variance A so that $B_{\text{freq}}(\mathbf{x})$ equals one at the usual acceptance point for a hypothesis test. If Andrews' acceptance point is put at critical level .90 then we need $A = 11.0$ in (3.5).

In our previous language this amounts to giving the prior 1/11th of the data's information content, reasonably close to our value of 1/15th. The difference comes from Andrews' use of proper Gaussian-like priors rather than the improper Jeffreys' priors used here. Jeffreys' priors give a better match to objective Bayes methods like the BIC (which was not Andrews' purpose of course), via the lemma of Section 2.2, while avoiding most of the asymptotics.

3.2 One-Sided Testing

We can consider the one-sided version of the one-dimensional Gaussian problem (1.1)-(1.2) by changing the bigger model to

$$\mathcal{M}:\ \theta > 0. \tag{3.6}$$

The improper prior $g(\theta) = c$ on $\mathcal{M}$ gives Bayes factor,

$$B(x) = c\ \Phi(x)/\varphi(x) \tag{3.7}$$

where $\Phi(x)$ and $\varphi(x)$ are the standard normal cdf and density. For y a breakeven point, $B(y) = 1$, we can write

$$\log\ B(x) = \log\ \{B(x)/B(y)\} = \frac{x^2 - y^2}{2} + \log \frac{\Phi(x)}{\Phi(y)}. \tag{3.8}$$

The frequentist breakeven point is now $y = 1.282$, the .90 critical level of the one-sided test statistic $S(x) = x$, giving

$$\log\ B_{\text{freq}}(x) = \frac{x^2 - y^2}{2} + \log\ \frac{\Phi(x)}{\Phi(y)} \qquad (y = 1.282) \tag{3.9}$$

as the one-sided version of (3.1). This can also be derived from a version of the lemma at (2.17) which takes account of the fact that in the one-sided case $\mathcal{M}_o$ is an extreme point of $\mathcal{M}$ rather than an interior point. (This changes (2.18), the Laplace approximation for $f(\mathbf{x})$.) Andrews (1994) gives similar formulas, for example at his equation (2.7).

Table 6 is the equivalent of Table 5 for the one-sided case. Notice that the critical levels of Fisher's scale, .90, .95, .975, ..., produce nearly the same Bayes factors in both tables. The aposteriori probabilities $\pi(x)$ (assuming $\pi/\pi_0 = 1$) are the same to two digits.

The Bayes factors in Table 6 are closely approximated by taking the prior density $g(\theta)$ to be uniform on $[0, 5.13]$. However a Bayesian who began with the uniform prior $\mathcal{U}[-4.85, 4.85]$ appropriate to the two-sided situation, and then decided that $\theta < 0$ was apriori impossible, would get Bayes factors $B(x)$ not much different than those in Table 6: Using 4.85 instead of 5.13 as the one-sided upper limit for $g(\theta)$ gives an excellent match to the version of (3.8) having breakeven point at $y = \Phi^{-1}(.893)$ instead of $\Phi^{-1}(.90)$.

Tables 5 and 6 show that a frequentist going from a two-sided to a one-sided Gaussian testing problem does so in reasonably coherent Bayesian fashion, essentially by cutting off the negative half of the $\mathcal{U}[-4.85, 4.85]$ prior. We can also use the $\mathcal{U}[-4.85, 4.85]$ prior to investigate frequentist behavior in multiple testing situations, see Remark J in Section 6.

Critical level α	**.90**	**.95**	**.975**	**.99**	**.995**	**.999**
x:	1.282	1.645	1.96	2.33	2.58	3.09
$B_{\text{freq}}(x)$:	1	1.80	3.25	7.24	13.42	57.86
$\pi(x)$:	.50	.64	.76	.88	.93	.98

Table 6. Critical levels and Bayes factors (3.9) for one-sided testing, one-dimensional Gaussian case; breakeven point at $y = 1.282$, the .90 quantile of x under $\mathcal{M}_o$; $\pi(x)$ aposteriori probability of $\mathcal{M}$ assuming prior probability 1/2.

3.3 Multidimensional Gaussian Testing

The Bayesian justification for Fisher's scale of evidence is less satisfactory in higher dimensional testing problems. Suppose that we observe an m-dimensional Gaussian vector with unknown expectation vector θ and covariance matrix the identity

$$\mathbf{x} \sim N_m(\boldsymbol{\theta}, \mathbf{I}), \tag{3.10}$$

and that we wish to test

$$\mathcal{M}_o : \boldsymbol{\theta} = 0 \quad \text{versus} \quad \mathcal{M} : \boldsymbol{\theta} \in R^m . \tag{3.11}$$

We will denote $x = \|\mathbf{x}\|$, $\theta = \|\boldsymbol{\theta}\|$, and write $B(x)$ instead of $B(\mathbf{x})$ in the case where $B(\mathbf{x})$ depends only on x, etc. The likelihood ratio statistic $\widehat{B}(x)$ equals $\exp(x^2/2)$ so that (2.27) gives

$$\log B(x) \doteq (x^2 - y^2)/2 \quad \text{where} \quad y^2 = \chi_m^{2(\alpha_0)} . \tag{3.12}$$

Here α_0 is the frequentist breakeven critical level, $\alpha_0 = .90$ on Fisher's scale. The arguments of this section extend easily to the case where $\boldsymbol{\theta}$ is partitioned as $\boldsymbol{\theta} = (\boldsymbol{\theta_0}, \boldsymbol{\theta_1})$, and $\mathcal{M}_o$ is $\boldsymbol{\theta_1} = \mathbf{0}$. We can also take $\mathbf{x} \sim N_m(\boldsymbol{\theta}, \sigma^2\mathbf{I})$ with σ^2 estimated independently from $\widehat{\sigma}^2 \sim \sigma^2\chi^2$, see Remark E of Section 6.

We used (3.12) in the one-dimensional case, with $\alpha_0 = .90$, and showed that it agreed well with a proper Bayesian analysis, starting from $g(\theta)$ uniform on [-4.85, 4.85]. The trouble in higher dimensions is that if we choose $\alpha_0 = .90$, then $B(x)$ in (3.12) is not close to being a Bayes factor for any genuine prior $g(\theta)$.

To show this we begin with $g(\boldsymbol{\theta})$, the density of $\boldsymbol{\theta}$ given $\mathcal{M}$, uniform on a disk of radius u,

$$g(\boldsymbol{\theta}) = \frac{\Gamma(m/2+1)}{\pi^{m/2}u^m} \qquad \text{for} \qquad 0 \le \theta \le u . \tag{3.13}$$

The constant in (3.13) makes g integrate to 1. Using definitions (2.6)-(2.7), and remembering that $g_0(\boldsymbol{\theta})$ is a delta function at zero, it is easy to show that the resulting Bayes factor is

$$B_u(x) = \frac{c_m}{u^m} e^{x^2/2} F_m(u,x) \qquad [c_m = 2^{m/2}\Gamma(m/2+1)] , \tag{3.14}$$

where

$$F_m(u, x) = \text{prob}\{\chi^2_m(x^2) < u^2\} \ , \tag{3.15}$$

the probability that a non-central χ^2 variate with m degrees of freedom and non-centrality x^2 is less than u^2.

We would like (3.14) to match the objective Bayes formula (3.12) $B(x) = \exp(x^2 - y^2)/2$. Notice that

$$\frac{\partial}{\partial x} \log B_u(x) = \frac{\partial}{\partial x} \log B(x) + \frac{\partial}{\partial x} \log F_m(u, x) \ . \tag{3.16}$$

The last term is always negative so $\frac{\partial}{\partial x} \log B_u(x) < \frac{\partial}{\partial x} \log B(x)$.

The value $u = u_0$ that makes $y = [\chi^{2(\alpha_0)}_m]^{1/2}$ the breakeven point satisfies $B_{u_0}(y) = 1$, or according to (3.14),

$$u_0 = [c_m e^{y^2/2} F_m(u_0, y)]^{1/m} \ . \tag{3.17}$$

For $m = 1$ and $y = 1.645$ we get $u_0 = 4.85$, the value used in Section 3.1. The second term $\frac{\partial}{\partial x} \log F_1(u_0, x)|_y$ in (3.16) is only -.0025 in this case, accounting for the good match between $B(x)$ and $B_{4.85}(x)$ in figure 2. Numerical calculations show that this match breaks down as the dimension m increases: the last term in (3.16) grows large, spoiling the agreement between $B_u(x)$ and $B(x)$. We might hope to save things by choosing $g(\boldsymbol{\theta})$ differently, but the next analysis shows not.

If $B_g(x)$ is the Bayes factor (2.6)-(2.7) corresponding to a spherically symmetric prior $g(\boldsymbol{\theta})$, and $B_{\text{freq}}(x, \alpha_0) = \exp[(x^2 - y^2)/2]$ for $y^2 = \chi^{2(\alpha_0)}_m$ define

$$Q(g, \alpha_0) \doteq \int_0^{x_{.999}} |B_g(x)/B_{\text{freq}}(x; \alpha_0) - 1| dx \ / x_{.999} \ , \tag{3.18}$$

$x_{.999} = (\chi^{2(.999)}_m)^{1/2}$, so Q measures the average absolute deviation between $B_g(x)$ and $B_{\text{freq}}(x; \ \alpha_0)$ over the range of Fisher's scale in Table 1.

Linear programming techniques allow us to minimize $Q(g; \ \alpha_0)$ over all possible choices of a spherically symmetric prior distribution $g(\theta)$. The results are shown in Figure 3. We see for example that for $\alpha_0 = .90$ and dimension $m = 6$, the minimum possible value of Q is .17. In order to match $B_{\text{freq}}(x, \alpha_0)$ with accuracy $Q = .011$, the accuracy shown in Figure 2, we need to increase α_0 to .96. This raises the breakeven

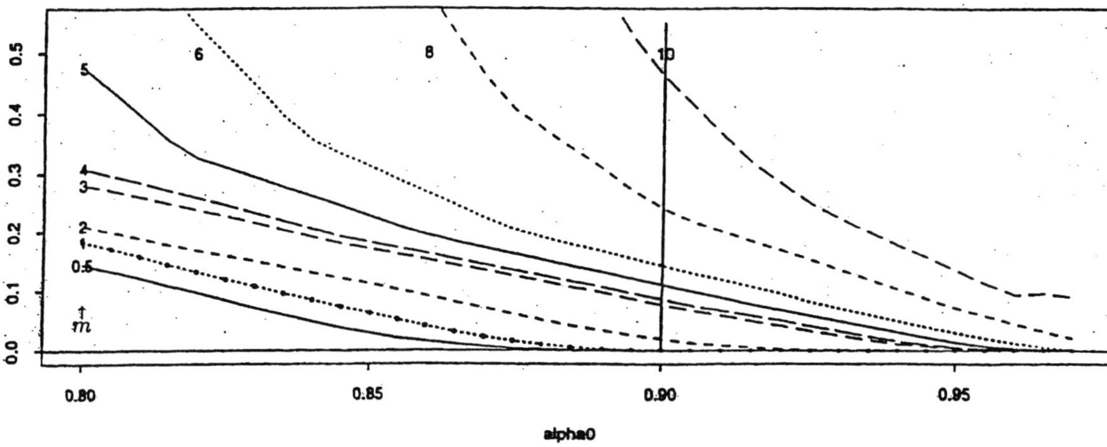

Figure 3. Minimum possible value of average absolute deviation (3.18) as a function of α_0 and dimension m; $m = .5$ is the one-sided one-dimensional problem.

point to $y = (\chi_6^{2(\alpha_0)})^{1/2} = 3.63$, rather than $(\chi_6^{2(.90)})^{1/2} = 3.26$, and decreases the Bayes factor $\exp((x^2 - y^2)/2)$ by a factor of 3.58. Some details of these calculations appear in the Appendix.

We used Fisher's scale of evidence to set the frequentist breakeven point at the .90 quantile of the test statistic. This definition turns out to be near the minimum possible for one-dimensional Gaussian testing problems: the $m = 1$ curve in Figure 3 shows that reducing α_0 to say .80 would make it impossible to get an accurate match with any genuine Bayes factor, the minimum attainable Q being .18. In higher dimensions $\alpha_0 = .90$ itself is too small. A frequentist who desires a proper Bayesian interpretation of p-values needs to set the breakeven quantile α_0 higher. A recipe for doing so appears in Remark G of Section 6.

4 Sample Size Coherency

The BIC assesses evidence in favor of the bigger model $\mathcal{M}$ more cautiously as the sample size n grows larger. The behavior is rooted in notions of coherent decision making, namely that the Bayesian specifications (2.3)-(2.4), whatever they may be, should stay the same for all sample sizes. This simple but important principle, which we have been calling "sample size coherency", causes most of the disagreement between Fisher's and Jeffreys' methods.

It is particularly easy to illustrate sample size coherency for the one-dimensional Gaussian example (1.6)-(1.7). Suppose that at sample size $n = 1$ we used density $g_1(\theta_1)$ for "g" in (2.4) (remembering that in this case g_0 puts all of its probability on zero.). Then, according to the coherency argument, we should still use $g_1(\theta_1)$ at sample size n. Making transformations (1.8),

$$x = \sqrt{n}\,\bar{x} \sim N(\theta, 1), \qquad \theta = \sqrt{n}\,\theta_1 \, , \tag{4.1}$$

restores us to situation (1.1), (1.2), now with

$$g(\theta) = g_1(\theta_1/\sqrt{n})/\sqrt{n} \, . \tag{4.2}$$

Equation (4.2) says that $g(\theta)$ is dilating at rate $\sqrt{n}$, and that the crucial value $g(0)$ is going to zero at rate $1/\sqrt{n}$. If $g_1(0)$ is positive and $g_1(\theta)$ is continuously differentiable at zero then it is easy to show that the breakeven point y occurs at

$$y \doteq [\log \frac{n}{2\pi g_1(0)^2}]^{\frac{1}{2}} \, . \tag{4.3}$$

Choosing $g_1(\theta)$ to be a standard $N(0,1)$ density gives $y \doteq \sqrt{\log(n)}$, which is the BIC breakeven point, (1.9). Variations of this argument are familiar in the Bayesian model selection literature, as in Section 3 of Kass and Wasserman (1995).

Frequentist model selection does not obey sample size coherency. Using Fisher's scale in the standard way amounts to using a fixed prior density no matter what the sample size may be, for instance $g(\theta) \doteq \mathcal{U}[\pm\, 4.85]$ in situation (1.6)-(1.8).

We have now reached a clear and unresolvable distinction between the Fisher and Jeffreys approaches to model selection. Which one is right? Here are some of the arguments that have been put forward:

Consistency Under quite general conditions BIC will select the correct model as $n \to \infty$, with $\log B_{\mathrm{BIC}}(\mathbf{x})$ going to $-\infty$ if $\theta \in \mathcal{M}_o$ and $+\infty$ if $\theta \notin \mathcal{M}_o$. The BIC penalty $-\frac{d}{2}\log(n)$ in (2.35) makes this happen (as would any other penalty function going to infinity at a rate slower than n.) The frequentist uses a fixed penalty $\doteq \chi_d^{2(.90)}/2$ and does not achieve consistency; even if $\theta \in \mathcal{M}_o$ there is always a fixed probability, .05 in the most common formulation, of selecting $\mathcal{M}$.

Power Consistency is not much of a comfort in a fixed sample size experiment. To use Neyman-Pearson terminology, what we really want is good size and good power too. Fisher's approach tends to aggressively maximize power by being satisfied with critical levels (i.e. sizes) in the .95 range. Jeffreys' methods are aimed at a more equitable balance. In using them the statistician risks settling for small power, as suggested by Figure 1.

Pre-experimental power calculations are an important part of frequentist model selection. A common prescription for the one-dimensional Gaussian situation (1.6)-(1.7) is to require sample size $n = (t_1/3.24)^2$, where t_1 is a preliminary guess for the treatment effect θ_1. The constant 3.24 results in .90 power for a two-sided .95 test, at $\theta_1 = t_1$. This kind of calculation, which has a strong Bayesian flavor, fits in well with the $\mathcal{U}[\pm\ 4.85]$ prior ascribed to the frequentist in Section 3.1. Selecting n on the basis of the prior reverses the BIC selection of the prior on the basis of n.

Sufficiency Transformations (1.8) restore problem (1.6), (1.7) to form (1.1), (1.2) no matter what n may be. Why should our assessment of evidence for $\mathcal{M}$ versus $\mathcal{M}_o$ depend in any way upon n? This could be the frequentist's riposte to the Bayesian argument for sample size coherency. It is strengthened by the difficulties of defining "n" in practical problems, see Section 5. The Bayesian argument is unassailable if we begin with a genuine prior but less so if "$g(\theta)$" expresses only a lack of prior knowledge.

Very Large Sample Sizes Raftery (1995) argues for the superiority of Jeffreys' scale and BIC model selection in the social sciences. His main example concerns a multinational social mobility study with data on $n = 113,556$ subjects. An appealingly simple sociometric model explains 99.7% of the deviance but still is rejected by the standard likelihood ratio test at a p-value of 10^{-120}. In this case the BIC penalty function is severe enough to give a Bayes factor in favor of the sociometric model, compared to a saturated model.

Many statisticians intuitively agree with Raftery that something goes wrong with Fisher's scale of evidence when n is very large, and that evidence against $\mathcal{M}_o$ should indeed be judged more cautiously in large samples. A counterargument says that Raftery is putting too much strain on the model selection paradigm: standard hypothesis tests and confidence intervals would show that the sociometric model does not fit the data perfectly but that the deviations from the model are quite small. Gelman and Rubin argue along these lines in the discussion following Raftery (1995), see also Diaconis and Efron (1985).

The Role of $\mathcal{M}_o$ The smaller model $\mathcal{M}_o$ is usually a straw man in Fisher's program, not an interesting scientific theory in its own right, see Section 4 of Efron (1971). Jeffreys' scale shows more respect for $\mathcal{M}_o$, perhaps on the Occam's razor principle that simpler hypotheses are preferred whenever tenable. This shows up in Table 5 (where the distorting effect of sample size coherency has been avoided by use of the frequentist breakeven value $y = 1.645$). Comparing $B_{\text{freq}}(x)$ with Jeffreys' scale, Table 2, Fisher's "moderate" evidence is "not worth more than a brief mention" for Jeffreys', "strong" is only "positive", etc.

This is more a difference in scientific context than a fundamental Bayesian-frequentist disagreement. Fisher worked in an agricultural field station where sample sizes were small and the data were noisy. Jeffreys' hard-science background suggests more abundant data, better structured models, and a more stringent standard of evidence. It is conceivable that had Jeffreys worked at Rothamsted he would have adjusted his scale downward, conversely for Fisher in geophysics.

Perhaps no single scale of evidence can serve satisfactorily in all contexts. In practice Jeffreys' scale, as opposed to Fisher's, tends to favor $\mathcal{M}_o$, and if implemented by the BIC will do so with increasing force as n increases.

The BIC is not the last word in objective Bayes model selection, though it seems to be the most popular method. Kass and Wasserman (1996) review a variety of techniques developed mostly since 1990 under such names as uninformative priors, reference priors, intrinsic priors, and fractional Bayes factors. All of these methods come close to obeying sample size coherency, and demonstrate BIC-like behavior in large samples.

Sample size coherency is an appealing principle in situations where the statistician actually sees the data set growing in size. It is less compelling in the more common case of fixed n, and taking it literally can lead to the situation seen in Figure 1 where evidence at the .995 level is barely worth mentioning. The selenium experiment of Section 5 has aspects of both fixed and changing sample sizes.

5 The Selenium Experiment

This section uses our methods to produce a combined Bayesian-frequentist analysis of a small but inferentially challenging data set. Table 7 shows total cancer mortality for a double-blind randomized trial of the trace element selenium taken as a cancer preventative, Clark et al. (1996). The original purpose of the trial was to test selenium's ability to prevent the recurrence of carcinoma of the skin. 1312 subjects, all of whom had suffered previous skin cancers, were recruited beginning in 1983 and received either 200 mmg per day of selenium or an identical-looking placebo. The results from 1983 to 1989, labeled "1st Period" in Table 7, did not show any skin cancer reduction in the selenium group. However total cancer mortality, mainly from lung, prostate, and colorectal cancers, did suggest a moderately significant reduction. The p-value of .032 shown in Table 7 is the one-sided binomial probability of seeing 7 or less occurrences of

	Selenium	Placebo	Total	s/N	**p-value**
1st Period:	7	16	23	7/23	**.032**
2nd Period:	22	41	63	22/63	**.00843**
Combined:	29	67	86	29/86	**.00124**

Table 7 Total cancer mortality in the selenium experiment; p-values are one-sided, based on binomial distribution (5.1). Data from Table 5 of Clark et al. (1996).

$s \sim$ binomial $(N, 1/2)$, splitting the probability atom at 7,

$$.032 = \frac{1}{2^N}\left[\sum_{i=0}^{s-1}\binom{N}{i} + \frac{1}{2}\binom{N}{s}\right], \tag{5.1}$$

with $N = 23$ and $s = 7$.

New funding was obtained, allowing a second trial period from 1990-1993. At the beginning of this period total cancer mortality was officially listed as a "secondary endpoint". The primary endpoint remained skin cancer, but given the results of the 1st period it seems fair to assume that the investigators' attention was now focused on total cancer mortality. Lung, prostate, and colorectoral cancer incidence were also listed as secondary endpoints, see Remark A.

We now consider selecting between the models

$$\begin{aligned} \mathcal{M}_o &: \text{ selenium has no effect on total cancer mortality} \\ \mathcal{M} &: \text{ selenium has an effect on total cancer mortality.} \end{aligned} \tag{5.2}$$

By conditioning on N the total number of cancer deaths in both groups, $N = 23$ for the first period and 63 for the second period, we can model s, the number of deaths in the selenium group, by

$$s \sim \text{binomial}(N, \theta) \qquad \text{with} \begin{cases} \mathcal{M}_o : \ \theta = .5 \\ \text{versus} \\ \mathcal{M} : \ \theta \neq .5 \end{cases} \tag{5.3}$$

as the competing models. We will also consider the one-sided version $\mathcal{M} : \ \theta < .5$. In either case we are dealing with a one-dimensional problem, $\dim(\mathcal{M}) - \dim(\mathcal{M}_o) = 1$.

How strong is the evidence for $\mathcal{M}$ and against $\mathcal{M}_o$? The combined-data p-value of .00124, even doubled for two-sided testing, is between "very strong" and "overwhelming" on Fisher's scale. However, this ignores the data-mining aspects of the experiment, which used the first period's outcome to change the focus of interest. Restating the frequentist results in terms of Bayes factors helps clarify the strength of evidence for selenium's cancer-preventing ability.

The simplest approach considers only the 2nd period data since it was then that attention was focused on total cancer mortality. A very quick way of doing the calculations transforms the one-sided binomial p-value .00843, computed as in (5.1) into an approximate normal deviate $x = \Phi^{-1}(1 - .00843) = 2.39$, and then calculates the Bayes factor for $\mathcal{M}$ from the one-dimensional Gaussian approximation formula (3.1),

$$B_{\text{freq}}(s) \doteq \exp\{(2.39^2 - 1.645^2)/2\} = 4.49 \ . \tag{5.4}$$

Starting from the conventional prior odds ratio $\pi/\pi_0 = 1$, the Bayes rule results in aposteriori probability $\pi(s) = .82$ for $\mathcal{M}$. Remark C of Section 6 shows that this is nearly the same as the aposteriori probability for the event of actual interest $\{\theta < .5\}$, even though we began the analysis with $\mathcal{M} : \theta \neq .5$. Remark B shows that the Gaussian approximation (5.4) works quite well in this case.

We might instead begin our analysis with the one-sided model $\mathcal{M} : \theta < .5$, on the grounds that the 1st period results removed most of our apriori probability on $\theta > .5$. This entitles us to use the one-sided formula (3.9) for the Bayes factor, giving $B_{\text{freq}}(s) = 8.42$, nearly double (5.4), and $\pi(s) = .89$. These results are shown in Table 8.

Instead of focusing on the 2nd period we might consider the combined data for both periods. If so we need to adjust our inferences to account for the fact that total cancer

	$\hat{B}$	---- B_{freq}	----	---- B_{BIC}	----
		2-sided	1-sided	$n = N$	$n = 1312$
2nd Period:	18.36	4.49	8.42	2.31	0.51
$[\pi(s)]$:		[**.82**]	[**.89**]	[**.70**]	[**.34**]
Combined:	103.8	25.14		11.19	2.87
$[\pi(s)]$:		[**.86**]		[**.74**]	[**.42**]

Table 8 Approximate Bayes factors for the Selenium experiment, as explained in text; boldface numbers are aposteriori probabilities for selenium having an effect, assuming $\pi/\pi_o = 1$ for 2nd period, $\pi/\pi_o = 1/4$ for combined data.

mortality was not the original primary endpoint. We will do this by setting the prior odds ratio to be

$$\pi/\pi_0 = 1/4 \ . \tag{5.5}$$

This is rather arbitrary of course but it cannot be wildly optimistic: after the 1st period results, which yielded a Bayes factor of only 1.44, the investigators effectively raised total cancer mortality to the status of primary endpoint, presumably with odds ratio near the conventional value 1 we have been using.

The combined data has $s = 29$ and $N = 86$. The two-sided normal approximation used in (5.4) is now

$$B_{\text{freq}}(s) = 25.14 \ . \tag{5.6}$$

Bayes' rule with $\pi/\pi_0 = 1/4$ gives aposteriori probability $\pi(s) = .86$ for $\mathcal{M}$. One-sided testing gives $B(s) = 47.46$ and $\pi(s) = .92$, but now we lack scientific justification for a one-sided analysis.

All of our frequentist-cum-Bayesian analyses yielded aposteriori probabilities $\pi(s)$ in the range .82 to .89 for selenium being efficacious in reducing cancer deaths. Perhaps this seems disappointing given the striking p-values in Table 7, but as Table 5 shows this is what we get from "strong" evidence on Fisher's scale. As far as Jeffreys' scale is concerned, $B_{\text{freq}}(s)$ never gets stronger than "positive" (remembering to divide by 4 for the combined data).

BIC analysis is predictably more pessimistic about selenium's efficacy. The BIC Bayes factor (1.5) for the combined data is

$$B_{\text{BIC}}(s) = 103.8/\sqrt{n} \tag{5.7}$$

Taking $n = 86$, the number of deaths, gives $B_{\text{BIC}}(s) = 11.19$, and $\pi(s) = .74$ starting from $\pi/\pi_0 = 1/4$. If we take $n = 1312$, the number of subjects in the study then $\pi(s) = .42$. Raftery (1986) makes a good argument for preferring $n = 86$ to 1312, but in general there is not a firm prescription for "n". If the data was collected in pairs should "n" be $n/2$? Kass and Wasserman (1995) aptly characterize the sample size question as "subtle but important", see also Lauritzen's commentary on O'Hagan (1995). These difficulties are avoided in the frequentist formulation, at the expense of ignoring sample size coherency.

Sample size coherency is unimportant in the usual fixed size experiment where the frequentist approach operates to best advantage. The selenium experiment is somewhere intermediate between fixed sample size and sequential. If we think of it as occurring in two stages, then the sample size coherency argument suggests dividing the 2nd period Bayes factors by $\sqrt{2}$, giving $B_{\text{freq}}(s) = 5.95$ and $\pi(s) = .86$ for the one-sided analysis. This kind of Bayesian correction is not much different than the standard frequentist approach to multiple testing situations, see Remark J.

6 Remarks

Remark A. Lung, prostate, and colorectal cancer incidence rates were also flagged as important secondary endpoints for the 2nd period of the selenium trial. Incidence of all three together was 17 in the selenium group versus 29 in the placebo group during the 1st period, giving binomial significance level .040 according to (5.1). 2nd period incidences were 21 versus 56, significance level $2.79 \cdot 10^{-5}$. Now the one-sided 2nd period Bayes factor corresponding to $B_{\text{freq}}(s) = 8.42$ in Table 8 is 1642. The very strong 2nd period results are a reminder that the two periods differ in the amount of selenium experienced by the treatment group.

Remark B. We do not need to rely on the Gaussian approximation (3.1) for the selenium analysis. Let y be the .95 percentile point for a binomial$(N, .5)$ distribution, calculated by interpolation of the "split-atom" cdf as in (5.1), so y is a .90 breakeven point for two-sided testing. Then according to (2.27)

$$B_{\text{freq}}(s) \doteq \widehat{B}(s)/\widehat{B}(y) = \frac{s^s(N-s)^{N-s}}{y^y(N-y)^{N-y}}\ . \tag{6.1}$$

This gives 4.63 instead of 4.49 for the two-sided 2nd period value of B_{freq} and 26.80 instead of 25.14 for the combined data. We see that (3.1) works quite well here.

Remark C In the two-sided formulation (5.2), 2nd period data, we calculated $B_{\text{freq}}(s) = 4.49$ in favor of $\mathcal{M} : \theta \neq .5$ versus $\mathcal{M}_o : \theta = .5$. However, we are really interested in the one-sided alternative $\theta < .5$. To this end we can state the results as follows: the aposteriori probability of $\mathcal{M} : \theta \neq .5$ is $\pi(s) = .82$, and given that $\mathcal{M}$ is true, the aposteriori probability that $\theta < .5$, (using $g(\theta) = \mathcal{U}[\pm\ 4.85]$) is about $\Phi(2.39) = .992$. This gives .81 for the aposteriori probability of $\{\theta < .5\}$.

Remark D Suppose that in situation (1.1), (1.2) we observe $x = 1.96$ and wish to estimate $\gamma = \text{prob}\{\theta > 0 \mid x\}$. The prior distribution of Section 3.2 appropriate to one-sided model selection, $\pi/\pi_0 = 1$ and $g(\theta)$ uniform on $[0, 5.13]$, gives aposteriori Bayes estimate $\gamma_1 = .76$. Section 3.1's prior for two-sided model selection, $\pi/\pi_0 = 1$ and $g(\theta)$ uniform on $[-4.85, 4.85]$, gives $\gamma_2 = .64$. This decrease is reasonable since the second prior puts only half as much probability on the positive axis.

Both γ_1 and γ_2 are much smaller than the value $\gamma_3 = .975$ we get using the standard objective prior for estimation, which has $g(\theta)$ constant over $(-\infty, \infty)$, with no special treatment for $\theta = 0$. This is the difference between model selection, which puts a bump of probability on, or at least near, $\mathcal{M}_o : \theta = 0$, and estimation which does not. The estimation paradigm is often more appropriate than model selection. If we are trying to choose between constant, linear, or quadratic regression functions for prediction purposes, then there may not be any reason to assign bumps of prior probability to zero values of the regression coefficients $\beta_0, \beta_1, \beta_2$. Efron and Tibshirani (1997) consider discrete selection problems from the "smooth prior" point of view. See also Lindley and O'Hagan's disagreement in the discussion following O'Hagan (1995).

Remark E The multidimensional Gaussian situation (3.10), (3.11) can be generalized to its more common form where we observe

$$\mathbf{x} \sim N_m(\boldsymbol{\theta}, \sigma^2\mathbf{I}) \quad \text{independent of} \quad \widehat{\sigma}^2 \sim \sigma^2\chi_q^2/q\ , \tag{6.2}$$

$\boldsymbol{\theta} = (\boldsymbol{\theta}_0, \boldsymbol{\theta}_1)$, $\dim(\boldsymbol{\theta}_1) = d$, and wish to select between

$$\mathcal{M}_o : \boldsymbol{\theta}_1 = \mathbf{0} \quad \text{versus} \quad \mathcal{M} : \boldsymbol{\theta} \in R^m\ . \tag{6.3}$$

Applying the lemma (2.17) in form (2.27), (2.28) gives estimated Bayes factor

$$B(\mathbf{x}, \widehat{\sigma}^2) \doteq \left[\frac{1 + \frac{d}{q}F_{d,q}}{1 + \frac{d}{q}F_{d,q}^{(\alpha_0)}}\right]^{\frac{d+q}{2}}, \tag{6.4}$$

where $F_{d,q}$ is the usual F statistic for testing $\mathcal{M}_o$ versus $\mathcal{M}$, and α_0 is the breakeven quantile, $\alpha_0 = .90$ on Fisher's scale. The shortcut formula (2.30) amounts to replacing $F_{d,q}^{(\alpha_0)}$ with its limit as $q \to \infty$.

Remark F Is Fisher's scale of evidence too liberal, as charged in the Bayesian literature? The answer depends at least partly on scientific context, though much of the contextual effect can be mitigated by an honest choice of the prior odds ratio π/π_0. Biomedical research has employed Fisher's scale literally millions of times, with generally good results. In crucial situations the scale may be implicitly tightened, for instance by the F.D.A. requirement of *two* .95 significant studies to qualify a new drug.

We have argued that Fisherian hypothesis testing has a reasonable Bayesian interpretation, at least if one is willing to forego sample size coherency. Section (3.3) shows the Bayesian interpretation breaking down in multidimensional testing problems, with the suggestion that Fisher's choice of breakeven quantile $\alpha_0 = .90$ needs to be increased.

Remark G Here is another argument for using bigger values of α_0 in higher dimensions. The BIC penalty function $d \cdot (\log n)/2$, (2.35), is linear in the dimensional difference d. By considering a nested sequence of models it is easy to show that this must always be the case if the penalty function depends only on d and n. However, the frequentist penalty function $\chi_d^{2(\alpha_0)}/2$ is not linear in d.

We can enforce linearity by replacing .90 in approximation (2.30) with $\alpha_o(d)$, where $\alpha_0(d)$ satisfies

$$\chi_d^{2(\alpha_0(d))} = d \cdot \chi_1^{2(\alpha_0(1))} , \tag{6.5}$$

$\alpha_0(1)$ being the breakeven quantile in dimension 1. Doing so makes $\alpha_0(d)$ increase with d, as shown in Table 9. The choice $\alpha_0(1) = .86$ is the minimum that keeps the discrepancy measure Q, (3.18) reasonably small (about .05) for dimensions 1-6.

d:	1	2	3	4	5	6
$\alpha_0(1) = .86$:	.86	.887	.912	.931	.946	.958
$\alpha_0(1) = .90$:	.90	.933	.956	.971	.981	.987

Table 9 Quantile $\alpha_0(d)$ satisfying linearity relationship (6.5); for two choices of $\alpha_0(1)$.

Remark H In the Gaussian situation (1.6)-(1.7), and in more general problems too, the BIC depends on a prior density $g(\theta)$ having $1/n$ of the data's information for estimating θ. Jeffreys' original proposal used a slightly more diffuse prior with information about $1/(\frac{\pi}{2}n)$, see Section 3 of Kass and Wasserman (1995). Berger and Pericchi (1993) suggest the value $1/(n/n_0)$, where $n_0 \geq 1$ is a fixed small constant. O'Hagan's (1995) proposal leads to larger values of n_0, for robustness purposes, an idea criticized in Berger and Mortera's discussion.

Remark I The calculations leading to the lemma's result (2.17) can be carried out more accurately by using a higher-order version of Laplace's method as in the appendix of Tierney and Kadane (1986). We get

$$\frac{B(\mathbf{x})}{B(\mathbf{x}^{\dagger})} = \frac{\widehat{B}(\mathbf{x})}{\widehat{B}(\mathbf{x}^{\dagger})} [1 + \frac{a(\widehat{\theta}) - a(\widehat{\theta}^{\dagger})}{n} + O(n^{-2})] , \tag{6.6}$$

where $a(\theta)$ is $O(1)$. In applications like the logistic regression example (2.26), $\widehat{\theta}^{\dagger}$ is $O_p(n^{-\frac{1}{2}})$. The interesting uses of (6.6), the ones where $B(\mathbf{x})$ is of moderate size, also have $a(\widehat{\theta})$ of order $O_p(n^{-\frac{1}{2}})$, so $a(\widehat{\theta}) - a(\widehat{\theta}^{\dagger})$ is $O_p(n^{-\frac{1}{2}})$ in (6.6). This suggests that approximation (2.17) is actually third-order accurate, giving the kind of good small-sample results seen in (2.26). However, this argument fails for the standard asymptotic calculation where the true θ is a fixed point in $\mathcal{M}$, in which case $a(\widehat{\theta}) - a(\widehat{\theta}^{\dagger})$ is $O_p(1)$.

Remark J The $\mathcal{U}[\pm\ 4.85]$ prior density figuring in the discussion of Section 3.1 also gives a rough Bayesian justification for the standard frequentist approach to multiple testing. Suppose that we observe J independent Gaussian variates, each having a possibly different expectation,

$$x_i \overset{i.i.d.}{\sim} N(\theta_i, 1), \qquad i = 1, 2, \ldots, J , \tag{6.7}$$

and that we wish to test

$$\begin{aligned} &\mathcal{M}_o: \quad \text{all} \quad \theta_i = 0 \quad \text{versus} \\ &\mathcal{M}: \quad \text{one of the} \quad \theta_i \quad \text{not zero.} \end{aligned} \tag{6.8}$$

For the conditional prior distribution of $\boldsymbol{\theta} = (\theta_1, \theta_2, \ldots, \theta_J)$ given model $\mathcal{M}$, we take an equal mixture of the J distributions

$$g_i(\boldsymbol{\theta}): \quad \theta_i \sim \mathcal{U}(-4.85, 4.85) \quad \text{and} \quad \theta_j = 0 \quad \text{for} \quad j \neq i , \tag{6.9}$$

$i = 1, 2, \ldots, J$. If $J = 2$, $g(\boldsymbol{\theta})$ is a cross-shaped distribution.

It is easy to calculate the Bayes factor for $\mathcal{M}$ versus $\mathcal{M}_o$ if the observed data vector is

$$\mathbf{x} = (x_1, 0, 0, \ldots, 0) , \tag{6.10}$$

in which case

$$B(\mathbf{x}) \doteq \sqrt{2\pi}\, \frac{e^{x_1^2/2} + J - 1}{2J \cdot 4.85} . \tag{6.11}$$

J:	2	3	4	6	8	10
Bonferroni:	1.96	2.13	2.24	2.39	2.50	2.58
(6.11):	1.95	2.13	2.25	2.41	2.52	2.60

Table 10. Simultaneous testing; comparison of critical point for the simultaneous .90 Bonferroni test with the breakeven point (6.11) for the Bayesian analysis suggested by the $\mathcal{U}[\pm\, 4.85]$ distribution; J is the number of simultaneous tests.

Using this approximation, the breakeven point $\mathbf{y} = (y, 0, 0, \ldots, 0)$ occurs at

$$y = \pm\ \{2 \log\, [\frac{2J \cdot 4.85}{\sqrt{2\pi}} - (J-1)]\}^{1/2}\ . \tag{6.12}$$

Table 10 compares the Bayesian breakeven point y from (6.12) with the .90 critical point $y = \Phi^{-1}(1-.05/J)$ of the usual frequentist Bonferroni procedure for J simultaneous tests. Once again we see that the frequentist is behaving in a reasonably Bayesian fashion, although caution is warranted here because of the somewhat special choices made in (6.8)-(6.10).

Summary Frequentist hypothesis testing, as interpreted on Fisher's scale of evidence, is the most widely used model selection technique. Jeffreys' theory of Bayes factors, as implemented by objective formulas like the BIC, implies that Fisher's scale is badly biased against smaller models, especially in large samples. This paper compares the two theories by giving an interpretation of Fisher's scale in terms of Bayes factors, as far as that is possible, along these lines:

- An ideal form for objective Bayes factors is developed in Section 2, $B(\mathbf{x}) \doteq \widehat{B}(\mathbf{x})/\widehat{B}(\mathbf{y})$, with $\mathbf{y}$ a breakeven point: $B(\mathbf{y}) = 1$.

- Fisher's theory is interpreted as putting the breakeven point at the 90th percentile of the test statistic, leading to a prescription $B_{\text{freq}}(\mathbf{x})$ for the frequentists' implied Bayes factor, as in (3.1).

- For the one-dimensional Gaussian problem Figure 2 shows that $B_{\text{freq}}(\mathbf{x})$ is close to the actual Bayes factor if the prior density on the bigger model is uniform over $[-4.85, 4.85]$. This portrays the frequentist as a somewhat unobjective Bayesian. By contrast, the BIC amounts to choosing the prior density uniform over $\pm\ [\pi \cdot n/2]^{1/2}$, getting wider as the sample size n increases. Roughly speaking, the frequentist assigns the prior 1/15 of the data's information, compared to $1/n$ for the BIC.

• The BIC, and other objectivist Bayesian methods, behave this way because of "sample-size coherency", the principle that a Bayesian specification for the model selection problem should be consistent over all possible sample sizes. Section 4 discusses arguments for and against this principle.

• Sample size coherency causes most of the disagreement between Fisher's and Jeffreys' methods, but even after correcting for it, Jeffreys' scale remains somewhat more inclined toward the smaller hypothesis $\mathcal{M}_o$. This is more a matter of scientific context than frequentist/Bayesian dispute.

• The argument that $B_{\text{freq}}(\mathbf{x})$ is close to being a proper Bayes factor weakens in higher dimensional problems. To restore it we need to increase the breakeven point on Fisher's scale above the $\alpha_0 = .90$ quantile of the test statistic. Figure 3 suggests an α_0 of about .96 for a six-dimensional Gaussian testing problem. It also shows that in the one-dimensional case we could not choose α_0 much smaller than .90.

• $B_{\text{freq}}(\mathbf{x})$ provides a Bayesian interpretation for standard p-values, as exemplified in Section 5's analysis of the selenium data.

Appendix

Numerical methods were used in Section 3.3 to find the spherically symmetric prior g on the alternative of the m-dimensional Gaussian test (3.10)-(3.11) yielding a Bayes factor B_g most closely matching the effective frequentist Bayes factor, B_{freq}, defined in (2.30). We will describe the calculations in detail here. The one-sided test (labelled $m = 0.5$ in Figure 3) may be treated similarly, as will be shown later.

As in Section 3.3 we put $x = \|\mathbf{x}\|$ and $\theta = \|\boldsymbol{\theta}\|$, and write $B_g(x)$ instead of $B_g(\mathbf{x}), g(\theta)$ instead of $g(\boldsymbol{\theta})$, etc. In this notation, we define the following objective function, a measure of distance between the Bayes factors B_g and B_{freq}:

$$Q(g, \alpha) = \int_0^{x_{.999}} |B_g(x)/B_{\text{freq}}(x; \alpha_0) - 1| dx/x_{.999} \ . \tag{A.1}$$

Here $x_{.999} = (\chi_m^{2(.999)})^{\frac{1}{2}}$. Other objective functions are possible; see below. The function Q is convenient because the problem can be formulated as a linear program. Formally, we want to solve the following:

$$\underset{g}{\text{minimize}} \qquad Q(g, \alpha_0)$$

(O1) subject to 1. g is a density on $\mathcal{R}^m$, spherically symmetric around 0,
2. $B_g(y) = B_{\text{freq}}(y; \alpha_0) = 1$, where $y^2 = \chi_m^{2(\alpha_0)}$.

The second constraint restricts consideration to priors g which result in a Bayesian test with the same "breakeven point" y (see Section 2.3) as the frequentist test.

The g minimizing (O1) for dimension $m = 1$ and $\alpha_0 = .90$ is given in (3.3). The minimum $q(g, \alpha_0)$ over a range of m and α_0 are shown in Figure 3.

Writing the spherically symmetric density g as $g(\boldsymbol{\theta}) \equiv h(\theta)$ we have

$$\int_0^\infty h(\theta)\theta^{m-1} d\theta = \frac{\Gamma(m/2)}{2\pi^{m/2}} . \tag{A.2}$$

If $m = 1$, for example, $\int_0^\infty h(\theta) d\theta = 1/2$.

An expression for B_{freq} is given in (3.12). We can calculate

$$B_g(x) = \int_0^u s_x(\theta) h(\theta) d\theta , \tag{A.3}$$

where

$$s_x(\theta) = \theta^{m-1} e^{-\theta^2/2} H(x\theta), \ \theta \geq 0 , \tag{A.4}$$

$$H(z) = \frac{2\pi^{(m-1)/2}}{\Gamma((m-1)/2)} \int_0^\pi (\sin\theta)^{m-2} e^{z \cos\theta} d\theta \ z \geq 0 . \tag{A.5}$$

To solve (O1) numerically we will discretize over the ranges of both x and θ. Let $\theta_1, \ldots, \theta_{n_\theta}$ be n_θ equally-spaced values from 0 to u, and $x_1, \ldots, x_{n_x}$ be n_x equally-spaced values from 0 to $x_{.999}$. We will approximate h by an n_θ-vector γ, with $h(\theta_j) = \gamma_j, j = 1, \ldots, n_\theta$.

Define the n_θ-vectors $b_i, i = 1, \ldots, n_x$, to have entries

$$b_{ij} = s_{x_i}(\theta_j) / B_{\text{freq}}(x_i, \alpha_0), \ j = 1, \ldots, n_\theta , \tag{A.6}$$

evaluated using (3.12), (7.4), and (7.5). The function H in (A.5) can be evaluated using numerical integration. Then a discrete approximation to Q is

$$Q'(\gamma; \alpha_0) = \sum_{i=1}^{n_x} |b_i^T \gamma - 1| , \tag{A.7}$$

so a discrete version of (O1) is

$$
\text{(O2)} \qquad
\begin{array}{ll}
\underset{\gamma}{\text{minimize}} & Q'(\gamma;\alpha_0) \\
\text{subject to} & 1a.\ \gamma_j \geq 0, j = 1,\ldots,n_\theta\ . \\
& 1b.\ \sum_{j=1}^{n_\theta} \gamma_j \theta_j^{m-1} = \Gamma(m/2)/(2\pi^{m/2}) \\
& 2.\ \sum_{j=1}^{n_\theta} \gamma_j s_y(\theta_j) = 1, \text{where } y^2 = \chi_m^{2(\alpha_0)}.
\end{array}
$$

Constraint 1 in (O1) has been split into two parts; part 1b. follows from (A.2).

To express (O2) in the standard form of a linear program we can introduce dummy variables $t_1, \ldots, t_{n_x}$, and write (O2) as

$$
\text{(O3)} \qquad
\begin{array}{ll}
\underset{\gamma}{\text{minimize}} & \sum_{i=1}^{n_x} t_i \\
\text{subject to} & \text{1a., 1b., and 2. in (O2), and} \\
& 3.\ t_i > b_i^T\gamma - 1, t_i > 1 - b_i^T\gamma,\ i = 1,\ldots,n_x\ .
\end{array}
$$

One-sided testing

The optimization problem (O2) may be adapted to the one-sided testing case of Section 3.2 with the following modifications. We now have B_{freq} as in (3.9), so that the breakeven point becomes $y = \Phi^{-1}(\alpha_0)$, and the upper limit of integration in (A.1) is $x_{.999} = \Phi^{-1}(.999)$. The b_{ij} in (A.6) are calculated using (3.9) and

$$s_\theta(x) = e^{-\theta^2/2+x\theta}\ . \tag{A.8}$$

Finally, since g (and γ) in this case has mass 1 on the positive real line, we replace constraint 1b. of (O2) with

$$1b.\ \sum_{j=1}^{n_\theta} \gamma_j = 1\ .$$

Choice of the objective function:

Note that since B_{freq} is $O(e^{x^2})$ for large x, while B_g is only $O(e^x)$ for any g, the two functions can, at most, remain close up to some finite x. We therefore restrict the integration in (A.1) below $x_{.999}$, Table 1's upper bound on plausible values for x under the null model.

Other objective functions could be considered. The squared-error criterion

$$Q_1(g,\alpha) = \int_0^{x_{.999}} (B_g(x)/B_{\text{freq}}(x;\alpha_0) - 1)^2 dx/x_{.999} \tag{A.9}$$

for example results in a quadratic, rather than linear, program. Besides computational simplicity, we prefer Q to Q_1 because Q penalizes values of B_g far from B_{freq} less heavily, and so matches the two functions more closely near $x = y$ where their values are equal. This is the region of primary interest.

The objective function could compare the Bayes factors on the log scale, using, for example,

$$Q_2(g,\alpha) = \int_0^{x_{.999}} |\log B_g(x) - \log B_{\text{freq}}(x;\alpha_0)| dx/x_{.999} \, . \tag{A.10}$$

Minimization over g is far more difficult in this case though, since Q_2, unlike Q or Q_1, is not convex in g.

REFERENCES

Andrews, D. (1994). The large sample correspondence between classical hypothesis tests and Bayesian posterior odds tests. *Econometrica* **62**, 1207-1232.

Berger, J.O. (1985). *Statistical Decision Theory and Bayesian Analysis.* Springer-Verlag, New York.

Berger, J.O. and Sellke, T. (1987). Testing a point null hypothesis: the irreconciliability of p-values and evidence. *J. Amer. Statist. Asso.* **82**, 112-135.

Berger, J.O. and Pericchi, L.R. (1993). The intrinsic Bayes factor for model selection. Technical Report 93-43C, Dept. of Statistics, Purdue Univ.

Berger, J., Boukai, B., and Wang, T. (1997). Unified frequentist and Bayesian testing of a precise hypothesis. *Statist. Sci.* **12**, 133-160.

Casella, G. and Berger, R. (1987). Reconciling Bayesian and frequentist evidence in the one-sided testing problem. *J. Amer. Statist. Asso.* **82**, 106-111.

Clark, L.C. Combs, J.F., Turnbull, B.W., et al. (1996). Effects of Selenium supplementation for cancer prevention in patients with carcinoma of the skin. *Jour. Amer. Medical Assoc.* **276**, 1957-1963, Editorial p. 1964-1965.

Efron, B. (1971). Does an observed sequence of numbers follow a simple rule? (Another look at Bode's Law). *J. Amer. Statist. Asso.* **66**, 552-568.

Efron, B. and Tibshirani, R. (1997). The problem of regions. Technical Report, Dept. of Statistics and Biostatistics, Stanford Univ.

Efron, B. and Diaconis, P. (1985). Testing for independence in a two-way table: new interpretations of the chi-square statistic (with discussion and rejoinder). *Ann. Stat.* **13**, 845-913.

Finney, D. (1947). The estimation from individual records of the relationship between dose and quantal response. *Biometrika* **34**, 320-335.

Fisher, R.A. (1954). *Statistical Methods for Research by Workers*, (12th ed). Haffner, New York.

Good, I.J. (1947 and 1950). *Probability and the Weighing of Evidence.* Griffen, London.

Good, I.J. (1969). A subjective evaluation of Bode's law and an objective test for approximate numerical rationality. *J. Amer. Statist. Asso.* **64**, 23-66.

Jeffreys, H. (1935). Some tests of significance treated by the theory of probability. *Proc. Cambridge Philosophical Soc.* **31**, 203-222.

Jeffreys, H. (1961). *Theory of Probability*, 3rd ed. Oxford University Press.

Kass, R.E. and Raftery, A.E. (1995). Bayes factors. *J. Amer. Statist. Asso.* **90**, 773-795.

Kass, R.E. and Wasserman, L. (1995). A reference Bayesian test for nested hypotheses and its relationship to the Schwarz criterion. *J. Amer. Statist. Asso.* **90**, 928-934.

Kass, R.E. and Wasserman, L. (1996). The selection of prior distributions by formal rules. *J. Amer. Statist. Asso.* **91**, 1343-1377.

O'Hagan, A. (1995). Fractional Bayes factors for model comparison. *J. Roy. Statist. Soc. Ser. B* **57**, 99-138.

Pettit, L. (1992). Bayes factors for outlier models using the device of imaginary observations. *J. Amer. Statist. Asso.* **87**, 541-545.

Raftery, A.E. (1986). A note on Bayes factors for log-linear contingency table models with vague prior information. *J. Roy. Statist. Soc. Ser. B* **48**, 249-250.

Raftery, A.E. (1995). Bayesian model selection in social research. *Sociology Methodology, 1995* (P.V. Mauden, ed.) 111-195, Blackwells, Cambridge, Mass.

Raftery, A.E. (1996). Approximate Bayes factors and accounting for model uncertainty in generalized linear models. *Biometrika* **83**, 251-266.

Schwarz, G. (1978). Estimating the dimension of a model. *Ann. Statist.* **6**, 461-464.

Smith, A.F. and Spiegelhalter, D.J. (1980). Bayes factors and choice criteria for linear models. *J. Roy. Statist. Soc. Ser. B* **42**, 213-220.

Spiegelhalter, D.J. and Smith, A.F. (1982). Bayes factors for linear and log-linear models with vague prior information. *J. Roy. Statist. Soc. Ser. B* **44**, 377-387.

Tierney, L. and Kadane, J. (1986). Accurate approximation for posterior moments and marginal densities. *J. Amer. Statist. Asso.* **81**, 82-86.

Tierney, L., Kass, R., and Kadane, J. (1989). Fully exponential Laplace approximations to expectations and variances of nonpositive functions. *J. Amer. Statist. Asso.* **84**, 710-716.

DISCUSSION

R.E. Kass

Carnegie-Mellon University

At this moment in the history of statistics, there seems to be less interest in the great Bayesian/frequentist divide than there had been in the nineteen seventies and eighties, when Efron (1986) asked, "Why isn't everyone a Bayesian?" We are all eager to get on to solving the many challenges of contemporary data analysis. Yet, we have our foundational conscience speaking to us; it continues to prod, with occasional welcome reminders from papers such as this one by Efron and Gous. How can these two great paradigms co-exist in peace? Where are the resolutions? What conflicts are irresolvable? And where does this leave us?

To me, the issues raised in this paper continue to be interesting. I find the authors' discussion clear and their new results informative. On the other hand, there are those in the Bayesian camp who see little relevance of all this to things they care about. Nearly all statisticians I have come across, regardless of philosophical persuastion, freely admit to *thinking* Bayesianly. Among the converted, however, there is a kind of Cartesian credo: "I think Bayesianly, therefore I am Bayesian." The impatience of the true believers comes in part from their taking the next step: "I think Bayesianly, therefore I must place all of my statistical work within the Bayesian paradigm."

A second, equally fundamental difficulty many Bayesians (and some frequentists) have with the perspective articulated in this paper, is in the importance it places on hypothesis testing and model selection. As the authors note, a recent version of this dissenting point of view is in Gelman and Rubin's discussion of Raftery (1995).

One might say that a major practical goal of this paper is to dissect Jeffreys's remark that his methods and Fisher's would rarely lead to different conclusions (Jeffreys, 1961, p. 435): "In spite of the difference in principle between my tests and those based

R. E. Kass is Professor and the Head , Department of Statistics, Carnegie-Mellon University, Pittsburgh, PA 15213-3890, U.S.A; email: kass@stat.cmu.edu.

on the P integrals, ... it appears that there is not much difference in the practical recommendations." Qualitatively and roughly, Bayes factors may be brought to approximate agreement with frequentist tests when the prior on the alternative shrinks at rate $O(n^{-1/2})$ toward the null value. This was made precise by Andrews (1994). Efron and Gous show that, in fact, the prior may be chosen so as to obtain a very close agreement in one dimension, but the agreement is not so close in higher dimensions. They rightly put their finger on sample size coherency as the key restriction of the Bayesian approach to testing and model selection, and their discussion and remarks on this point are helpful. They present an example of hypothesis testing in which the conclusions to be drawn are important, and they apparently like the way their B_{freq} measures the evidence. What remains, however, is the deep and vexing problem of model selection.

I have recently come across a data set on prosopagnosia—a disorder in which patients have difficulty recognizing faces—where the goal is to better understand the cognitive deficiency. We developed a generalized ROC analysis in which probit regression is used with 7 explanatory variables defined by the experimental design, the goal being to make inferences about several contrasts. There are a total of just over 20,000 binary observations on roughly 40 subjects and to examine the contrasts of interest, something must be done to take account of the remaining explanatory variables. Thus, we have a familiar problem of "model selection," in quotes because one may well decide not to literally select a model.

My own preference would be to use BIC to sift through a wide variety of models (perhaps using MCMC), drawing conclusions based on the potentially many models that have substantial posterior probability according to BIC (i.e., prob $= \exp(BIC)/(1+\exp(BIC)))$. However, we run into a problem, which the authors allude to toward the end of their Section 5: What value of the sample size n should be used in the penalty term? The difficulty is most easily appreciated in the one-way ANOVA context with k replications in I groups. At first glance, there are two possibilities: n could be the total number of observations $n = I \cdot k$ or, if we view each replication as an observation on an I-dimensional vector, we might instead take $n = k$. In many examples, the results are strikingly different. The same problem arises in our probit experimental design. I think the choices $n = I \cdot k$ and $n = k$ are both intuitively reasonable. It may seem preferable to take $n = k$, but keep in mind that in linear regression one would usually take n to be the total number of observations. Thus, if we take $n = k$ for replicated designs we create a discontinuity: adding very small epsilons to the 1's and 0's in the design matrix for the linear regression version of ANOVA creates a non-replicated linear regression that is essentially identical to the ANOVA design, yet we would be using a very differenct value of n than in the ANOVA design and would thus get very different results.

One potential way out of this bind is to begin by regarding the sample size as being as large as possible; in the one-way ANOVA setting we would take $n = I \cdot k$. In the terminology of Kass and Wasserman (1995), this would correspond to using a "unit-information prior" with n units of information in the sample. We might then ask how many units of information the prior should contain. That is, we might consider the possibility that the prior might contain c/n units of information. In the one-way ANOVA setting, we could choose $c = 1$ to obtain a BIC-based analysis with penalty term depending on $\log(I \cdot k)$, or we could instead choose $c = I$ to obtain a penalty term involving $\log(k)$. Furthermore, there is now an additional possibility of estimating c from the data—finding the value of c that provides the best fit. An interesting version of this approach to model selection (albeit more involved than I am describing here) has been proposed by George and Foster (2000).

This is a somewhat lengthy background to a simple question for the authors: Would they extend their B_{freq} to more complicated contexts? Can, for instance, their formula (2.30) be regarded as a general model-selection criterion, and would they apply it even in non-nested situations (which is crucial for applications of the kind I have mentioned)?

I'd like to come back to Jeffreys' remark (about the garden-variety small-sample problems most commonly treated) that his methods and Fisher's might not differ much in practice, and to a point that continues to bother me, at least a little. There is a pretty substantial body of work comparing Bayes factors to significance tests, following from Jeffreys and Savage (especially Edwards, Lindman, and Savage, 1963), including work by Berger and various colleagues. To me, the most important practical take-home message is that .05 really doesn't represent much evidence against the null. The authors provide a nice quote from Fisher that seems to give his perspective pretty clearly on the use of .05 as an appropriate conventional cutoff value, as well as setting the stage for their use of $\alpha_0 = .9$ as the breakeven point. Another relevant reference, however, is Fisher's famous description (in the introduction to his *Experimental Design*) of "A Lady Tasting Tea." There, he rejects the design involving 6 cups of tea, prefering one involving 8, because the p-value from exact identification with 6 cups would be $p = .05$ and this he considered insufficient evidence. In a similar vein, Gosset wrote (Pearson, 1938):

> [A test] doesn't in itself necessarily prove that the sample is not drawn randomly from the population even if the chance is very small, say .00001: what it does is to show that if there is any alternative hypothesis which will explain the occurrence of the sample with a more reasonable probability, say .05 ... you will be very much more inclined to consider that the original hypothsis is not true."

The point here is that Gosset picked the .05 level to correspond to a "reasonable" tail-area for an *acceptable* hypothesis. In my elementary statistics classes we can and sometimes do get into extended discussions about sample size, and the arbitrariness of any cutoff value, as well as the virtues of estimation rather than testing. However, statistics students—like everyone else—need to have clear and concise guidelines they can remember. I tell students to interpret $p > .05$ as "no evidence against H_0" and $p < .05$ as "some evidence against H_0" but I frequently emphasize that a p near .05 would be pretty shaky. In doing this I think I am offering a bit more conservative advice than is traditional, but it seems to me to be good advice that is more or less consistent with its Gosset/Fisher origins while being enlightened by subsequent analysis. So, my last question is this: How has any of this reconsideration of the foundations affected their thinking, and what scale of evidence do they tell their students to use?

Let me now, with great pleasure, thank the authors for a stimulating article.

ADDITIONAL REFERENCES

Edwards, W., Lindman, H., and Savage, L.J. (1963). Bayesian statistical inference for psychological research. *Psychol. Rev.* **70**, 193-242.

Efron, B. (1986). Why isn't everyone a Bayesian? *Amer. Statist.* **40**, 1-11.

George, E.I. and Foster, D.P. (2000). Calibration and empirical Bayes variable selection. *Biometrika* **87**, 731-748.

Pearson, E.S. (1938) "Student" as statistician, in William Sealy Gosset, 1876-1937. *Biometrika* **30**, 210-250.

G. S. Datta and P.Lahiri

University of Georgia and University of Nebraska-Lincoln

Starting from a simple testing hypothesis problem, the authors have very successfully demonstrated how one can possibly reconcile two apparently different approaches (Fisher's and Jeffreys') to model selection. In the process, they have proposed a new frequentist Bayes factor for model selection.

G. S. Datta is Professor, Department of Statisitcs, University of Georgia, Athens, GA 30602, U.S.A; email: gauri@stat.uga.edu. P. Lahiri is Milton Mohr Distinguished Professor of Statisitcs, Department of Mathematics and Statistics, University of Nebraska-Lincoln, Lincoln, NE 68588-0323, U.S.A; email: plahiri@math.unl.edu.

Datta was partially supported by NSF grants SBR-9705145 and DMS-0071642 and by NSA. Lahiri was supported in part by NSF grant SBR-9978145. The authors thank Mr. Shijie Chen for computing help.

In this discussion, we will consider the important problem of selection between a random effect model and a fixed effect model. Such a problem is encountered in many applications including animal breeding and small-area estimation. For the simplicity of exposition, let us consider the following one-way balanced random effects model:

$$y_{ij} = \mu + a_i + e_{ij}, \qquad i = 1, ..., m; \qquad j = 1, ..., n_0,$$

where a_i's and e_{ij}'s are independent with $a_i \stackrel{\text{i.i.d}}{\sim} N(0, \sigma_a^2)$ and $e_{ij} \stackrel{\text{i.i.d}}{\sim} N(0, \sigma_e^2)$. Let us call this model M.

Note that this model belongs to the exponential family considered by the authors. The model selection in question can be equivalently viewed as the following one-sided testing hypothesis problem:

$$H_0 : \sigma_a^2 = 0 \qquad \text{vs} \qquad H_a : \sigma_a^2 > 0.$$

Under this null hypothesis model M reduces to model M_0: $y_{ij} \stackrel{\text{i.i.d}}{\sim} N(\mu, \sigma_e^2)$. We consider this model selection problem for two reasons. First, unlike the testing of μ considered by the authors, our example distinguishes among different types of noninformative priors available in the literature. Secondly, the parameter value specified by the null hypothesis falls on the boundary of the parameter space, a case not covered by the authors. Is the approximation formula given in Lemma (section 2.2) valid in this situation? We try to investigate this question by numerical examples. Do the authors have any comments here?

Define $S = \Sigma\Sigma(y_{ij} - \bar{y}_i)^2$, $SSB = n_0\Sigma(\bar{y}_i - \bar{y})^2$ and $SST = \sum_i \sum_j (y_{ij} - \bar{y})^2$. We also define $\eta = \sigma_a^2/\sigma_e^2$. To calculate the Bayes factor in favor of M against M_0, we need to calculate the marginal densities $f_M(y)$ and $f_{M_0}(y)$, where

$$f_M(y) = \int L(\eta, \sigma_e^2, \mu)\pi(\eta, \sigma_e^2, \mu) d\mu d\eta d\sigma_e^2, \tag{1}$$

and

$$f_{M_0}(y) = \int L(0, \sigma_e^2, \mu)\pi_0(\sigma_e^2, \mu) d\mu d\sigma_e^2, \tag{2}$$

where the unrestricted likelihood $L(\eta, \sigma_e^2, \mu)$ and the likelihood under the null hypothesis $L(0, \sigma_e^2, \mu)$ are given by

$$\begin{aligned} L(\eta, \sigma_e^2, \mu) &= (2\pi)^{-\frac{n_0 m}{2}} (\sigma_e^2)^{-\frac{mn_0}{2}} (1 + n_0\eta)^{-\frac{m}{2}} \times \exp[-\frac{1}{2\sigma_e^2}\{s + \frac{n_0}{1 + n_0\eta}\sum_{i=1}^{m}(\bar{y}_i - \mu)^2\}], \\ L(0, \sigma_e^2, \mu) &= (2\pi)^{-\frac{mn_0}{2}} (\sigma_e^2)^{-\frac{mn_0}{2}} \exp[-\frac{s + n_0\sum_{i=1}^{m}(\bar{y}_i - \mu)^2}{2\sigma_e^2}]. \end{aligned} \tag{3}$$

To consider authors' suggestion in calculating Bayes factor based on Jeffreys' priors (under M and M_0), we use a general class of priors. Under the model M, we use

$$\pi(\eta, \sigma_e^2, \mu) = (\sigma_e^2)^{\{a_1+a_2+a_3+2a_4+1\}} \eta^{a_1} (1+n_0\eta)^{a_3} [(n_0-1)(1+n_0\eta)^2+1]^{a_4}. \qquad (4)$$

Under $H_0 : \sigma_a^2 = 0$, the class of priors to be considered is

$$\pi_0(\mu, \sigma_e^2) = (\sigma_e^2)^{a_2^*}. \qquad (5)$$

It can be shown (see Datta and Lahiri 2000) that the Bayes factor B in favor of M against M_0 is given by

$$B = \frac{f_M(y)}{f_{M_0}(y)} = 2^{-(a_1+a_2+a_3+2a_4+1-a_2^*)} [SSB+s]^{1/2(mn_0-2a_2^*-3)}$$

$$\times \Gamma(\frac{mn_0-2a_1-2a_2-2a_3-4a_4-5}{2}) \{\Gamma(\frac{mn_0-2a_2^*-3}{2})\}^{-1}$$

$$\times \int_0^\infty \eta^{a_1}(1+n_0\eta)^{1/2\{m(n_0-1)-2a_1-2a_2-4a_4-4\}} [SSB+s(1+n_0\eta)]^{-1/2(mn_0-2a_1-2a_2-2a_3-4a_4-5)} d\eta. \qquad (6)$$

In order that B in (6) remains free from the unit of measurement we must have

$$a_1 + a_2 + a_3 + 2a_4 - a_2^* + 1 = 0. \qquad (7)$$

Thus, we must be careful in choosing the noninformative priors for the model M and M_0 so that the resulting Bayes factor is unit-free. We will consider the following pairs of priors from the classes (4) and (5) described above in calculating the Bayes factor B.

(a) Jeffreys' priors: Here $a_2^* = a_3 = -3/2$.

Prior under $M = \pi_{JM}(\mu, \sigma_a^2, \sigma_e^2) = (\sigma_e^2)^{-1}(\sigma_e^2 + n_0\sigma_a^2)^{-3/2}$,

Prior under $M_0 = \pi_{JM_0}(\mu, \sigma_e^2) = (\sigma_e^2)^{-3/2}$.

(b) Matching prior for η vs. Jeffrey's recommended prior: Here $a_2^* = a_3 = -1$.

Prior under $M = \pi_{\eta M}(\mu, \sigma_a^2, \sigma_e^2) = \{\sigma_e^2(\sigma_e^2 + n_0\sigma_a^2)\}^{-1}$

Prior under $M_0 = \pi_{M_0}(\mu, \sigma_e^2) = (\sigma_e^2)^{-1}$.

(c) Matching prior for σ_a^2 in the full model: Here $a_2^* = a_3 = -2$.

Prior under $M = \pi_{\sigma_a^2 M}(\mu, \sigma_a^2, \sigma_e^2) = \pi_M^{(2)}(\mu, \sigma_a^2, \sigma_e^2) = \frac{[(n_0-1)(\sigma_e^2+n_0\sigma_a^2)^2+\sigma_e^4]^{1/2}}{[\sigma_e^2(\sigma_e^2+n_0\sigma_a^2)]^2}$.

Prior under $M_0 = \pi(\mu, \sigma_e^2) = (\sigma_e^2)^{-2}$.

Datta and Lahiri (2000) showed that for the prior-pair (a)-(c), B further simplifies to

$$B = \frac{1}{n_0} \frac{\int_b^1 w^{\frac{1}{2}\{m(n_0-1)\}-1}(1-w)^{\frac{1}{2}(m-2a_3-3)-1}dw}{b^{\frac{1}{2}(n_0-1)m}(1-b)^{\frac{1}{2}(m-2a_3-3)}} = B(a_3) \text{ (say)}, \tag{8}$$

where $b = S/SST$. Note that for $b < w < 1$ and for $a_3'' < a_3'$ it is easy to see that $(1-w)^{-a_3'} > (1-w)^{-a_3''}(1-b)^{a_3''-a_3'}$. Using this it can be shown from (8) that $B(a_3') > B(a_3'')$. Thus, B is increasing in a_3. Hence, B for prior pair $(b) > B$ for prior pair $(a) > B$ for prior pair (c). In other words, the matching prior for η is the least favorable for M_0, the matching prior for σ_a^2 is the most favorable for M_0, and Jeffreys' prior (the recommended prior of the authors) falls in between.

The frequentist Bayes factor as given by the authors involves the log-likelihood ratio statistic

$$\begin{aligned}\hat{B}_l(y) &= 2\{\sup_{\mu,\sigma_a^2,\sigma_e^2} \log\ L(\sigma_a^2,\sigma_e^2,\mu) - \sup_{\mu,\sigma_e^2}\log\ L(0,\sigma_e^2,\mu)\}\\ &= 2\{\sup_{\sigma_a^2,\sigma_e^2} \log\ L(\sigma_a^2,\sigma_e^2,\bar{y}) - \sup_{\sigma_e^2}\log\ L(0,\sigma_e^2,\bar{y})\}. \end{aligned} \tag{9}$$

It can be shown that

$$\hat{B}_l(y) \equiv \hat{B}_l(b) = m\{h(\tilde{b}) - h(b)\}I(b < \tilde{b}),$$

where $\tilde{b} = (n_0-1)/n_0$ and $h(x) = (n_0-1)\log\ x + \log(1-x), \quad 0 < x < 1.$

Then the authors' approximation to frequentist Bayes factor $\hat{B}_{Freq}(b)$ is given by

$$\hat{B}_{Freq}(b) = \exp\{\frac{1}{2}\hat{B}_l(b)\}/2.2687, \tag{10}$$

where 2.2687 is based on the cut-off point at level 0.10 of $\hat{B}_l(b)$. Since asymptotically (see Chernoff 1954) $P_{M_0}[\hat{B}_l(b) > x] \approx \frac{1}{2}P[\chi_1^2 > x]$, we get at level 0.10, $x = (1.28)^2$ and $\exp\{x/2\} = 2.2687$.

Consider the dyestuff data given in Box and Tiao (1973, Sec. 5.1.2). Five samples (n_0=5) from each of the six (m=6) randomly chosen batches of raw materials were taken and yield of dyestuff of standard color for each sample was determined. Table 1 reports the values of Bayes factors for three different choices of the noninformative priors and the frequentist Bayes factor. It is interesting to note that the frequentist Bayes factor and the Bayes factors under the noninformative priors (a) - (c) all provide "positive" support for random batch effect.

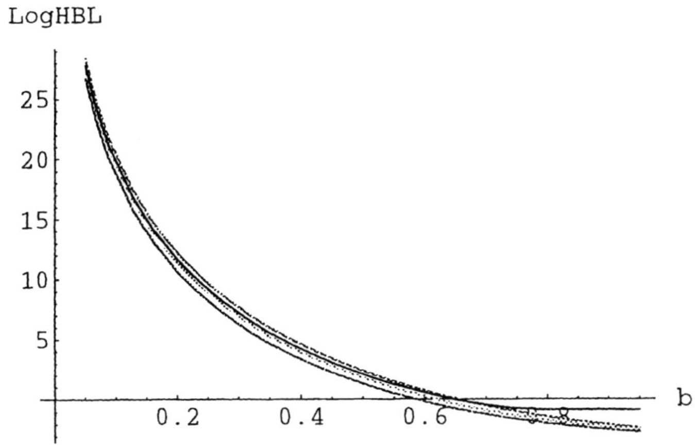

Figure 1: *Logarithm of frequentist Bayes factor and various noninformative Bayes factors for dyestuff data (six batches) of Box and Tiao.*

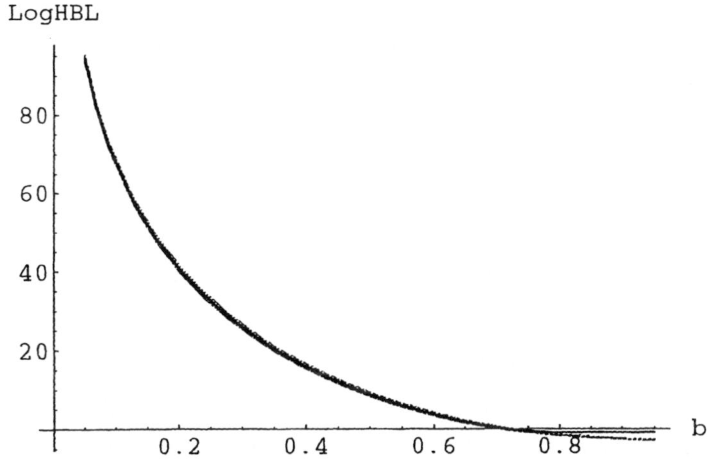

Figure 2: *Logarithm of frequentist Bayes factor and various noninformative Bayes factors for dyestuff data (20 batches, simulated) of Box and Tiao*

Table 1. Frequentist Bayes factor and the Bayes factors under priors (a)-(c)for dyestuff data.

Frequentist	(a)	(b)	(c)
6.568	4.92424	8.67469	3.02869

Note that all the three Bayes factors constructed using noninformative priors (a)-(c) and the frequentist Bayes factor is a function of b. Figures 1 and 2 plot logarithm of Bayes factors against b for $m = 6$ and $m = 20$ (in each case $n_0 = 5$). It is clear that there is very good reconciliation of the Bayes factors under noninformative priors (a)-(c).

ADDITIONAL REFERENCES

Box, G. E. P. and Tiao, G. C. (1973). *Bayesian Inference in Statistical Analysis.* Wiley, New York.

Chernoff, H. (1954). On the distribution of the likelihood ratio, *J. Amer. Statist. Asso.* **25**, 573-578.

Datta, G.S. and Lahiri, P. (2000). A comparison of the Bayes factors and frequentist Bayes factor for a balanced one-way random effects model, *unpublished manuscript.*

REJOINDER

Bradley Efron and Alan Gous

This article was written under the following rule of thumb: no method that's been heavily used in serious statistical practice can be entirely wrong. The rule certainly applies to Fisherian hypothesis testing, but it also applies to Jeffreys and the BIC, leaving us to worry about Figure 1. The two scales of evidence seem to be giving radically different answers, even for sample sizes as small as $n = 100$.

Our paper localizes the disagreement to coherency, in this case sample size coherency, the key distinguishing feature of modern Bayesian philosophy. The BIC, along with any other methodology that acts coherently across different sample sizes, must share Figure 1's behavior, treating the smaller hypothesis M_0 ever more favorably as n increases. Fisher's theory, which is usually presented with the sample size fixed, eschews sample size coherency in favor of a more aggressive demand for statistical power.

Professor Kass' good-natured commentary seems to share our ecumenical rule of thumb. His approach to the prosopagnosia data set, using BIC to screen possible working models before confronting the key inferential questions, is a reasonable beginning point. Predictably, it will be slanted toward smaller working models, with more of the nuisance variables deemed unimportant, than a similar screening based on standard hypothesis tests. In fact with $n = 20,000$, BIC could easily prefer *very* small models. Our own experience with the selenium analysis did not inspire confidence in the "biggest possible choice of n" rule, see the $n = 1312$ column of Table 8.

There is no question that hypothesis testing is greatly overused in statistical practice, and that many problems would be better phrased in terms of estimation, confidence intervals, or their Bayesian equivalents. There remains nevertheless a core of situations where hypothesis testing conveys the correct scientific attitude. The selenium experiment is a good example. The null hypothesis point $\theta = .5$ in (5.3) commands attention, and cannot be smoothly incorporated with its neighbors into a continuous prior density. Maybe we don't need a perfect delta function of prior belief at $\theta = .5$, a very narrow Gaussian curve serving just as well, but in any case the overall prior for θ must incorporate a bump near $\theta = .5$. Bumps make for mathematical difficulties, and neither Fisher's nor Jeffreys' model-selection methodology is as convincing as their estimation theories, but the hypothesis testing problem won't disappear just because it is awkward.

Professors Datta and Lahiri try various uninformative priors on an interesting components-of-variance example, partly to see how well our "frequentist Bayes factor" $B_{\text{freq}}(x)$, (2.30), works in a different setting. The big difference here is that the null hypothesis occurs at the extreme of the larger hypothesis instead of inside. We considered one such situation in Section 3.2, the one-sided hypothesis testing case. It is important to note that the approximation $B(\mathbf{x}) \doteq \widehat{B}(\mathbf{x})/\widehat{B}(\mathbf{y})$ that leads to $B_{\text{freq}}(x)$ must be modified in one-sided cases, the change going back to a one-sided application of Laplace's method in the Lemma of Section 2.2. The difference can be seen in the additional term $\log \Phi(x)/\Phi(y)$ in (3.9). Even without this correction, $B_{\text{freq}}(x)$ seems to be giving quite reasonable Bayes factors in Datta & Lahiri's two figures.

Kass also questions us about $B_{\text{freq}}(x)$ and its application to more complicated situations. One strong caveat: in higher dimensional problems, $d > 1$, Fisher's .90 choice for the breakeven quantile α_o is too low. This is the message of Figure 3, where for $d > 1$ we can't reconcile p-values with *any* prior unless α_o is increased. Table 9 suggests reasonable values of α_o for increasing dimension d. With this improvement to (2.30) kept in mind, $B_{\text{freq}}(x)$ seems quite useful. We haven't tried to extend it to nonnested testing situations, which are not easily handled by any frequentist methodology.

At its heart the Bayesian-frequentist controversy is an argument about principles, not so much which principles as whether or not to use them at all. The Bayesian guiding principle is focussed on consistent decision-making across different frames of reference, sample-size coherency being a classic example. Examples of frequentist inconsistency, in which the Bayesian model-selection literature abounds, are apt to fall on deaf ears, frequentists being more focussed on just the problem at hand. Everyone uses Bayes rule, as Kass observes, but not everybody agrees on how it should be used in practice. Jeffreys' use in the model-selection problem leads to sample size coherency, while Fisher's use does not.

That brings us to Professor Kass' last and most difficult question: which model-selection paradigm do we teach our students? Fisher's scale seems perfectly suited to the common situation of fixed sample size and a straw-man null hypothesis that the investigators wish to disprove. However it is less satisfactory for more complicated problems involving multiple comparisons, data-mining, null hypotheses of genuine interest (as in the bioequivalence problem), or sequential decision making. Even slightly more complicated situations, like the selenium example, made us grateful for some Bayesian guidance in the form of $B_{\text{freq}}(x)$.

We are grateful to the discussants and the editors, both for producing this volume and allowing our participation. A preliminary version of this article was presented as the 1996 Morris DeGroot memorial lecture at Carnegie-Mellon University. Morris was a good-natured but always effective exponent of Bayesian thought, and this paper is dedicated to him.

IMS Lecture Notes—Monograph Series
NSF-CBMS Regional Conference Series in Probability & Statistics
Order online at: http://www.imstat.org/publications/imslect/index.phtml

Vol	Title	Authors or Editors	Member price	General price
	IMS LECTURE NOTES-MONOGRAPH SERIES			
1	*Essays on the Prediction Process*	FB Knight	$9	$15
2	*Survival Analysis*	J Crowley & RA Johnson	$15	$25
3	*Empirical Processes*	P Gaenssler	$12	$20
4	*Zonal Polynomials*	A Takemura	$9	$15
5	*Inequalities in Statistics & Probability*	YL Tong	$15	$25
6	*The Likelihood Principle*	JO Berger & RL Wolpert	$15	$25
7	*Approximate Computation of Expectations*	C Stein	$12	$20
8	*Adaptive Statistical Procedures & Related Topics*	J Van Ryzin	$24	$40
9	*Fundamentals of Statistical Exponential Families*	LD Brown	$15	$25
10	*Differential Geometry in Statistical Inference*	SI Amari, OE Barndorff-Nielsen, RE Kass, SL Lauritzen & CR Rao	$15	$25
11	*Group Representations in Probability & Statistics*	PW Diaconis	$18	$30
12	*An Introduction to Continuity, Extrema & Related Topics for General Gaussian Processes*	RJ Adler	$15	$25
13	*Small Sample Asymptotics*	CA Field & E Ronchetti	$15	$25
14	*Invariant Measures on Groups & Their Use in Statistics*	RA Wijsman	$18	$30
15	*Analytic Statistical Models*	IM Skovgaard	$15	$25
16	*Topics in Statistical Dependence*	HW Block, AR Sampson & TH Savits	$27	$45
17	*Current Issues in Statistical Inference: Essays in Honor of D. Basu*	M Ghosh & PK Pathak	$15	$25
18	*Selected Proceedings of the Sheffield Symposium on Applied Probability*	IV Basawa & RL Taylor	$9	$15
19	*Stochastic Orders & Decision Under Risk*	K Mosler & M Scarsini	$18	$30
20	*Spatial Statistics & Imaging*	A Possolo	$21	$35
21	*Weighted Empiricals & Linear Models*	HL Koul	$18	$30
22	*Stochastic Inequalities*	M Shaked & YL Tong	$24	$40
23	*Change-point Problems*	HG Mueller & D Siegmund	$26	$45
24	*Multivariate Analysis & Its Applications*	TW Anderson, KT Fang & I Olkin	$26	$45
25	*Adaptive Designs*	N Flournoy & WF Rosenberger	$24	$40
26	*Stochastic Differential Equations in Infinite Dimensional Spaces*	G Kallianpur & J Xiong	$24	$40
27	*Analysis of Censored Data*	HL Koul & JV Deshpande	$18	$30
28	*Distributions with Fixed Marginals & Related Topics*	L Ruschendorf, B Schweizer & MD Taylor	$31	$52
29	*Bayesian Robustness*	JO Berger, B Betro, E Moreno, LR Pericchi, F Ruggeri, G Salinetti & L Wasserman	$29	$49
30	*Statistics, Probability & Game Theory: Papers in Honor of D. Blackwell*	TS Ferguson, LS Shapley & JB MacQueen	$32	$54
31	*L_1-Statistical Procedures & Related Topics*	Y Dodge	$41	$69
32	*Selected Proceedings of the Symposium on Estimating Functions*	IV Basawa, VP Godambe & RL Taylor	$42	$69
33	*Statistics in Molecular Biology & Genetics*	F. Seillier-Moiseiwitsch	$36	$45
34	*New Developments & Applications in Experimental Design*	N Flournoy, WF Rosenberger & WK Wong	$21	$35
35	*Game Theory, Optimal Stopping, Probability & Statistics: Papers in Honor of Thomas S. Ferguson*	FT Bruss & L Le Cam	$39	$66
36	*State of the Art in Probability & Statistics: Festschrift for Willem R. van Zwet*	M de Gunst, C Klaassen & A van der Vaart	$79	$120
37	*Selected Proceedings of the Symposium on Inference for Stochastic Processes*	IV Basawa, CC Heyde & RL Taylor		
38	*Model Selection*	P Lahiri		
	NSF-CBMS REGIONAL CONFERENCE SERIES			
1	*Group Invariance Applications in Statistics*	ML Eaton	$15	$25
2	*Empirical Processes: Theory & Applications*	D Pollard	$12	$20
3	*Stochastic Curve Estimation*	M Rosenblatt	$15	$25
4	*Higher Order Asymptotics*	JK Ghosh	$15	$25
5	*Mixture Models: Theory, Geometry & Applications*	BG Lindsay	$15	$25
6	*Statistical Inference from Genetic Data on Pedigrees*	E Thompson	$18	$30